Lecture Notes in Computer Science 11321

Commenced Publication in 1973
Founding and Former Series Editors:
Gerhard Goos, Juris Hartmanis, and Jan van Leeuwen

More information about this series at http://www.springer.com/series/7407

Lilya Budaghyan
Francisco Rodríguez-Henríquez (Eds.)

Arithmetic of Finite Fields

7th International Workshop, WAIFI 2018
Bergen, Norway, June 14–16, 2018
Revised Selected Papers

 Springer

Editors
Lilya Budaghyan
Department of Informatics
University of Bergen
Bergen, Norway

Francisco Rodríguez-Henríquez
Centro de Investigación y de Estudios
Avanzados del Instituto Politécnico
Nacional
Mexico, Mexico

ISSN 0302-9743   ISSN 1611-3349   (electronic)
Lecture Notes in Computer Science
ISBN 978-3-030-05152-5   ISBN 978-3-030-05153-2   (eBook)
https://doi.org/10.1007/978-3-030-05153-2

Library of Congress Control Number: 2018962540

LNCS Sublibrary: SL1 – Theoretical Computer Science and General Issues

This Springer imprint is published by the registered company Springer Nature Switzerland AG
The registered company address is: Gewerbestrasse 11, 6330 Cham, Switzerland

Preface

These are the proceedings of WAIFI 2018, the 7th International Workshop on the Arithmetic of Finite Fields, held in Bergen, Norway, during June 14–16, 2016. The six previous editions of this workshop were held in Madrid, Spain (WAIFI 2007), Siena, Italy (WAIFI 2008), Istanbul, Turkey (WAIFI 2010), Bochum, Germany (WAIFI 2012), Gebze, Turkey (WAIFI 2014), and Ghent, Belgium (WAIFI 2016). Springer has published all previous volumes of the WAIFI proceedings in the LNCS series. Since 2008, WAIFI has been held every even year, bringing together mathematicians, computer scientists, engineers, and physicists who conduct research in different areas of finite field arithmetic.

The program consisted of six invited talks and 14 contributed papers. The invited speakers were Simon Blackburn (Royal Holloway University of London, UK), Anwar Hasan (University of Waterloo, Canada), Daniel Panario (Carleton University, Canada), Daniel Katz (California State University, USA), Ferruh Özbudak (Middle East Technical University, Turkey), and Benjamin Smith (Inria, France). The papers supporting the three last invited talks were also included in the proceedings. The contributed talks were selected from 26 submissions, each of which was assigned to at least three committee members or external reviewers chosen by the members. Additionally, the Program Committee had a significant online discussion phase for several days. Two of the papers were selected for the best paper award: "On Symmetry and Differential Properties of Generalized Boolean Functions" by Thor Martinsen, Wilfried Medil, Alexander Pott, Pantelimon Stanica, and "A New Family of Pairing-Friendly Elliptic Curves" by Michael Scott and Aurore Guillevic.

We are very grateful to the members of the Program Committee for their dedication, professionalism, and careful work with the review and selection process. We also sincerely thank the external reviewers who contributed with their special expertise to review papers for this workshop. We deeply thank the general co-chairs, Lilya Budaghyan and Tor Helleseth, for their support of the Program Committee and their hard work in leading the overall organization of the workshop helped by the Organizing Committee. We would also like to sincerely thank members of the Steering Committee of the workshop series for their constant support and encouragement in our efforts to create a stimulating scientific program leading to this volume. Furthermore, we are very grateful to José Luis Imaña for his valuable help in publicity and for diligently maintaining the workshop website. As with the previous volumes, Springer agreed to publish the revised and expanded versions of the WAIFI 2016 papers as an LNCS volume. We thank Alfred Hoffman from Springer for making this possible. We would like to acknowledge that the submission and selection of papers were done using the EasyChair conference management system. We would also like to thank Bergen Research Foundation for sponsoring the workshop. Finally, but most importantly, we deeply thank all the authors who submitted their papers to the workshop and the participants from all over the world who chose to honor us with their attendance.

June 2018

Lilya Budaghyan
Francisco Rodríguez-Henríquez

Organization

International Workshop on the Arithmetic of Finite Fields

Bergen, Norway
June 14–16, 2018

Organized by
University of Bergen

Steering Committee

Claude Carlet	University of Paris 8, France
Anwar Hasan	University of Waterloo, Canada
José Luis Imaña	Complutense University of Madrid, Spain
Çetin Kaya Koç	University of California Santa Barbara, USA
Sihem Mesnager	University of Paris 8, France
Ferruh Özbudak	Middle East Technical University, Turkey
Christof Paar	Ruhr-Universität Bochum, Germany
Francisco Rodríguez-Henríquez	CINVESTAV-IPN, Mexico
Erkay Savas	Sabanci University, Turkey
Berk Sunar	Worcester Polytechnic Institute, USA
Gustavo Sutter	Autonomous University of Madrid, Spain

General Chairs

Lilya Budaghyan	University of Bergen, Norway
Tor Helleseth	University of Bergen, Norway

Program Chairs

Lilya Budaghyan	University of Bergen, Norway
Francisco Rodríguez-Henríquez	CINVESTAV-IPN, México

Program Committee

Diego F. Aranha	University of Campinas, Brazil
Daniel Augot	Inria, France
Angela Barbero	University of Valladolid, Spain
Marco Calderini	University of Bergen, Norway
Claude Carlet	University of Paris 8, France
Craig Costello	Microsoft Research, USA

Robert Coulter	University of Delaware, USA
Luca De Feo	Université Paris Saclay, UVSQ and Inria, France
Cunsheng Ding	Hong Kong University of Science and Technology, SAR China
Huseyin Hisil	Yasar University, Turkey
Koray Karabina	Florida Atlantic University, USA
Alla Levina	ITMO University, Russia
Chunlei Li	University of Bergen, Norway
Nian Li	Hubei University, China
Oleg Logachev	Moscow State University, Russia
Florian Luca	Wits University, South Africa
Matthew Parker	University of Bergen, Norway
Léo Perrin	Inria, France
Arash Reyhani-Masoleh	Western University, Canada
Igor Semaev	University of Bergen, Norway
Max Sala	University of Trento, Italy
Erkay Savas	Sabanci University, Turkey
Patrick Solé	Télécom ParisTech, France
Natalia Tokareva	Novosibirsk State University, Russia
Øyvind Ytrehus	Simula, Norway

Additional Reviewers

Daniele Bartoli	University of Perugia, Italy
Massimo Giulietti	University of Perugia, Italy
Bernhard Heim	German University of Technology, Oman
Brandon Langenberg	PQSecure Technologies, USA
Michael Naehrig	Microsoft Research, USA
Erdinc Ozturk	Sabanci University, Turkey
Federico Pintore	University of Oxford, UK
Joost Renes	University of Nijmegen, The Netherlands
Éric Schost	University of Waterloo, Canada
Deng Tang	Southwest Jiaotong University, China
Frederik Vercauteren	Katholieke Universiteit Leuven, Belgium
Vanessa Vitse	Université de Grenoble I, France
Arne Winterhof	Österreichische Akademie der Wissenschaften, Austria

Organizing Committee

Marco Calderini	University of Bergen, Norway
Nikolay Kaleyski	University of Bergen, Norway
Chunlei Li	University of Bergen, Norway

Sponsoring Institutions

Bergen Research Foundation

Contents

Invited Talk 1

Pre- and Post-quantum Diffie–Hellman from Groups, Actions, and Isogenies

Benjamin Smith[(✉)]

Inria and Laboratoire d'Informatique de l'École polytechnique (LIX),
Université Paris–Saclay, Palaiseau, France
smith@lix.polytechnique.fr

Abstract. Diffie–Hellman key exchange is at the foundations of public-key cryptography, but conventional group-based Diffie–Hellman is vulnerable to Shor's quantum algorithm. A range of "post-quantum Diffie–Hellman" protocols have been proposed to mitigate this threat, including the Couveignes, Rostovtsev–Stolbunov, SIDH, and CSIDH schemes, all based on the combinatorial and number-theoretic structures formed by isogenies of elliptic curves. Pre- and post-quantum Diffie–Hellman schemes resemble each other at the highest level, but the further down we dive, the more differences emerge—differences that are critical when we use Diffie–Hellman as a basic component in more complicated constructions. In this survey we compare and contrast pre- and post-quantum Diffie–Hellman algorithms, highlighting some important subtleties.

1 Introduction

The Diffie–Hellman key-exchange protocol is, both literally and figuratively, at the foundation of public-key cryptography. The goal is for two parties, Alice and Bob, to derive a shared secret from each other's public keys and their own private keys. Diffie and Hellman's original solution [51] is beautifully and brutally simple: given a fixed prime p and a primitive element g in the finite field $\mathbb{F}_p$ (that is, a generator of the multiplicative group $\mathbb{F}_p^\times$), Alice and Bob choose secret keys a and b, respectively, in $\mathbb{Z}/(p-1)\mathbb{Z}$. Alice computes and publishes her public key $A = g^a$, and Bob his public key $B = g^b$; the shared secret value is $S = g^{ab}$, which Alice computes as $S = B^a$, Bob as $S = A^b$.

The security of the shared secret depends on the hardness of the *Computational Diffie–Hellman Problem* (CDHP), which is to compute S given only A, B, and the public data of the structures that they belong to. For finite-field Diffie–Hellman, this means computing g^{ab} given only g, g^a, and g^b (mod p). The principal approach to solving the CDHP is to solve the *Discrete Logarithm Problem* (DLP), which is to compute x from g and g^x. We thus recover a from $A = g^a$ (or, equivalently, b from $B = g^b$), then power B by a (or A by b) to recover S. Attacking the DLP means directly attacking one of the public keys, regardless of any particular shared secret they may be used to derive.

© Springer Nature Switzerland AG 2018
L. Budaghyan and F. Rodríguez-Henríquez (Eds.): WAIFI 2018, LNCS 11321, pp. 3–40, 2018.
https://doi.org/10.1007/978-3-030-05153-2_1

Over the past four decades, the Diffie–Hellman protocol has been generalized from multiplicative groups of finite fields to a range of other algebraic groups, most notably elliptic curves. Partly motivated by this cryptographic application, there has been great progress in discrete logarithm algorithms for some groups, most notably Barbulescu, Gaudry, Joux, and Thomé's quasipolynomial algorithm for discrete logarithms in finite fields of fixed tiny characteristic [11].

The most stunning development in discrete logarithm algorithms came with the rise of the quantum computation paradigm: Shor's quantum algorithm [123] solves the discrete logarithm problem—and thus breaks Diffie–Hellman—in any group in polynomial time and space on a quantum computer.[1] The development of quantum computers of even modest capacity capable of running Shor's algorithm remains an epic challenge in experimental physics: at the time of writing, the largest implementation of Shor's algorithm was used to factor the integer 21, so there is some way to go [94]. But in anticipation, cryptographic research has already bent itself to the construction of *post-quantum* cryptosystems, designed to be used on conventional computers while resisting known quantum attacks.

Nowadays, Diffie–Hellman is often an elementary component of a more complicated protocol, rather than the entire protocol itself. For example, the TLS protocol for establishing secure internet connections includes an ephemeral Diffie–Hellman [116]. But to give a more interesting example, the X3DH protocol [93] used to establish connections in Signal and WhatsApp includes *four* simple Diffie–Hellmans between various short and long-term keypairs. The common use of Diffie–Hellman as a component makes the search for a drop-in post-quantum replacement for classical Diffie–Hellman particularly relevant today.

While many promising post-quantum candidates for public-key encryption and signatures have been developed —the first round of the NIST post-quantum standardization process [107] saw 59 encryption/KEM schemes and 23 signature schemes submitted— finding a simple post-quantum drop-in replacement for Diffie–Hellman (as opposed to a KEM) has proven to be surprisingly complicated. Some interesting post-quantum "noisy Diffie–Hellman" key exchange protocols based on hard problems in codes, lattices, and Ring-LWE have been put forward over the years (including [4,6,24,52,53,109], and [49]), but these typically require a reconciliation phase to ensure that Alice and Bob have the same shared secret value (as opposed to an approximate shared secret with acceptable noise on each side); we will not discuss these protocols further here. Perhaps surprisingly, given the loudly trumpeted quantum destruction of elliptic curve cryptography by Shor's algorithm, the most serious candidates for post-quantum Diffie–Hellman come from isogeny-based cryptography, which is founded in the deeper theory of elliptic curves.

The key idea in moving from conventional elliptic-curve cryptography to isogeny-based cryptography is that points on curves are replaced with entire curves, and relationships between points (scalars and discrete logarithms) are

---

[1] More generally, Armknecht, Gagliardoni, Katzenbeisser, and Peter have shown that no group-homomorphic cryptosystem can be secure against a quantum adversary, essentially because of the existence of Shor's algorithm [8].

replaced with relationships between curves (isogenies). Isogeny classes have just enough algebraic structure to define efficient asymmetric cryptosystems, but not enough to make them vulnerable to Shor's algorithm.

But what should a "post-quantum Diffie–Hellman" scheme be, and how closely should it match classical Diffie–Hellman functionalities and semantics? To what extent can the intuition and theoretical lore built up over decades of classical Diffie–Hellman carry over to these new protocols? This survey is an attempt to begin addressing these questions. The aim is to help cryptographers, mathematicians, and computer scientists to understand the similarities and differences between classical Diffie–Hellman and the new post-quantum protocols.

The Plan. We begin with a quick survey of classical group-based Diffie–Hellman in Sects. 2–5. We discuss modern elliptic-curve Diffie–Hellman in Sect. 7; this dispenses with the underlying group on some levels, and thus forms a pivot for moving towards post-quantum Diffie–Hellman. We review Couveignes' *hard homogeneous spaces* framework in Sects. 8 and 9, before introducing HHS cryptosystems in the abstract in Sect. 10; we go deeper into the underlying hard problems in Sects. 11 and 12. Moving into the concrete, we recall basic facts about isogenies in Sect. 13, before discussing commutative isogeny-based key exchange in Sect. 14 and the stranger SIDH scheme in Sect. 15. Our focus is mostly constructive, and our discussion of quantum cryptanalysis will be purely asymptotic, for reasons discussed in Sect. 6.

Limiting Scope. The basic Diffie–Hellman scheme is completely unauthenticated: it is obviously vulnerable to a man-in-the-middle attack where Eve impersonates Bob to Alice, and Alice to Bob. Alice and Bob must therefore authenticate each other outside the Diffie–Hellman protocol, but we do not discuss authentication mechanisms here. We also essentially ignore the provable-security aspects of these protocols, beyond noting that each has been proven session-key secure in Canetti and Krawczyk's adversarial authenticated-links model [33] (see [59, Sect. 5.3] for a proof for commutative isogeny key exchange, and [45, Sect. 6] for SIDH). As we noted above, we do not discuss noisy Diffie–Hellman schemes here, mostly for lack of time and space, but also because these are further from being drop-in replacements for classical Diffie–Hellman. Finally, we must pass over *decision* Diffie–Hellman-based protocols in silence, partly for lack of space, but mostly because at this early stage it seems hard to say anything nontrivial about decision Diffie–Hellman in the post-quantum setting. This is with some reluctance: as Boneh declared in [20], "the decision Diffie–Hellman assumption is a gold mine" (at least for theoretical cryptographers). Revisiting [20] in the post-quantum setting would be highly interesting, but this is neither the time nor the place for that investigation.

Notation. We will use abelian groups written additively and multiplicatively, depending on the context. To minimise confusion, we adopt these typographical conventions for groups and elements:

- $\mathcal{G}$ always denotes an abelian group written **additively**, with group operation $(P, Q) \mapsto P + Q$, inverse $P \mapsto -P$, and identity element 0.
- $\mathfrak{G}$ always denotes an abelian group written **multiplicatively**, with group operation $(\mathfrak{p}, \mathfrak{q}) \mapsto \mathfrak{p}\mathfrak{q}$, inverse $\mathfrak{p} \mapsto \mathfrak{p}^{-1}$, and identity element 1.

2   Abstract Groups and Discrete Logarithms

Diffie and Hellman presented their key exchange in the multiplicative group of a finite field, but nothing in their protocol requires the field structure. We will restate the protocol in the setting of a general finite abelian group in Sect. 3; but first, we recall some basic facts about abstract groups and discrete logarithms.

Let $\mathcal{G}$ be a finite abelian group of order N (written additively, following the convention above). We can assume $\mathcal{G}$ is cyclic. For every integer $m \pmod{N}$ we have an exponentiation endomorphism $[m] : \mathcal{G} \to \mathcal{G}$, called *scalar multiplication*, defined for non-negative m by

$$[m] : P \longmapsto \underbrace{P + \cdots + P}_{m \text{ copies}}$$

and for negative m by $[m]P = [-m](-P)$. We can compute $[m]$ in $O(\log m)$ $\mathcal{G}$-operations using a variety of addition chains; typically $m \sim \#\mathcal{G} = N$.

The fundamental hard algorithmic problem in $\mathcal{G}$ is computing the inverse of the scalar multiplication operation: that is, computing discrete logarithms.

Definition 1 (DLP). *The* Discrete Logarithm Problem *in $\mathcal{G}$ is, given P and Q in $\langle P \rangle \subseteq \mathcal{G}$, compute an x such that $Q = [x]P$.*

Any DLP instance in any $\mathcal{G}$ can always be solved using $O(\sqrt{N})$ operations in $\mathcal{G}$, using (for example) Shanks' baby-step giant-step algorithm (BSGS), which also requires $O(\sqrt{N})$ space [122]; Pollard's ρ algorithm reduces the space requirement to $O(1)$ [113]. If N is composite and its (partial) factorization is known, then we can do better using the Pohlig–Hellman–Silver algorithm [112], which solves the DLP by reducing to the DLP in subgroups of $\mathcal{G}$ (see Sect. 12 below).

The DLP enjoys random self-reducibility: if we have an algorithm that solves DLPs for a large fraction $1/M$ of all possible inputs, then we can solve DLPs for all possible inputs after an expected M random attempts. Suppose we want to solve an arbitrary DLP instance $Q = [x]P$. We choose a random integer r, try to solve $Q' = Q + [r]P = [x + r]P$ for $x + r$, and if we succeed then we recover $x = (x + r) - r$. After M randomizations, we expect to find an r for which Q' lands in the set of inputs to which the algorithm applies.

In the pure abstract, we consider $\mathcal{G}$ as a black-box group: operations are performed by oracles, and elements are identified by (essentially random) bit-strings. This models the absence of useful information that we could derive from any concrete representation of $\mathcal{G}$. In this setting, Shoup [124] has proven that the DLP is not merely in $O(\sqrt{N})$, but in $\Theta(\sqrt{N})$. But in the real world, we do not have black-box groups; every group has a specific concrete element representation and an explicit algorithmic group law. The difficulty of the DLP then varies with the representation, as we will se in Sect. 5.

3 Pre-quantum Diffie–Hellman

Now let us consider Diffie–Hellman in the abstract. Let $\mathcal{G}$ be a cyclic group, and fix a public generator P of $\mathcal{G}$. Public keys are elements of $\mathcal{G}$; private keys are bitstrings, interpreted as elements of $\mathbb{Z}/N\mathbb{Z}$. Each (public,private)-keypair $(Q = [x]P, x)$ presents a DLP instance in $\mathcal{G}$.

The Diffie–Hellman protocol takes place in into two logical phases, which in practice may be separated by a significant period of time. In the first phase, the parties generate their keypairs using Algorithm 1 (KeyPair):

– Alice generates her keypair as (A, a) ←KeyPair() and publishes A;
– Bob generates his as (B, b) ←KeyPair() and publishes B.

In the second phase, they compute the shared secret S with Algorithm 2 (DH):

– Alice computes $S \leftarrow \text{DH}(B, a)$;
– Bob computes $S \leftarrow \text{DH}(A, b)$.

Alice and Bob have the same value S because $S = [a]B = [b]A = [ab]P$.

Algorithm 1. Keypair generation for textbook Diffie–Hellman in a group $\mathcal{G} = \langle P \rangle$ of order N.

Input: ()
Output: A pair (Q, x) in $\mathcal{G} \times \mathbb{Z}/N\mathbb{Z}$ such that $Q = [x]P$
1 **function** KeyPair()
2   $\quad x \xleftarrow{\$} \mathbb{Z}/N\mathbb{Z}$
3   $\quad Q \leftarrow [x]P$
4   $\quad$ **return** (Q, x)

Algorithm 2. Textbook Diffie–Hellman key exchange in $\mathcal{G}$

Input: A public key R in $\mathcal{G}$, and a private key x in $\mathbb{Z}/N\mathbb{Z}$
Output: A shared secret $S \in \mathcal{G}$
1 **function** DH(R, x)
2   $\quad S \leftarrow [x]R$
3   $\quad$ **return** S // To be input to a KDF

The security of the (entire) shared secret depends on the hardness of the *Computational Diffie–Hellman Problem* (CDHP) in $\mathcal{G}$.

Definition 2 (CDHP). *The* Computational Diffie–Hellman Problem *in $\mathcal{G}$ is, given P, $A = [a]P$, and $B = [b]P$ in $\mathcal{G}$, to compute $S = [ab]P$.*

While it is obvious that an algorithm that solves the DLP in $\mathcal{G}$ can solve the CDHP in $\mathcal{G}$, constructing a reduction in the other direction—that is, efficiently solving DLP instances given access to an oracle solving CDHP instances—is a much more subtle matter. It is now generally accepted, following the work of Den Boer [48], Maurer and Wolf [95, 96], Muzereau, Smart, and Vercauteren [105], Bentahar [13], and Boneh and Lipton [21] that the DLP and CDHP are equivalent for the kinds of $\mathcal{G}$ that cryptographers use in practice. Since solving DLP instances is the only way we know to solve CDHP instances, Diffie–Hellman is generally considered to be a member of the DLP-based family of cryptosystems.

The shared secret S is *not* suitable for use as a cryptographic key[2]; rather, it should be treated with a Key Derivation Function (KDF) to produce a proper secret cryptographic key K. This essentially hashes the secret S, spreading the entropy of S uniformly throughout K, so deriving any information about K requires computing the whole of S, hence solving a CDHP in $\mathcal{G}$. The indistinguishability of S, and hence its security as a key, depends on the weaker *Decisional* Diffie–Hellman Problem, which is beyond the scope of this article.

The lifespan of keypairs is crucial in Diffie–Hellman-based cryptosystems. In *ephemeral* Diffie–Hellman, Alice and Bob's keypairs are unique to each execution of the protocol. Ephemeral Diffie–Hellman is therefore essentially interactive. In contrast, *static* Diffie–Hellman uses long-term keypairs across many sessions. Alice may obtain Bob's long-term public key and complete a Diffie–Hellman key exchange with him—and start using the shared secret—without any active involvement on his part. Static Diffie–Hellman is therefore an important example of a *Non-Interactive Key Exchange* (NIKE) protocol [61].

Efficient *public-key validation*—that is, checking that a public key was honestly generated—is an important, if often overlooked, requirement for many Diffie–Hellman systems, particularly those where keys are re-used. Suppose Alice derives a shared secret key K from a Diffie–Hellman exchange with Bob's public key B, and then uses K to communicate with Bob. A malicious Bob might construct an invalid public key B in such a way that K reveals information about Alice's secret key a. If (a, A) is ephemeral then Bob has learned nothing useful about a, since it will never be used again; but the keypair (A, a) is to be reused, as in static Diffie–Hellman, then secret information has been leaked, and Alice thus becomes vulnerable to active attacks (see e.g. [91] for an example). Public key validation is simple in a finite field: it usually suffices to check the order of the element. Antipa, Brown, Menezes, Struik, and Vanstone describe the process for elliptic-curve public keys [7], and their methods extended to most curve-based algebraic groups without serious difficulty. We will see that this is a more serious problem in post-quantum systems.

---

[2] Some protocols do use the shared secret S as a key, most notably the textbook ElGamal encryption presented at the start of Sect. 4.

4 Encryption and Key Encapsulation

The classic ElGamal public-key encryption scheme [55] is closely related to Diffie–Hellman key exchange. Its key feature is that messages are viewed as elements of the group $\mathcal{G}$, so adding a random-looking element of $\mathcal{G}$ to a message in $\mathcal{G}$ acts as encryption.

Algorithm 3 lets Alice encrypt a message to Bob. Alice first generates an ephemeral keypair (E, e), completes the Diffie–Hellman on her side using Bob's static public key B to compute a shared secret S, which she uses as a (symmetric) key to encrypt an element of $\mathcal{G}$ via $M \mapsto C = M + S$ (with corresponding decryption $C \mapsto M = C - S$). Sending her ephemeral public key E together with the ciphertext allows Bob to compute S and decrypt with Algorithm 4. Since the secret key here is the bare shared secret S, untreated by a KDF, the security of this protocol depends not on the CDHP but rather on the (easier) decisional Diffie–Hellman problem in $\mathcal{G}$.

Algorithm 3. Classic ElGamal encryption: Alice encrypts to Bob

Input: Bob's public key B and a message $M \in \mathcal{G}$
Output: An element $E \in \mathcal{G}$ and a ciphertext $C \in \mathcal{G}$

```
1 function ElGamalEncrypt(B, M)
2      (E, e) ← KeyPair()                        // E = [e]P
3      S ← DH(B, e)                // B = [b]P ⟹ S = [eb]P
4      C ← M + S                              // C = M + [eb]P
5      return (E, C)
```

Algorithm 4. Classic ElGamal decryption: Bob decrypts from Alice

Input: An element $E \in \mathcal{G}$, a ciphertext $M \in \mathcal{G}$, and Bob's private key b
Output: A plaintext messge $M \in \mathcal{G}$

```
1 function ElGamalDecrypt((E, C, b))
2      S ← DH(E, b)             // E = [e]P ⟹ S = [eb]P
3      M ← C - S                          // C = M + [eb]P
4      return M
```

It is important to note that this cryptosystem does not provide semantic security. For example, if (E_1, C_1) and (E_2, C_2) are encryptions of M_1 and M_2, respectively, then $(E_1 + E_2, C_1 + C_2)$ is a legitimate encryption of $M_1 + M_2$. While this property is desirable for certain applications (such as E-Voting [12]), in most contexts textbook ElGamal cannot be safely used for public-key encryption.

The homomorphic nature of the scheme is due to the fact that the group law $+$ is being used as the encryption and decryption algorithm. But even if this behaviour is actually desired, requiring the message to be encoded as an element of $\mathcal{G}$ creates two further problems. First, it imposes a hard and inconvenient limit on the size of the message space; second, it requires an efficient encoding of messages to group elements. At first glance, this second requirement seems straightforward for ElGamal instantiated in $\mathbb{F}_p^{\times}$, since bitstrings of length $\leq \log_2 p$ can be immediately interpreted as integers in $\mathbb{Z}/p\mathbb{Z}$, and hence elements of $\mathbb{F}_p$; but this embedding does not map us into the prime-order subgroups where the protocol typically operates. This complication is worse in the elliptic curve setting, where the limit on message length can be even more restrictive.

The modern, semantically secure version of ElGamal encryption is an example of hybrid encryption, best approached through the more general setting of Key Encapsulation Mechanisms (KEMs) and Data Encryption Mechanisms (DEMs) [44,77]. We establish encryption keys using an asymmetric system (the KEM), before switching to symmetric encryption for data transport (the DEM).

Algorithms 5 and 6 illustrate a simple Diffie–Hellman-based KEM. In Algorithm 5, Bob has already generated a long-term keypair (B, b) and published B. Alice takes B, generates an ephemeral keypair (E, e), completes the Diffie–Hellman on her side, and derives a cryptographic key K from the shared secret S. She can use K to encrypt messages to Bob[3], while E encapsulates K for transport. To decapsulate E and decrypt messages from Alice, Bob follows Algorithm 6, completing the Diffie–Hellman on his side and deriving the cryptographic key K from the shared secret S.

Algorithm 5. DH-based KEM: Alice encapsulating to Bob.

Input: Bob's public key $B \in \mathcal{G}$
Output: A symmetric encryption key $K \in \{0,1\}^n$ and an encapsulation $E \in \mathcal{G}$ of K under B

1 **function** DHEncapsulate(B)
2     $\quad (E, e) \leftarrow$ KeyPair() $\qquad$ // $E = [e]P$
3     $\quad S \leftarrow$ DH(B, e) $\qquad$ // $S = [eb]P$
4     $\quad K \leftarrow$ KDF($E \parallel S$) $\qquad$ // $K =$ KDF($E \parallel [eb]P$)
5     $\quad$ **return** (K, E)

Algorithm 6. DH-based KEM: Bob decapsulating from Alice.

Input: An encapsulation $E \in \mathcal{G}$ of a symmetric key $K \in \{0,1\}^n$ under Bob's public key $B \in \mathcal{G}$, and Bob's private key $b \in \mathbb{Z}$
Output: A symmetric encryption key $K \in \{0,1\}^n$

1 **function** DHDecapsulate(E, b)
2     $\quad S \leftarrow$ DH(E, b) $\qquad$ // $E = [e]P \implies S = [eb]P$
3     $\quad K \leftarrow$ KDF($E \parallel S$) $\qquad$ // $E = [e]P \implies K =$ KDF($[e]P \parallel [eb]P$)
4     $\quad$ **return** K

---

[3] If Alice immediately encrypts a message under K and sends the ciphertext to Bob with E, then this is "hashed ElGamal" encryption (see [1] for a full encryption scheme in this style).

Remark 1. While KEMs provide a convenient API and formalism for key establishment, that they cannot always be used as a replacement for plain-old Diffie–Hellman, especially as a component in more complicated protocols.

5   Concrete Groups and Discrete Logarithms

So far, everything has been presented in the abstract; but if we want to use any of these schemes in practice, then we need to choose a concrete group $\mathcal{G}$. As we noted in Sect. 2, the hardness of the DLP (and hence the CDHP) varies according to the representation of $\mathcal{G}$, and may fall far short of the $O(\sqrt{N})$ ideal. Here we give a very brief overview of the main candidate groups for group-based Diffie–Hellman, and DLP-based cryptography in general. We refer the reader to Guillevic and Morain's excellent survey [75] for further detail on discrete logarithm algorithms.

The DLP in prime finite fields, as used by Diffie and Hellman, is subexponential: the General Number Field Sieve [89] solves DLP instances in $\mathbb{F}_p$ in time $L_p[1/3, (64/9)^{1/3}]$.[4] In extension fields of large characteristic, or when the characteristic has a special form, the complexity is lower, while still subexponential (see [75]); in the extreme case of extension fields of tiny characteristic, the DLP is quasipolynomial in the field size [11]. These algorithms can also be used to attack DLPs in algebraic tori, which are compact representations of subgroups of $\mathbb{F}_q^\times$ which offer smaller key sizes and efficient arithmetic [90, 120].

Elliptic curves have long been recognised by number theorists as a generalization of the multiplicative group (indeed, both the multiplicative and additive groups can be seen as degenerate elliptic curves; see e.g. [34, Sect. 9]). Once Diffie and Hellman had proposed their protocol in a multiplicative group, then, it was perhaps only a matter of time before number theorists proposed elliptic-curve Diffie–Hellman; and within a decade Miller [100] and Koblitz [81] did just this, independently and almost simultaneously. The subexponential finite-field DLP algorithms do not apply to general elliptic curves, and so far we know of no algorithm with complexity better than $O(\sqrt{N})$ for the DLP in a general prime-order elliptic curve. Indeed, the only way we know to make use of the geometric structure is to run generic $O(\sqrt{N})$ algorithms on equivalence classes modulo ± 1, but this only improves the running time by a factor of roughly $\sqrt{2}$ [18]. We can do better for elliptic curves defined over some extension fields [72], and for some small special classes of curves [10, 62, 98, 126] (notably pairing-friendly curves); but in the more than thirty years since Miller and Koblitz introduced elliptic curve cryptography, this $\sqrt{2}$ speedup represents the only real non-quantum algorithmic improvement for the general elliptic-curve DLP.[5]

---

[4] Recall that $L_X[\alpha, c] = \exp((c + o(1))(\log X)^\alpha (\log \log X)^{1-\alpha})$.

[5] At least, it is the only improvement as far as asymptotic complexity is concerned: implementation and distribution have improved substantially. It is, nevertheless, quite dumbfounding that in over thirty years of cryptographically-motivated research, we have only scraped a tiny constant factor away from the classical asymptotic complexity of the DLP in a generic prime-order elliptic curve over a prime finite field.

Going beyond elliptic curves, a range of other algebraic groups have been proposed for use in cryptography. Koblitz proposed cryptosystems in Jacobians of hyperelliptic curves as an obvious generalization of elliptic curves [82]. Others have since suggested Jacobians of general algebraic curves, and abelian varieties [104,117]; but as the genus of the curve (or the dimension of the abelian variety) grows, index-calculus algorithms become more effective, quickly outperforming generic DLP algorithms. At best, the DLP for curves of fixed genus ≥ 3 is exponential, but easier than $O(\sqrt{N})$ [50,71,73,127]; at worst, as the genus and field size both tend to infinity, the DLP becomes subexponential [56]. Déchàne proposed generalized Jacobians as a bridge between elliptic-curve and finite-field cryptography [46], but these offer no constructive advantage [68].

The groups mentioned above are all *algebraic groups*: elements are represented by tuples of field elements, and group operations are computed using polynomial formulæ. Algebraic groups are well-suited to efficient computation on real-world computer architectures, but they are not the only such groups: another kind consists of class groups of number fields. Buchmann and Williams proposed Diffie–Hellman schemes in class groups of quadratic imaginary orders [32], leading to a series of DLP-based cryptosystems set in more general rings (see [31] for a survey); but ultimately these are all vulnerable to subexponential index-calculus attacks.

In the classical world, therefore, elliptic curves over $\mathbb{F}_p$ and $\mathbb{F}_{p^2}$ and genus-2 Jacobians over $\mathbb{F}_{p^2}$ present the hardest known DLP instances with respect to group size (and hence key size). Elliptic curves over prime fields have become, in a sense, the gold standard to which all other groups are compared.

6   Concrete Hardness and Security Levels

It is important to note that our understanding of DLP hardness is not just a matter of theory and sterile asymptotics; all of the algorithms above are backed by a tradition of experimental work. Recently discrete logarithms have been computed in 768-bit general and 1024-bit special prime fields [63,80], and in 112-bit and 117-bit binary elliptic curve groups [17,136]. A table of various record finite field discrete logarithm computations can be found at [74].

These computations give us confidence in defining cryptographic parameters targeting concrete security levels against classical adversaries. For example, it is generally accepted that DLP-based cryptosystems in $\mathbb{F}_p^\times$ with $\log_2 p \approx 3072$ or in $\mathcal{E}(\mathbb{F}_p)$ with $\log_2 p \approx 256$ for a well-chosen $\mathcal{E}$ should meet a classical approximate 128-bit security level: that is, a classical adversary equipped with current algorithms should spend around 2^{128} computational resources to break the system with non-negligable probability.

For quantum adversaries, we know that DLPs can be solved in polynomial time—but we still know relatively little about the concrete difficulty and cost of mounting quantum attacks against DLP-based cryptosystems, let alone against candidate post-quantum systems. For example, we mentioned above that the current record for Shor's factoring algorithm is 21; but to our knowledge, Shor's algorithm for discrete logarithms has never been implemented. Roetteler, Naehrig,

Svore and Lauter have estimated the quantum resources required to compute elliptic-curve discrete logarithms [118], which is an important first step.

The situation becomes even murkier for the quantum algorithms we will meet below, including the Kuperberg, Regev, Tani, and Childs–Jao–Soukharev algorithms. We have asymptotic estimates, but no concrete estimates (or real data points, for that matter). It is not clear what the most useful performance metrics are for these algorithms, or how to combine those metrics with classical ones to estimate overall problem difficulty.

For this reason, we will refrain from giving any concrete security estimates or recommendations for key-lengths for the postquantum systems in the second half of this article. We look forward to detailed theoretical estimates along the lines of [118], and to the eventual development of quantum computers sufficiently large to implement these algorithms and get some actual cryptanalysis done.

7 Modern Elliptic-Curve Diffie–Hellman

At first glance, elliptic-curve cryptography is just finite-field cryptography with a different algebraic group seamlessly swapped in[6], and no theoretical modification. But Miller's original article [100] ends with an interesting observation that departs from the multiplicative group perspective:

> Finally, it should be remarked, that even though we have phrased everything in terms of points on an elliptic curve, that, for the key exchange protocol (and other uses as one-way functions), that only the x-coordinate needs to be transmitted. The formulas for multiples of a point cited in the first section make it clear that the x-coordinate of a multiple depends only on the x-coordinate of the original point.

Miller is talking about elliptic curves in Weierstrass models $y^2 = x^3 + ax + b$, where $-(x, y) = (x, -y)$, so x-coordinates correspond to group elements modulo sign. The mapping $(m, x(P)) \mapsto x([m]P)$ is mathematically well-defined, because every $[m]$ commutes with $[-1]$; but it can also be computed efficiently.

In Diffie–Hellman, then, instead of using

$$A = [a]P, \qquad\qquad B = [b]P, \qquad\qquad S = [ab]P$$

Miller is proposing that we use

$$A = \pm[a]P \qquad\qquad B = \pm[b]P \qquad\qquad S = \pm[ab]P$$
$$= x([a]P), \qquad\qquad = x([b]P), \qquad\qquad = x([ab]P)$$

Clearly, we lose nothing in terms of security by doing this: the x-coordinate CDHP reduces immmediately to the CDHP in the elliptic curve. Given a CDHP

---

[6] Not entirely seamlessly: some operations, like hashing into $\mathcal{G}$, become slightly more complicated when we pass from finite fields to elliptic curves (see [108]).

oracle for $\mathcal{E}$, we can compute $\pm[ab]P$ from $(\pm P, \pm[a]P, \pm[b]P)$ by choosing arbitrary lifts to signed points on $\mathcal{E}$ and calling the oracle there; conversely, given an x-coordinate CDHP oracle, we can solve CDHP instances on $\mathcal{E}$ by projecting to the x-line, calling the oracle there, and then guessing the sign on S.

The idea of transmitting only the x-coordinates may seem advantageous in terms of reducing bandwidth, but in reality, where elliptic curve point keys are systematically compressed to an x-coordinate plus a single bit to indicate the "sign" of the corresponding y-coordinate, there is little to be gained here beyond avoiding the small effort required for compression and decompression. The real practical value in Miller's idea is that working with only x-coordinates is faster, and requires less memory: $x([a]P)$ can be computed from a and $x(P)$ using fewer field operations than would be required to compute $[a]P$ from a and P.

This advantage was convincingly demonstrated by Bernstein's Curve25519 software [14], which put Miller's idea into practice using carefully selected curve parameters optimized for Montgomery's ladder algorithm, which computes the pseudo-scalar multiplications $x(P) \mapsto x([m]P)$ using a particularly efficient and regular differential addition chain [42,101]. The result was not only a clear speed record for Diffie–Hellman at the 128-bit security level, but a new benchmark in design for key exchange software. Curve25519 is now the Diffie–Hellman to which all others are compared in practice.

The elliptic curve cryptosystems that were standardized in the 1990s and early 2000s, such as the so-called NIST [106] and Brainpool [92] curves, are based on full elliptic curve arithmetic and are not optimized for x-only arithmetic. More recently, Curve25519 and similar systems have been standardized for future internet applications [88]. These systems are also preferred in newer practical applications, such as the Double Ratchet algorithm used for key management within the Signal protocol for end-to-end encrypted messaging [110].

In theory, Miller's idea of working modulo signs (or, more generally, automorphisms) extends to any algebraic group.[7] For example, the quotients of Jacobians of genus-2 curves by ± 1 are Kummer surfaces. Under suitable parametrizations, these have highly efficient pseudo-scalar multiplications [70], which have been used in high-speed Diffie–Hellman implementations [16,115].

While the x-only approach to elliptic curve Diffie–Hellman is particularly useful in practice, it also highlights an important theoretical point: on a formal level, Diffie–Hellman does *not* require a group structure.[8] By this, we mean that

---

[7] While quotienting by ± 1 is useful in curve-based cryptosystems, it is counterproductive in multiplicative groups of finite fields. There, the pseudo-scalar multiplication is $(m, P + 1/P) \mapsto P^m + 1/P^m$; computing this is slightly slower than computing simple exponentiations, and saves no space at any point.

[8] Buchmann, Scheidler, and Williams later proposed what they claimed was the first group-less key exchange in the infrastructure of real quadratic fields [29]. Mireles Morales investigated the infrastructure in the analogous even-degree hyperelliptic function field case [102], relating it to a subset of the class group of the field; in view of his work, it is more appropriate to describe infrastructure key exchange as group-based. In any case, coming nearly a decade after Miller, this would not have been the first non-group Diffie–Hellman.

the group law never explicitly appears in the protocol—and this is precisely why Diffie–Hellman works on elliptic x-coordinates, where there is no group law (but where there *are* maps induced by scalar multiplication).

The group structure is still lurking behind the scenes, of course. It plays several important roles:

1. *Correctness.* The group law gives an easy proof that the pseudo-scalar multiplication operations $(m, x(P)) \mapsto x([m]P)$ exist and commute.
2. *Efficiency.* The group law induces biquadratic relations on x-coordinates that we use to efficiently compute pseudo-scalar multiplications using suitable differential addition chains [15].
3. *Security.* The hardness of the CDHP in the full group underwrites the hardness of the x-coordinate CDHP.

Remark 2. Can we do without a group entirely? Heading into the pure abstract, we can consider a Diffie–Hellman protocol with minimal algebraic structure. For example, we could take a set $\mathcal{X}$ of public keys in place of the group $\mathcal{G}$, and sample private keys from a set $\mathcal{F}$ of functions $\mathcal{X} \to \mathcal{X}$ defined by the property[9]

$$(a \circ b)(P) = (b \circ a)(P) \quad \text{for all} \quad a, b \in \mathcal{F}.$$

The associated Diffie–Hellman protocol is then defined by

$$A = a(P), \qquad B = b(P), \qquad S = a(b(P)) = b(a(P)).$$

We need $\mathcal{F}$ to be large enough to prevent brute force attacks on S; we must be able to efficiently sample functions from $\mathcal{F}$, and evaluate them at elements of $\mathcal{X}$; and we need to justify the hardness of the associated CDHP. An algebraic structure on $\mathcal{F}$ may not be strictly necessary to ensure all of this, but it certainly makes life easier.

8 Principal Homogeneous Spaces

At the time of writing, the closest thing we have to a post-quantum analogue of Diffie–Hellman comes from *isogeny-based cryptography*, whose origins go back to Couveignes' "Hard Homogeneous Spaces" manuscript [43]. This went unpublished for ten years, before appearing online more or less at the same time as its ideas were independently rediscovered by Rostovtsev and Stolbunov [119].

Couveignes' work is a convenient framework for reasoning about isogeny-based cryptosystems: the hard detail on class groups and isogeny classes is abstracted away into groups acting on sets. We warn the reader that from now on we will mostly be working with abelian groups written *multiplicatively*, which we denote by $\mathfrak{G}$ in accordance with the convention from Sect. 1.

---

[9] If we require this property to hold for *all* P in $\mathcal{X}$, then $\mathcal{F}$ is a *commutative magma*. Diffie–Hellman protocols where $\mathcal{F}$ is equipped with a semigroup or semiring structure have been investigated [97], though the results are only of theoretical interest.

Recall that a (left) *action* of a group $\mathfrak{G}$ on a set $\mathcal{X}$ is a mapping $\mathfrak{G} \times \mathcal{X} \to \mathcal{X}$, written $(\mathfrak{g}, P) \mapsto \mathfrak{g} \cdot P$, compatible with the group operation in $\mathfrak{G}$: that is,

$$\mathfrak{g}_1 \cdot (\mathfrak{g}_2 \cdot P) = (\mathfrak{g}_1 \mathfrak{g}_2) \cdot P \quad \text{for all } \mathfrak{g}_1, \mathfrak{g}_2 \in \mathfrak{G} \text{ and } P \in \mathcal{X}.$$

In our case $\mathfrak{G}$ is abelian, so $\mathfrak{g}_1 \cdot (\mathfrak{g}_2 \cdot P) = \mathfrak{g}_2 \cdot (\mathfrak{g}_1 \cdot P)$ for all $\mathfrak{g}_1, \mathfrak{g}_2$, and P.

Definition 3 (PHS). *A principal homogeneous space (PHS) for an abelian group $\mathfrak{G}$ is a set $\mathcal{X}$ equipped with a simple, transitive action of $\mathfrak{G}$: that is, for any P and Q in $\mathcal{X}$, there is a unique $\mathfrak{g}$ in $\mathfrak{G}$ such that $Q = \mathfrak{g} \cdot P$. Equivalently, for every P in $\mathcal{X}$, the map $\varphi_P : \mathfrak{G} \to \mathcal{X}$ defined by $\mathfrak{g} \mapsto \mathfrak{g} \cdot P$ is a bijection.*

Example 1. The trivial example of a PHS is a group acting on itself via its own group operation: that is, $\mathcal{X} = \mathfrak{G}$, with $\mathfrak{g} \cdot \mathfrak{a} = \mathfrak{g} \mathfrak{a}$.

Example 2. The classic first example of a nontrivial PHS is a vector space acting by translation on its underlying affine space.

Example 2 illustrates a classic informal definition: a PHS is a group whose identity element has been forgotten (mislaid, not omitted). Affine spaces have no distinguished "origin"; on the other hand, as soon as one is (arbitrarily) chosen, then each point defines an implicit displacement vector, and we can identify the affine space with a vector space. More generally, for each P in $\mathcal{X}$, if we define $\varphi_P(\mathfrak{g}_1)\varphi_P(\mathfrak{g}_2) := \varphi_P(\mathfrak{g}_1 \mathfrak{g}_2)$ then we get a well-defined group structure on $\mathcal{X}$; in fact, we get a different group structure for each choice of P. The idea therefore is not so much that the identity element has been forgotten, yet might still be remembered; it is rather that every single element is an equally plausible identity.

Example 3. Let $\mathcal{X}$ be the set of points on a curve $\mathcal{C}$ of genus 1, and let $\mathfrak{G} = \mathrm{Pic}^0(\mathcal{C})$ be the group of degree-0 divisor classes on $\mathcal{C}$. By the Riemann–Roch theorem, for every class $[D]$ in $\mathrm{Pic}^0(\mathcal{C})$ and point P on $\mathcal{C}$, there exists a unique P_D on $\mathcal{C}$ such that $[D] = [P_D - P]$. We therefore have an explicit action of $\mathfrak{G}$ on $\mathcal{X}$, defined by $[D] \cdot P = P_D$. If we fix a choice of distinguished "base point" O in $\mathcal{X}$, then we can identify each class $[D]$ with the point $[D] \cdot O$, and thus, by transport of structure, we get a group law on $\mathcal{X}$. (Cognoscenti will recognise the definition of the group law on an arbitrary elliptic curve via the Picard group.)

Our final example of a PHS is fundamental in isogeny-based cryptography. It is far more complicated to define than Examples 1, 2, and 3; we will give an extremely brief description here, returning to it in greater detail in Sects. 13 and 14.

Example 4. Let q be a prime power and t an integer with $|t| \leq 2\sqrt{q}$; let $\mathcal{O}_K$ be the ring of integers of the imaginary quadratic field $K = \mathbb{Q}(\sqrt{\Delta})$, where $\Delta := t^2 - 4q$. Let $\mathcal{X}$ be the set of $\mathbb{F}_q$-isomorphism classes of elliptic curves $\mathcal{E}/\mathbb{F}_q$ whose $\mathbb{F}_q$-endomorphism ring is isomorphic to $\mathcal{O}_K$ (and where the image of the Frobenius endomorphism of $\mathcal{E}$ in $\mathcal{O}_K$ has trace t). Then $\mathcal{X}$ is a PHS under the ideal class group $\mathfrak{G} = \mathrm{Cl}(\mathcal{O}_K)$ of $\mathcal{O}_K$, with ideals acting by $\mathfrak{a} \cdot \mathcal{E} = \mathcal{E}/\mathcal{E}[\mathfrak{a}]$, where

$\mathcal{E}[\mathfrak{a}]$ is the intersection of the kernels of the endomorphisms in $\mathfrak{a}$. This PHS is central to the theory of Complex Multiplication; there is also a well-developed algorithmic theory for it, used to compute fundamental number-theoretic objects such as modular and Hilbert class polynomials (see e.g. [28]).

Example 4 highlights another view of PHSes: we can consider $\mathcal{X}$ as a version of $\mathfrak{G}$ whose structure is hidden by the maps φ_P. In this case, the elements of $\mathcal{X}$ are j-invariants, and (when the class group is sufficiently large) look like random elements of $\mathbb{F}_q$; the class group itself has no such encoding.

9 Hard Homogeneous Spaces

Let $\mathcal{X}$ be a PHS under $\mathfrak{G}$. From now on we assume we can efficiently compute group operations, evaluate actions, test equality, and hash elements of $\mathfrak{G}$ and $\mathcal{X}$. We also assume we can uniformly randomly sample elements of $\mathfrak{G}$. Figure 1 illustrates the two interesting computational problems in this setting, which Couveignes called *vectorization* (Definition 4) and *parallelization* (Definition 5).

Definition 4 (Vectorization). *The* vectorization *problem in a PHS $\mathcal{X}$ under $\mathfrak{G}$ is, given P and Q in $\mathcal{X}$, to compute the unique $\mathfrak{g}$ in $\mathfrak{G}$ such that $Q = \mathfrak{g} \cdot x$.*

Definition 5 (Parellelization). *The* parallelization *problem in a PHS $\mathcal{X}$ under $\mathfrak{G}$ is, given P, A, and B in $\mathcal{X}$, to compute the unique S in $\mathcal{X}$ such that $S = (\mathfrak{a}\mathfrak{b}) \cdot P$ where $A = \mathfrak{a} \cdot P$ and $B = \mathfrak{b} \cdot P$. (Note that then $S = \mathfrak{a} \cdot B = \mathfrak{b} \cdot A$.)*

Definition 6 (HHS). *Let $\mathcal{X}$ be a PHS under $\mathfrak{G}$. We say $\mathcal{X}$ is a* hard homogeneous space *(HHS) if the action of $\mathfrak{G}$ on $\mathcal{X}$ is efficiently computable, but the vectorization and parallelization problems are computationally infeasible.*

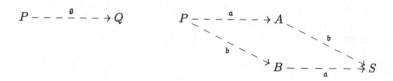

Fig. 1. Vectorization (left: finding the unique $\mathfrak{g}$ such that $Q = \mathfrak{g} \cdot P$) and parallelization (right: computing the unique S such that $S = \mathfrak{g}_B \cdot A = \mathfrak{g}_A \cdot B = (\mathfrak{g}_A \mathfrak{g}_B) \cdot P$). The dashed arrows denote actions of the unknown group elements $\mathfrak{g}$, $\mathfrak{g}_A$, and $\mathfrak{g}_B$.

The names "vectorization" and "parallelization" are intuitive in the context of Example 2: vectorization is computing the displacement vector between points P and Q in the space $\mathcal{X}$, while parallelization is completing the parallelogram with vertices P, A, and B. These are routine operations in vector and affine spaces, so the PHS of Example 2 is typically not something that we would consider an HHS. Similarly, the PHS of Example 3 is not an HHS, because we can always vectorize by formally subtracting points to form a degree-0 divisor. Couveignes suggested that the PHS of Example 4 might be an HHS—and with the current state of classical and quantum class group and isogeny algorithms, it is.

10   Cryptography in Hard Homogeneous Spaces

On a purely symbolic level, the vectorization problem $(P, \mathfrak{g} \cdot P) \mapsto \mathfrak{g}$ in a PHS bears an obvious formal resemblance to the DLP $(P, [x]P) \mapsto x$ in a group, just as the parallelization problem $(P, \mathfrak{a} \cdot P, \mathfrak{b} \cdot P) \mapsto \mathfrak{ab} \cdot P$ resembles the CDHP $(P, [a]P, [b]P) \mapsto [ab]P$. The presence of abelian groups in each problem suggests deeper connections—connections that do not necessarily exist. Indeed, we saw above that parallelization in a PHS is an implicit computation of the group law, while the Diffie–Hellman operation in a group is something completely different.

But irrespective of the relationship between parallelizations and CDHPs, this syntactical resemblance allows us to define a cryptosystem analogous to Diffie–Hellman. Algorithms 7 and 8 define Couveignes' key exchange in the HHS setting, with security depending on the hardness of parallelization.

Algorithm 7. Key generation for cryptosystems in an HHS $\mathcal{X}$ under $\mathfrak{G}$, with a fixed base point P in $\mathcal{X}$

Input: ()
Output: A private-public keypair $(Q, \mathfrak{g}) \in \mathfrak{G} \times \mathcal{X}$ s.t. $Q = \mathfrak{g} \cdot P$
1 **function** KeyPair()
2     $\mathfrak{g} \leftarrow$ Random($\mathfrak{G}$)   // $\mathfrak{g}$ is sampled uniformly at random from $\mathfrak{G}$
3     $Q \leftarrow \mathfrak{g} \cdot P$
4   **return** $(Q, \mathfrak{g})$

Algorithm 8. Diffie–Hellman in an HHS $\mathcal{X}$ under a group $\mathfrak{G}$

Input: A private key $\mathfrak{g}_A \in \mathfrak{G}$ and a public key $Q_B \in \mathcal{X}$, each generated by calls to Algorithm 7
Output: A shared secret value $S \in \mathcal{X}$
1 **function** DH($Q_B, \mathfrak{g}_A$)
2     $S \leftarrow \mathfrak{g}_A \cdot Q_B$
3   **return** S

Algorithms 7 and 8 immediately raise some important restrictions on the kinds of $\mathfrak{G}$ and $\mathcal{X}$ that we can use. The first is that we need some kind of canonical representation for elements of $\mathcal{X}$, to ensure that Alice and Bob can derive equal shared cryptographic keys from the shared secret S. This property is also important in settings where public keys are required to be unique for a given private key. We also need to be able to efficiently draw uniformly random samples from $\mathfrak{G}$; and then, given a random element of $\mathfrak{G}$, we need to be able to efficiently compute its action on arbitrary elements of $\mathcal{X}$. An alternative approach is to repeatedly randomly sample from a subset of efficiently-computable elements of $\mathfrak{G}$, building a sequence of such elements to be used as the secret key, with the action of the key being the composition of the action of its components. This approach requires an argument that the distribution of these compositions is sufficiently close to the uniform distribution on the whole of $\mathfrak{G}$. Both approaches are relevant in isogeny-based cryptography, as we will see in Sect. 14.

Many CDHP-based cryptosystems have obvious HHS analogues. We can define an HHS-based KEM (and implicitly, a hashed-ElGamal-type public key encryption scheme) along the lines of Algorithms 5 and 6, by replacing the calls to Algorithms 1 and 2 with calls to Algorithms 7 and 8, respectively. But not all DLP-based protocols have HHS-based analogues: for example, the obvious HHS analogue of Schnorr's signature scheme [121] would appear to require an efficient (decisional) parallelization algorithm in order to verify signatures.

HHS-Diffie–Hellman is *not* a natural generalization of group-Diffie–Hellman. As we noted in Sect. 7, in group-DH, we have a ring (integers modulo N) acting on the group $\mathcal{G}$; the composition operation at the heart of DH is ring multiplication, but the ring only forms a group under addition. Formally, in group-DH we only use the fact that the scalars form a commutative magma; but algorithmically, we exploit the fact that the elements form an abelian group and the scalars form a commutative ring, mapping addition in the ring onto the group law, in order to efficiently evaluate scalar multiplications using addition chains.

More concretely, we noted in Sect. 8 that the maps $\varphi_P : \mathfrak{g} \mapsto \mathfrak{g} \cdot P$ can be seen as hiding the group $\mathfrak{G}$ in $\mathcal{X}$. The parallelization $(P, \mathfrak{a} \cdot P, \mathfrak{b} \cdot P) \mapsto \mathfrak{a}\mathfrak{b} \cdot P$ can be written as $(\varphi_P(1), \varphi_P(\mathfrak{a}), \varphi_P(\mathfrak{b})) \mapsto \varphi_P(\mathfrak{a}\mathfrak{b})$; that is, parallelization computes the group law in the hidden representation of $\mathfrak{G}$ in $\mathcal{X}$ corresponding to P. From this perspective, HHS-Diffie–Hellman is a hidden version of the ridiculous key exchange where the shared secret is the product of the two public keys: obviously, without the hiding, this offers no security whatsoever.

11   Vectorization and Parallelization

To argue about the security of the schemes in Sect. 10, we must address the following questions: how hard are vectorization and parallelization? What is the relationship between these problems, and to what extent does our common intuition relating the DLP and CDHP carry over to vectorization and parallelization in the PHS setting?

It might seem excessive to require the simple and transitive action of a PHS in all this: we could relax and set up the same cryptosystems with a group $\mathfrak{G}$ acting on a set $\mathcal{X}$ in any old way. While we might lose uniqueness and/or existence of vectorizations and parallelizations, many of the arguments in this section would still go through. However, using PHSes instead of general group actions makes one thing particularly simple: the proof of random self-reducibility for vectorization and parallelization is identical to the usual arguments for groups, which we sketched in Sect. 5. More precisely: if an algorithm successfully solves vectorizations in $(\mathfrak{G}, \mathcal{X})$ with a probability of $1/M$, then we can solve *any* vectorization in $(\mathfrak{G}, \mathcal{X})$ with an expected M calls to the algorithm. Given a target vectorization $(P, Q = \mathfrak{g} \cdot P) \mapsto \mathfrak{g}$, we attempt to solve $(\mathfrak{a} \cdot P, \mathfrak{b} \cdot Q) \mapsto \mathfrak{a}\mathfrak{b}\mathfrak{g}$ for randomly chosen $\mathfrak{a}$ and $\mathfrak{b}$ in $\mathfrak{G}$; we expect to land in the subset of inputs where the algorithm succeeds within M attempts, and then recovering $\mathfrak{g}$ from $(\mathfrak{a}, \mathfrak{b}, \mathfrak{abg})$ is trivial. This means that the average- and worst-case difficulties for vectorization are equivalent; a similar argument yields the same result for parallelization.

Now, consider the relationship between vectorization and parallelization. If we can solve vectorizations $(P, \mathfrak{g} \cdot P) \mapsto \mathfrak{g}$, then we can solve parallelizations $(P, \mathfrak{a} \cdot P, \mathfrak{b} \cdot P) \mapsto \mathfrak{ab} \cdot P$, so parallelization is notionally easier than vectorization.

As we have seen, the parallelization operation $(P, \mathfrak{a} \cdot P, \mathfrak{b} \cdot P) \mapsto \mathfrak{ab} \cdot P$ acts as the group law induced on $\mathcal{X}$ by $\mathfrak{G}$ when elements are hidden by $\varphi_P : \mathfrak{g} \mapsto \mathfrak{g} \cdot P$. Given a parallelization oracle for a PHS $(\mathfrak{G}, \mathcal{X})$ with respect to one fixed base point P (call this P-*parallelization*), we can view $\mathcal{X}$ as an efficiently computable group, and thus apply any black-box group algorithm to $\mathcal{X}$. Further, given an efficient P-parallelization algorithm, the map φ_P becomes an efficiently computable group homomorphism. Even further, if we have a P-parallelization oracle for *all* P in $\mathcal{X}$, then the mapping $(\mathfrak{g}, P) \mapsto \mathfrak{g} \cdot P$ becomes an efficiently computable bilinear pairing $\mathfrak{G} \times \mathcal{X} \to \mathcal{X}$ (viewing $\mathcal{X}$ as a version of $\mathfrak{G}$ hidden by one φ_O).

The efficient homomorphism $\varphi_P : \mathfrak{G} \to \mathcal{X}$ implied by a P-parallelization oracle is of course an isomorphism (because $\#\mathcal{X} = \#\mathfrak{G}$), but its inverse is not necessarily efficient[10]: if it were, then we could solve vectorizations $(P, \mathfrak{g} \cdot P) \mapsto \mathfrak{g}$ because $\mathfrak{g} = \varphi_P^{-1}(\mathfrak{g} \cdot P)$. Conversely, if we can vectorize with respect to P, then we can invert φ_P: the preimage $\varphi_P^{-1}(T)$ of any T in $\mathcal{X}$ is the vectorization of T with respect to $\varphi_P(1)$. Parallelization therefore yields an efficient isomorphism in one direction, while vectorization yields the inverse as well.

In the case where a group $\mathcal{G}$ has prime order, we can use a CDHP oracle to view $\mathcal{G}$ as a *black-box field* in the sense of Boneh and Lipton [21]. Given a base point P in $\mathcal{G}$, each element $[a]P$ of $\mathcal{G}$ is implicitly identified with its discrete logarithm a. The Diffie–Hellman operation $(P, [a]P, [b]P) \mapsto [ab]P$ becomes an implicit multiplication, allowing us to view $\mathcal{G}$ as a model of $\mathbb{F}_N$, and thus to apply various subexponential and polynomial-time reductions from the DLP to the CDHP in $\mathcal{G}$ (as we noted in Sect. 3). A parallelization oracle for $(\mathfrak{G}, \mathcal{X})$, however, only allows us to view $\mathcal{X}$ as a black-box group, not a black-box field; we therefore have no equivalent of the den Boer or Maurer reductions in the HHS setting.

The separation between vectorization and parallelization therefore seems more substantial than the somewhat thin and rubbery separation between the DLP and CDHP. However, we would still like to have some upper bounds for the hardness of these problems. For vectorization, we can give some algorithms.

In the classical setting, Couveignes noted that Shanks' baby-step giant-step (BSGS) and Pollard's probabilistic algorithms for DLPs in groups extend to vectorizations in PHSes. Algorithm 9 is a BSGS analogue for a PHS $\mathcal{X}$ under $\mathfrak{G}$.[11] Given P and $Q = \mathfrak{g} \cdot P$ in $\mathcal{X}$ and a generator $\mathfrak{e}$ for (a subgroup of) $\mathfrak{G}$, Algorithm 9

---

[10] The term *one-way group action* is used for the HHS framework in [35] and [23]. This hints at a more general setting, where actions are not necessarily simple or transitive.

[11] Algorithm 9 becomes the usual BSGS for DLPs in $\mathfrak{G} = \langle \mathfrak{e} \rangle$ if we let $\mathcal{X} = \mathfrak{G}$ (with the group operation as the action), let $P = 1$, and let Q be the discrete log target.

computes an exponent x such that $\mathfrak{g} = \mathfrak{e}^x$, if it exists (if $\mathfrak{g}$ lies outside the subgroup generated by $\mathfrak{e}$, then the algorithm will fail and return $\bot$).

Algorithm 9. BSGS for a PHS $\mathcal{X}$ under (a subgroup of) a group $\mathfrak{G}$.

Input: Elements P and Q in $\mathcal{X}$, and an element $\mathfrak{e}$ in $\mathfrak{G}$
Output: x such that $Q = \mathfrak{e}^x \cdot P$, or $\bot$

1 **function** BSGS($P, Q, \mathfrak{e}$)
2     $\beta \leftarrow \lceil \sqrt{\#\langle \mathfrak{e} \rangle} \rceil$ // May be replaced with an estimate if $\#\langle \mathfrak{e} \rangle$ not known
3     $S \leftarrow \{\}$ // Hash table: keys in $\mathcal{X}$, values in $\mathbb{Z}/N\mathbb{Z}$
4     $T \leftarrow P$
5   **for** i in $[0, \ldots, \beta]$ **do**
6        $S[T] \leftarrow i$
7        $T \leftarrow \mathfrak{e} \cdot T$
8     $(T, \mathfrak{c}) \leftarrow (Q, \mathfrak{e}^\beta)$
9   **for** j in $[0, \ldots, \beta]$ **do**
10   **if** $T \in S$ **then**
11           $i \leftarrow S[T]$
12   **return** $i - j\beta$ // $\mathfrak{e}^{j\beta} \cdot Q = \mathfrak{e}^i \cdot P$
13        $T \leftarrow \mathfrak{c} \cdot T$
14   **return** $\bot$

Vectorization in a PHS $\mathcal{X}$ under $\mathfrak{G}$ can always be solved classically in time and space $\tilde{O}(\sqrt{\#\mathfrak{G}})$ using Algorithm 9 and random self-reducibility, provided a generator of a polynomial-index subgroup of $\mathfrak{G}$ is known. Algorithm 9 does more than what is required: it returns not just the desired vectorization $\mathfrak{g}$, but the discrete logarithm of $\mathfrak{g}$ with respect to $\mathfrak{e}$. This betrays the fact that Algorithm 9 is just a black-box group algorithm operating on a hidden version of $\mathfrak{G}$.

Pollard's algorithms also generalize easily to the HHS setting, because we can compute the pseudorandom walks using only translations, or "shifts", by group elements. These translations in the group setting can be replaced by actions by the same elements in the HHS setting. The space complexity of vectorization can thus be reduced to as little as $O(1)$ for the same time complexity.

Moving to the quantum setting: despite its resemblance to the DLP, vectorization cannot be solved with Shor's algorithm. In fact, vectorization is an instance of the *abelian hidden shift problem* [134]: given functions f and g such that $f(x \cdot s) = g(x)$ for all x and some "shift" s, find s. The hidden shift instance corresponding to the vectorization instance $(P, Q = \mathfrak{g} \cdot P)$ has $f = \varphi_P : \mathfrak{G} \to \mathcal{X}$, $g = \varphi_Q : \mathfrak{G} \to \mathcal{X}$, and $s = \mathfrak{g}$. Kuperberg reduces the abelian hidden shift problem to an instance of the dihedral hidden subgroup problem, which is then solved with a quantum algorithm with a query complexity of $L_N[1/2, c]$, where $c = \sqrt{2}$ according to [37]. Kuperberg's original algorithm [86] uses subexponential space; Regev's simpler algorithm [114] uses polynomial quantum space; and

Kuperberg's most recent work [87] uses linear quantum space, but subexponential classical space. More detailed perspectives on these algorithms in the context of the isogeny class HHS appear in [23,35,59].

12   The Difficulty of Exploiting Subgroup Structures

Moving back to the abstract: when we think about DLPs, in black-box or in concrete groups, we implicitly and systematically apply the Pohlig–Hellman–Silver algorithm to reduce to the prime-order case. It is interesting to note that for PHSes, no such reduction is known: it appears difficult to exploit the subgroup structure of $\mathfrak{G}$ when solving vectorization problems in $\mathcal{X}$.[12]

Algorithm 10 presents the Pohlig–Hellman–Silver algorithm for discrete logarithms in a group $\mathfrak{G}$ whose order has known factorization $N = \prod_i N_i^{e_i}$. Line 9 applies a DLP-solving algorithm (like BSGS, Pollard ρ, or a specialized algorithm for a concrete group) in the order-N_i subgroup of $\mathfrak{G}$. If the factorization is complete and the N_i are prime, then the global DLP is reduced to a polynomial number smaller prime-order sub-DLPs.

Algorithm 10. Pohlig–Hellman–Silver for a group $\mathfrak{G}$ whose order has known (partial) factorization.

Input: An element $\mathfrak{e}$ of $\mathfrak{G}$, a $\mathfrak{g}$ in $\langle \mathfrak{e} \rangle$, and $((N_1, e_1), \ldots, (N_n, e_n))$ such that
$\#\mathfrak{G} = N = \prod_{i=1}^{n} N_i^{e_i}$, with the N_i pairwise coprime and the $e_i > 0$.
Output: x such that $\mathfrak{g} = \mathfrak{e}^x$

1 **function** PohligHellman($\mathfrak{e}, \mathfrak{g}, ((N_1, e_1), \ldots, (N_n, e_n))$)
2   **for** $1 \leq i \leq n$ **do**
3        $\mathfrak{e}_i \leftarrow \mathfrak{e}^{N/N_i^{e_i}}$
4        $\mathfrak{g}_i \leftarrow \mathfrak{g}^{N/N_i^{e_i}}$
5        $x_i \leftarrow 0$
6   **for** j *in* $(e_i - 1, \ldots, 0)$ **do**
7           $\mathfrak{s} \leftarrow \mathfrak{e}_i^{(N_i)^j}$ // $\mathfrak{s}$ is in the order-N_i subgroup
8           $t \leftarrow \mathfrak{s}^{-x_i} \cdot \mathfrak{g}_i^{(N_i)^j}$ // t is in the order-N_i subgroup
9           $y \leftarrow \log_{\mathfrak{s}}(t)$ // Use e.g. baby-step giant-step
10          $x_i \leftarrow x_i N_i + y$
11       $x \leftarrow$ CRT($\{(x_i, N_i^{e_i}) : 1 \leq i \leq n\}$)
12   **return** x

The key steps involve producing the subgroup DLPs. Lines 3 and 4 project the DLP instance $(\mathfrak{g}, \mathfrak{h})$ into the order-$N_i^{e_i}$ subgroup of $\langle \mathfrak{g} \rangle$. Lines 7 and 8 then

---

[12] It might seem odd that some black-box group algorithms like BSGS and Pollard ρ adapt easily to PHSes, but not others like Pohlig–Hellman. But looking closer, BSGS and Pollard ρ in groups only require translations, and not a full group law. We can therefore see BSGS and Pollard ρ not as black-box group algorithms, but rather as black-box PHS algorithms that are traditionally applied with $\mathcal{X} = \mathfrak{G}$.

produce a DLP instance in the order-N_i subgroup. This is always done by exponentiation by $N/N_i^{e_i}$ and N_i; indeed, this is the only way that the factors N_i are used in the algorithm.

In the PHS setting, subgroup DLPs should be replaced with subgroup vectorizations. Line 9 could be replaced with a call to Algorithm 9, using $\mathfrak{e}^{N_i^{e_i-1}}$ as the subgroup generator; the problem is to produce a vectorization instance in a sub-PHS acted on by the corresponding subgroup. We cannot naively concatenate "$\cdot P$" (or "$\cdot Q$") to most lines of the algorithm to turn group elements and operations into PHS elements and operations: Line 4, for example, would require computing $\mathfrak{g}^{N/N_i^{e_i}} \cdot P$ from $Q = \mathfrak{g} \cdot P$, P, and $\mathfrak{g}$, but this amounts to an iterated parallelization— and parallelization is supposed to be hard in an HHS.

A thorough investigation of the possibility and difficulty of exploiting subgroup structures for vectorization and parallelization would require working with subgroup actions on quotient spaces; we do not have room to discuss this here.[13] We note, however, that in some protocols a limited number of exploitable parallelizations are provided by the protocol itself, as in the group-based protocols subject to Cheon's attack [36], and this should have some consequences for the security of any HHS analogues of these protocols.

13   A Quick Introduction to Isogenies

This section provides enough background on isogenies and endomorphisms of elliptic curves to make sense of the HHS from Example 4 before we describe cryptosystems based on it in Sect. 14. We also want to fill in some background on supersingular curves before we need it in Sect. 15. We assume a basic familiarity with the arithmetic of elliptic curves; readers familiar with isogenies and isogeny graphs can safely skip this section. As a mathematical reference, we suggest [125] and [84]; for greater detail focused on the cryptographic use case, see [58].

We want to talk about relationships between elliptic curves over a fixed finite field $\mathbb{F}_q$, where q is a power of some prime p. We can assume that $p \neq 2$ or 3, to simplify, though the theory applies to those cases as well. We will work with elliptic curves as short Weierstrass models $\mathcal{E} : y^2 = x^3 + ax + b$, with a and b in $\mathbb{F}_q$: in practice we might compute using other curve models (many isogeny-based cryptosystem implementations have preferred Montgomery arithmetic [42]), but since we end up working with curves up to $\mathbb{F}_q$-isomorphism, and every curve is $\mathbb{F}_q$-isomorphic to a short Weierstrass model, we lose nothing in restricting to this simple and universal curve shape in this article. The m-torsion $\mathcal{E}[m]$ of $\mathcal{E}$ is the subgroup of points P such that $[m]P = 0_{\mathcal{E}}$.

---

[13] Biasse, Jacobson, and Iezzi [19] have made some preliminary steps in this direction, and claim a classical complexity of $O(\sqrt{N/M})$ for vectorization when $\mathfrak{G}$ contains a subgroup of order M. However, their algorithm assumes we can correctly guess the subgroup orbits of the vectorization targets—and this is a problem for which we have no solution that improves on exhaustive search (or vectorization). When run as a probabilistic vectorization algorithm, therefore, their algorithm runs in time $O(\sqrt{MN})$, which is actually a factor-of-$\sqrt{M}$ slowdown over BSGS.

A *homomorphism* $\phi : \mathcal{E} \to \mathcal{E}'$ is, by definition[14], a rational map such that $\phi(0_\mathcal{E}) = 0_{\mathcal{E}'}$. Homomorphisms induce homomorphisms on groups of points [125, Sect. III.4], but not every homomorphism of groups of points is induced by a homomorphism of curves. An $\mathbb{F}_q$-homomorphism is one that is defined over $\mathbb{F}_q$: that is, the rational functions defining it as a rational map have coefficients in $\mathbb{F}_q$. Every homomorphism here will be defined over $\mathbb{F}_q$, unless explicitly stated otherwise.

Isogenies are nonzero homomorphisms of elliptic curves. If there is an isogeny from $\mathcal{E}$ to $\mathcal{E}'$, then we say that $\mathcal{E}$ and $\mathcal{E}'$ are *isogenous*. We will see below that for each isogeny $\mathcal{E} \to \mathcal{E}'$ there is a dual isogeny $\mathcal{E}' \to \mathcal{E}$, so isogeny is an equivalence relation on elliptic curves.

Isomorphisms are invertible homomorphisms. The j-invariant of a curve $\mathcal{E}$: $y^2 = x^3 + ax + b$ is $j(\mathcal{E}) = 1728 \cdot \frac{4a^3}{4a^3 + 27b^2}$; two curves $\mathcal{E}$ and $\mathcal{E}'$ are $\overline{\mathbb{F}}_q$-isomorphic if and only if $j(\mathcal{E}) = j(\mathcal{E}')$. We need to work with the stronger notion of $\mathbb{F}_q$-isomorphism, where the j-invariant does not tell the whole story. Curves that are $\overline{\mathbb{F}}_q$-isomorphic but not $\mathbb{F}_q$-isomorphic are called *twists*. The most important example is the *quadratic twist*, which is isomorphic over $\mathbb{F}_{q^2}$ but not $\mathbb{F}_q$: the quadratic twist of $\mathcal{E} : y^2 = x^3 + ax + b$ is $\mathcal{E}' : v^2 = u^3 + \mu^2 au + \mu^3$, where μ is any nonsquare in $\mathbb{F}_q$ (the choice of nonsquare makes no difference up to $\mathbb{F}_q$-isomorphism, which is why we say *the* rather than *a* quadratic twist). The isomorphism $\tau : \mathcal{E} \to \mathcal{E}'$ is defined by $(x, y) \mapsto (u, v) = (\mu x, \mu^{3/2} y)$; this is clearly not defined over $\mathbb{F}_q$, yet $j(\mathcal{E}) = j(\mathcal{E}')$. The quadratic twist of a curve $\mathcal{E}$ is its only twist, up to $\mathbb{F}_q$-isomorphism, unless $j(\mathcal{E}) = 0$ or 1728 (in which case there may be four or two more twists, respectively). Specifying an $\mathbb{F}_q$-isomorphism class therefore comes down to specifying a j-invariant and a choice of twist.

Endomorphisms are homomorphisms from a curve to itself. The endomorphisms of a given curve $\mathcal{E}$ form a ring $\mathrm{End}(\mathcal{E})$, with the group law on $\mathcal{E}$ inducing addition of endomorphisms and composition of endomorphisms corresponding with multiplication. The structure of the set of isogenies from $\mathcal{E}$ to other curves is deeply connected to the structure of $\mathrm{End}(\mathcal{E})$, and vice versa.

The scalar multiplication maps $[m]$ are endomorphisms, so $\mathrm{End}(\mathcal{E})$ always contains a copy of $\mathbb{Z}$. It also includes the *Frobenius* endomorphism $\pi : (x, y) \mapsto (x^q, y^q)$, which satisfies the quadratic equation $P(X) := X^2 - tX + q = 0$ for some integer t in the *Hasse interval* $[-2\sqrt{q}, 2\sqrt{q}]$; we call t the *trace* of Frobenius (and of $\mathcal{E}$). Since points in $\mathcal{E}(\mathbb{F}_q)$ are precisely the points fixed by π, we have $\#\mathcal{E}(\mathbb{F}_q) = P(1) = q + 1 - t$. If $\mathcal{E}'$ is the quadratic twist of $\mathcal{E}$ and we pull back the Frobenius on $\mathcal{E}'$ to an endomorphism on $\mathcal{E}$ via the twisting isomorphism, then the result is $-\pi$, so the trace of $\mathcal{E}'$ is the negative of the trace of $\mathcal{E}$.

Now consider the set of all elliptic curves over $\mathbb{F}_q$. Tate's theorem tells us that two elliptic curves are $\mathbb{F}_q$-isogenous if and only if they have the same trace (and

---

[14] An elliptic curve is by definition a pair $(\mathcal{E}, 0_\mathcal{E})$, where $\mathcal{E}$ is a curve of genus 1 and $0_\mathcal{E}$ is a distinguished point on $\mathcal{E}$ (which becomes the identity element of the group of points; cf. Example 3); so it makes sense that a morphism $(\mathcal{E}, 0_\mathcal{E}) \to (\mathcal{E}', 0_{\mathcal{E}'})$ in the category of elliptic curves should be a mapping of algebraic curves $\mathcal{E} \to \mathcal{E}'$ preserving the distinguished points, that is, mapping $0_\mathcal{E}$ onto $0_{\mathcal{E}'}$.

hence the same number of rational points). This means that the set of all elliptic curves is partitioned into $\mathbb{F}_q$-isogeny classes, indexed by the integers in the Hasse interval (via the trace). Since the trace of a curve over $\mathbb{F}_q$ is the negative of the trace of its quadratic twist, and the quadratic twist is generally the only twist, we can use the j-invariant to uniquely identify elements of the isogeny class of trace $t \neq 0$ up to $\mathbb{F}_q$-isomorphism, even though j normally only classifies curves up to $\overline{\mathbb{F}}_q$-isomorphism. We can handle $j = 0$ and 1728 as rare special cases, but for the case $t = 0$ we need to be more careful.

Now let us focus on a single $\mathbb{F}_q$-isogeny class. The isogeny class immediately breaks up into a union of $\mathbb{F}_q$-isomorphism classes. The modern way of looking at an $\mathbb{F}_q$-isogeny class is as a graph, with $\mathbb{F}_q$-isomorphism classes of curves for vertices, and $\mathbb{F}_q$-isomorphism classes of isogenies for edges (isogenies $\phi_1 : \mathcal{E}_1 \to \mathcal{E}_1'$ and $\phi_2 : \mathcal{E}_2 \to \mathcal{E}_2'$ are isomorphic if there are isomorphisms $\iota : \mathcal{E}_1 \to \mathcal{E}_2$ and $\mathcal{E}_1' \to \mathcal{E}_2'$ such that $\phi_2 \circ \iota = \iota' \circ \phi_1$).

Tate's theorem is not constructive, so we generally don't know how to get from one point to another in an isogeny graph. The difficulty of computing a path representing an unknown isogeny between given elliptic curves in the same isogeny class is the source of most hard problems in isogeny-based cryptography.

To take a closer look at the structure of isogeny graphs we need to classify isogenies, and to break them down to into fundamental components. Our main tool for this is the *degree*. Since an isogeny $\phi : \mathcal{E} \to \mathcal{E}'$ is defined by nonconstant rational maps, it induces an extension $\phi^\# : \mathbb{F}_q(\mathcal{E}') \hookrightarrow \mathbb{F}_q(\mathcal{E})$ of function fields; the degree $\deg(\phi)$ of ϕ is defined to be the degree of that extension. (We extend the definition of degree to homomorphisms by defining the degree of zero maps to be 0.) If $\phi : \mathcal{E} \to \mathcal{E}'$ and $\phi' : \mathcal{E}' \to \mathcal{E}''$ are isogenies, then $\deg(\phi' \circ \phi) = \deg \phi \cdot \deg \phi'$. Two examples are particularly important: $\deg[m] = m^2$, and $\deg \pi = q$. If $\deg \phi = d$, then we say that ϕ is a *d-isogeny*.

Another important quality of isogenies is *(in)separability*, which we define according to the (in)separability of the corresponding function field extension. For our purposes, the purely inseparable isogenies are all iterated compositions of p-powering $(x, y) \mapsto (x^p, y^p)$ (such as Frobenius); these can be factored out of any other isogeny, and then what remains is separable.

Suppose S is a finite subgroup of $\mathcal{E}(\overline{\mathbb{F}}_q)$. Now S must include 0, and it is also fixed by $[-1]$; so S is determined precisely by the x-coordinates of its nonzero elements. We can therefore encode S as a polynomial $F_S(X) = \prod_P (X - x(P))$, where the product runs over the nonzero points P of S in such a way that P is included iff $-P$ is not. The subgroup S is defined over $\mathbb{F}_q$ if and only if the polynomial F_S has coefficients in $\mathbb{F}_q$.

Being homomorphisms, isogenies have kernels. The kernel of an isogeny $\phi : \mathcal{E} \to \mathcal{E}'$ is always a finite subgroup of $\mathcal{E}$. If ϕ is separable, then $\#\ker \phi = \deg \phi$. The points of $\ker \phi$ are generally defined over an extension of $\mathbb{F}_q$, but $\ker \phi$ can be encoded as the *kernel polynomial* $F_{\ker \phi}$, which is defined over $\mathbb{F}_q$. Separable isogenies are defined by their kernels, up to isomorphism.

Going in the other direction, given a finite subgroup S of $\mathcal{E}$ defined over $\mathbb{F}_q$, there exists a separable *quotient isogeny* $\phi : \mathcal{E} \to \mathcal{E}/S$ with $\ker \phi = S$. The isogeny and the curve $\mathcal{E}/S$ are both defined up to $\mathbb{F}_q$-isomorphism; they can be computed using *Vélu's formulæ* [135]. (If S is encoded as the polynomial F_S, then we compute ϕ_S using the symmetric version of Vélu's formulæ in [84, Sect. 2.4].)

Given an ideal $\mathfrak{a} \subset \mathrm{End}(\mathcal{E})$, we can consider the subgroup $\mathcal{E}[\mathfrak{a}] := \cap_{\psi \in \mathfrak{a}} \ker \psi$. This is the kernel of an isogeny $\phi : \mathcal{E} \to \mathcal{E}/\mathcal{E}[\mathfrak{a}]$; the isogenies that arise in this way are central to the key exchange of Sect. 14. The degree of ϕ is the norm of $\mathfrak{a}$ in $\mathbb{Z} \subset \mathrm{End}(\mathcal{E})$. If $\mathfrak{a} = (\psi)$ is principal, then ϕ is isomorphic to ψ.

Given any d-isogeny $\phi : \mathcal{E} \to \mathcal{E}'$, we can compute the subgroup $S = \phi(\mathcal{E}[d]) \subset \mathcal{E}'$, and then the quotient $\phi_S : \mathcal{E}' \to \mathcal{E}'/S$ is a d-isogeny such that $\phi_S \circ \phi$ has kernel $\mathcal{E}[d]$; hence, ϕ_S is isomorphic to a d-isogeny $\phi^\dagger : \mathcal{E}' \to \mathcal{E}$ such that $\phi^\dagger \circ \phi = [d]$ on $\mathcal{E}$ (and $\phi \circ \phi^\dagger = [d]$ on $\mathcal{E}'$). We call $\phi^\dagger$ the *dual* of ϕ. The upshot is that every d-isogeny $\mathcal{E} \to \mathcal{E}'$ has a corresponding d-isogeny $\mathcal{E}' \to \mathcal{E}$.

Every isogeny can be factored into a composition of isogenies of prime degree, though there are two important caveats: factorization is not unique, and generally a factorization may only exist over some extension field. For example, if $\ell \neq p$ is prime, then $\mathcal{E}(\overline{\mathbb{F}}_q) \cong (\mathbb{Z}/\ell\mathbb{Z})^2$, so there are $\ell + 1$ order-ℓ subgroups $S \subset \mathcal{E}(\overline{\mathbb{F}}_q)[\ell]$, each the kernel of a different isogeny $\phi_S : \mathcal{E} \to \mathcal{E}/S$, and then the dual isogeny gives us a factorization $[\ell] = \phi_S^\dagger \circ \phi_S$. Each decomposition is only defined over the field of definition of the associated subgroup S.

Just as we decompose isogenies into ℓ-isogenies, so consider the subgraphs formed by ℓ-isogenies. The structures of ℓ-isogeny graphs depend strongly on the endomorphism rings of curves in the isogeny class, as we will see.

A curve $\mathcal{E}$ is *supersingular* if p divides its trace (over $\mathbb{F}_p$, this implies the trace is 0). If $\mathcal{E}$ is not supersingular, then it is *ordinary*. The j-invariant of any supersingular curve is in $\mathbb{F}_p$ or $\mathbb{F}_{p^2}$, so any supersingular curve is isomorphic to one defined over $\mathbb{F}_p$ or $\mathbb{F}_{p^2}$. There are roughly $\lfloor p/12 \rfloor$ supersingular j-invariants in $\mathbb{F}_{p^2}$, of which $O(\sqrt{p})$ are in $\mathbb{F}_p$ (more precisely, this number is the class number of $\mathbb{Q}(\sqrt{-p})$). Since supersingularity is defined by the trace, either all of the curves in an isogeny class are supersingular, or all of them are ordinary; the two kinds of curves do not mix.

There are two possibilities for the general structure of the endomorphism ring of an elliptic curve over a finite field:

commutative $\mathrm{End}(\mathcal{E})$ is isomorphic to an order in a quadratic imaginary field;
  or
noncommutative $\mathrm{End}(\mathcal{E})$ is isomorphic to an order in a quaternion algebra.

All ordinary curves have commutative endomorphism rings. If a supersingular curve is defined over $\mathbb{F}_p$, then its endomorphism ring is commutative[15]; if it is defined over $\mathbb{F}_{p^2}$, then its endomorphism ring is noncommutative.

The commutative case is relatively simple: each $\mathrm{End}(\mathcal{E})$ is an order in $K = \mathbb{Q}(\pi)$ containing the quadratic ring $\mathbb{Z}[\pi]$. The discriminant of $\mathbb{Z}[\pi]$ is $\Delta_\pi := t^2 - 4q = m^2 \Delta_K$, where Δ_K is the fundamental discriminant of K. The algorithmic exploration of ordinary isogeny graphs begins with Kohel's thesis [84, Chap. 4];

---

[15] If we consider endomorphisms defined over $\mathbb{F}_{p^2}$, then the ring is noncommutative.

these graphs are now mainstream computational tools in the arithmetic of elliptic curves and elliptic curve cryptography [60, 65, 66, 79]. The analogous theory for supersingular curves over $\mathbb{F}_p$, whose endomorphism rings are commutative and thus behave like ordinary curves, was explored by Delfs and Galbraith [47].

If $\phi : \mathcal{E} \rightarrow \mathcal{E}'$ is an ℓ-isogeny of endomorphism rings with commutative endomorphism rings (with ℓ prime), then there are three possibilities: $\mathrm{End}(\mathcal{E}) \cong \mathrm{End}(\mathcal{E}')$ (we say ϕ is *horizontal*), $\mathrm{End}(\mathcal{E}) \subset \mathrm{End}(\mathcal{E}')$ with index ℓ (we say ϕ is *ascending*), or $\mathrm{End}(\mathcal{E}) \supset \mathrm{End}(\mathcal{E}')$ with index ℓ (we say ϕ is *descending*). An ℓ-isogeny can only be ascending or descending if ℓ divides the conductor m of $\mathbb{Z}[\pi]$ in O_K, and an ℓ-isogeny $\phi : \mathcal{E} \rightarrow \mathcal{E}'$ can only be horizontal if $\mathrm{End}(\mathcal{E})$ and $\mathrm{End}(\mathcal{E}')$ are isomorphic to the maximal order O_K of K. The ℓ-isogenies of ordinary curves thus form "volcano" structures: cycles of horizontal isogenies link the curves $\mathcal{E}$ with $\mathrm{End}(\mathcal{E}) \cong O_K$, and from each of these curves a regular tree grows downwards, with its leaves in the curves with $\mathrm{End}(\mathcal{E}) \cong \mathbb{Z}[\pi]$ (which is minimal). The vertices with $\mathrm{End}(\mathcal{E}) \cong O_K$ have two horizontal ℓ-isogenies (or one or zero, if the cycle is degenerate), and $\ell - 1$ descending isogenies; each other vertex has one ascending and ℓ descending isogenies, except for the minimal vertices, which have no further descending isogenies.

From our perspective, what is most interesting about the commutative case is that the (isomorphism classes) of curves $\mathcal{E}$ with $\mathrm{End}(\mathcal{E}) \cong O_K$ form a PHS under the action of the class group of $\mathrm{Cl}(O_K)$. We met this PHS in Example 4.

The noncommutative case is much more complicated, and we will be much more brief here. The algorithmic applications of the full supersingular isogeny graph go back to Mestre and Oesterlé [99], and more detail appears in the second half of Kohel's thesis [84, Chap. 7]. In the non-commutative case, the ℓ-isogeny graph is $(\ell + 1)$-regular and connected, and it is an expander graph.

14   Commutative Isogeny-Based Key Exchange

Recall the PHS space from Example 4, which Couveignes conjectured was an HHS: fix a prime power q and an integer t with $|t| \leq 2\sqrt{q}$, set $\Delta := t^2 - 4q$, and let O_K be the maximal order (the ring of integers) of the quadratic imaginary field $K := \mathbb{Q}(\sqrt{\Delta})$. For this HHS,

- the space $\mathcal{X}$ is the set of isomorphism classes of elliptic curves over $\mathbb{F}_q$ whose endomorphism rings are isomorphic to O_K;
- the group $\mathfrak{G}$ is the ideal class group $\mathrm{Cl}(O_K)$ of O_K; and
- the action $(\mathfrak{a}, \mathcal{E}) \mapsto \mathfrak{a} \cdot \mathcal{E}$ is evaluated by computing the isogeny $\phi : \mathcal{E} \rightarrow \mathcal{E}/\mathcal{E}[\mathfrak{a}]$, and taking $\mathfrak{a} \cdot \mathcal{E}$ to be the isomorphism class of $\mathcal{E}/\mathcal{E}[\mathfrak{a}]$.

The cardinality N of $\mathfrak{G}$ is the class number of O_K, which is roughly $\sqrt{\Delta_K}$, where Δ_K is the discriminant of K (essentially the squarefree part of Δ). There is no point in not maximising N with respect to q, so we should use t such that Δ is already a fundamental discriminant; this forces all curves in the isogeny class to have $\mathbb{Z}[\pi] = \mathrm{End}(\mathcal{E}) = O_K$, and then $N = \#\mathfrak{G} \sim \sqrt{q}$.

The vectorization and parallelization problems in this HHS are expressed concretely in terms of computing paths in isogeny graphs. The fastest known classical algorithms for vectorization and parallelization in this HHS are the generic square-root algorithms, which run in time $O(\sqrt[4]{q})$. In the quantum world, Childs, Jao, and Soukharev have defined a subexponential quantum isogeny evaluation algorithm [37], which in combination with Kuperberg's algorithm gives a full subexponential quantum algorithm for solving vectorization in this HHS. This applies identically to the ordinary and commutative-supersingular cases. Further analysis of this approach can be found in [23].

Couveignes defined a key exchange (essentially Algorithms 7 and 8) and an identification protocol in this HHS in [43]. These protocols were essentially unknown outside the French community until Rostovtsev and Stolbunov independently proposed a public key encryption scheme based on the same HHS [119]. Stolbunov [128] then derived more protocols, including an interactive key exchange scheme similar to Algorithms 7 and 8. The only real difference between Couveignes' and Stolbunov's cryptosystems is in the sampling of private keys, each representing one of the two approaches mentioned in Sect. 10. Couveignes uses a true uniform random sampling over the whole of the keyspace, then applies a lattice reduction-based algorithm to produce an equivalent key whose action is efficiently computable. Rostovtsev and Stolbunov sample keys from a subset of efficiently computable keys whose distribution they conjecture to be close enough to the uniform distribution on the entire group.

One particularly nice aspect of these schemes is that key validation can be made simple and efficient (see [59, Sect. 5.4]), so we can safely use the scheme for static key exchange. Since the group action is simple and transitive, every element of the space is a legitimate public key. To validate a given x in $\mathbb{F}_q$ as a public key, therefore, it suffices to check that x is the j-invariant of a curve with endomorphism ring O_K. We immediately construct a curve $\mathcal{E}$ with j-invariant x, and check that it has the right trace (which amounts to checking that $\mathcal{E}(\mathbb{F}_q)$ has the claimed cardinality), switching to the quadratic twist if necessary. This ensures that $\mathrm{End}(\mathcal{E}) \subseteq O_K$; if t is chosen such that $\mathbb{Z}[\pi] = O_K$, then we are already done; otherwise, we check $\mathrm{End}(\mathcal{E}) = O_K$ using Kohel's algorithm [84].

Regardless of how the private key ideals are sampled, by the time we want to use them in the group action they are presented as factored ideals

$$\mathfrak{a} = \prod_{i=1}^{r} \mathfrak{l}_i^{e_i} \quad \text{with} \quad -B_i \leq e_i \leq B_i,$$

where the $\mathfrak{l}_i$ are distinguished prime ideals whose corresponding ℓ_i-isogenies can be evaluated very quickly. If the cost of evaluating an isogeny associated with kernel $\mathcal{E}[\mathfrak{l}_i]$ (for a random $\mathcal{E}$ in the isogeny class) is C_i, then the exponent bounds B_i should be chosen in such a way that the cost of evaluation $\sum_{i=1}^{r} B_i C_i$ is minimised while keeping the number of private keys $\prod_{i=1}^{r}(2B_i + 1)$ big enough.

Suppose then that want to compute an ℓ-isogeny from an elliptic curve $\mathcal{E}/\mathbb{F}_q$ for some prime $\ell \neq p$. We consider two methods of computing ℓ-isogenies here. The classic approach is based on modular polynomials. An alternative approach based on Vélu's formulæ was originally proposed in [59], and subsequently used in [35]. Both approaches are discussed in greater detail in [59, Sect. 3.2].

First, consider the "modular" approach. Recall that the ℓ-th (classical[16]) modular polynomial $\Phi_\ell(J_1, J_2)$ is defined over $\mathbb{Z}$, is monic of degree $\ell + 1$ in both J_1 and J_2, and the roots of $\Phi_\ell(j(\mathcal{E}), X)$ in $\mathbb{F}_q$ are the j-invariants of the curves ℓ-isogenous to $\mathcal{E}$ over $\mathbb{F}_q$. In fact, the $\mathbb{F}_q$-irreducible factors correspond to Galois orbits of ℓ-isogenies (or, equivalently, to Galois orbits of order-ℓ subgroups of $\mathcal{E}$). To compute the ℓ-isogenous curves up to isomorphism, therefore, we (pre)compute Φ_ℓ and reduce it modulo p; then we evaluate one variable at $j(\mathcal{E})$, and compute the roots in $\mathbb{F}_q$ of the resulting univariate polynomial.

There are $\ell + 1$ curves ℓ-isogenous to $\mathcal{E}$ over $\overline{\mathbb{F}}_q$. Of these isogenies, at most two can preserve the endomorphism ring. If we choose $\mathcal{E}$ such that $\mathrm{End}(\mathcal{E})$ is the maximal order and equal to $\mathbb{Z}[\pi]$ (or at least such that ℓ does not divide the conductor of $\mathbb{Z}[\pi]$ in O_K, so $\mathbb{Z}[\pi]$ is locally maximal) then $\Phi_\ell(j(\mathcal{E}), X)$ will have only two roots, corresponding precisely to these horizontal isogenies. If this is the first step in a walk of ℓ-isogenies then we must determine which of the two is in the "correct" direction, corresponding to the ideal; we can do this by checking the eigenvalue of Frobenius on the kernel, for example. But if we have already started walking, then there is no need to do this: we know that the "wrong" isogeny is the dual of the preceding step, so we just ignore the j-invariant of that curve. The total cost of this approach is dominated by finding the roots of $\Phi_\ell(j(\mathcal{E}), \mathbb{F}_q)$, which is $O(\ell \log q)$ $\mathbb{F}_q$-operations.

The alternative "Vélu" approach is to construct isogeny kernels explicitly, and compute the corresponding isogeny steps using Vélu's formulæ [135]. The idea is simple: suppose $\mathcal{E}(\mathbb{F}_q)$ contains a point P_ℓ of order ℓ; we want to compute the isogeny $\mathcal{E} \to \mathcal{E}/\langle P_\ell \rangle$ with kernel generated by P_ℓ. We can compute the kernel polynomial $F(X) = \prod_{i=1}^{(\ell-1)/2}(X - [i]P_\ell)$ in $\widetilde{O}(\ell)$ $\mathbb{F}_q$-operations; if P_ℓ is defined over a small extension of $\mathbb{F}_q$, say $\mathbb{F}_{q^k}$, then the cost is $\widetilde{O}(k^2\ell)$ $\mathbb{F}_q$-operations. We can then apply Vélu's formulæ (as in [135] or [84, Sect. 2.4]) to compute an equation for $\mathcal{E}/\langle P_\ell \rangle$; we do not need an expression for the isogeny itself. The total cost is dominated by the cost of computing F, which is $\widetilde{O}(k^2\ell)$ $\mathbb{F}_q$-operations.

The Vélu approach is much faster than the modular approach when $k^2 \ll \log q$, but it requires us to use isogeny classes of curves with many small-order subgroups over very low-degree extensions. Such curves are rare, and hard to find by exhaustive search: constructing them presents similar challenges to the construction of pairing-friendly curves (though here we want many small primes dividing the order over a degree-k extension, rather than one big prime). We might try to do better by using the CM method [3,129], which constructs elliptic curves with a specified group order—but the CM method only works when the discriminant Δ_K of the maximal order (and hence the class group of the maximal order) is very small, because $\#\mathrm{Cl}(O_K) \sim \sqrt{|-\Delta_K|}$. This means that if we use the CM method to generate parameters, then the private key space is far too

---

[16] We use classical modular polynomials here for simplicity, but alternative modular polynomials such as Atkin's, which have smaller degree, are better in practice. These degrees are still in $O(\ell)$, so the asymptotic efficiency of this approach does not change.

small for these cryptosystems to be secure. In [59] curve parameters are selected by running an extensive search to maximise the number of primes ℓ with points in $\mathcal{E}[\ell](\mathbb{F}_{q^k})$ for smallish k, using the Vélu approach for these primes and the modular approach for the others. This gives a significant improvement over the pure modular approach, but the result is still far from truly practical.

CSIDH [35] steps around this obstruction in an extremely neat way, by switching to supersingular curves over $\mathbb{F}_p$. Since their endomorphism rings are commutative quadratic orders, these curves behave like ordinary curves, and the Couveignes–Rostovtsev–Stolbunov protocol carries over without modification. However, the fact that these curves necessarily have order $p + 1$ makes it extremely simple to control their group structure and class group size by appropriately choosing p from within the desired range. This close control means that we can force all of the small primes to be "Vélu" with $k = 1$, which results in a speedup that beats ordinary-curve constructions like that of [59] by orders of magnitude. Key validation is also simpler for these curves. We are unaware of any impact on security, negative or positive, stemming from the use of supersingular curves as opposed to ordinary curves; so far, each attack described as targeting either CRS or CSIDH (e.g. [23]) applies equally to the other. CSIDH therefore represents a practical post-quantum Diffie–Hellman replacement, though the development of efficient side-channel-aware implementations of commutative isogeny protocols remains an open problem.

15   Supersingular Isogeny Diffie–Hellman

We conclude with a brief discussion of Jao and De Feo's supersingular isogeny Diffie–Hellman, known as SIDH [45, 78]. On the surface, SIDH resembles the commutative isogeny key exchange of Sect. 14: Alice and Bob each compute a sequence of isogenies to arrive at their public keys, and later the shared secret. However, the differences are striking.

The most fundamental difference is that the endomorphism rings in SIDH are noncommutative, so the group acting on the isogeny class is not abelian: SIDH falls squarely outside the HHS framework. In particular, Alice and Bob's isogeny walks do not automatically commute; some extra data must be passed around to correctly orient their walks for the second phase of the key exchange.

The second crucial difference with commutative isogeny key exchange is that the underlying ℓ-isogeny graphs are no longer cycles; rather, each is $(\ell + 1)$-regular, connected, and an expander graph. Since there are $O(\sqrt{p})$ vertices, computing a random sequence of $O(\log p)$ ℓ-isogenies from a given base curve takes us to a curve that is as good as randomly sampled from the isogeny class.

To define the protocol, we fix distinct primes ℓ_A and ℓ_B (these will be very small, typically 2 and 3), and exponents n_A and n_B, respectively, and let p be a prime such that $p = c \cdot \ell_A^{n_A} \cdot \ell_B^{n_B} \pm 1$ for some very small c. We want to choose ℓ_A and ℓ_B such that $\ell_A^{n_A}$ and $\ell_B^{n_B}$ are roughly the same size; ideally, $\ell_A^{n_A} \sim \ell_B^{n_B} \sim \sqrt{p}$.

Now consider the supersingular isogeny class over $\mathbb{F}_{p^2}$: every curve $\mathcal{E}$ in it has $\mathcal{E}[\ell_A^{n_A}] \cong (\mathbb{Z}/\ell_A^{n_A}\mathbb{Z})^2$ and $\mathcal{E}[\ell_B^{n_B}] \cong (\mathbb{Z}/\ell_B^{n_B}\mathbb{Z})^2$. Fix a base curve $\mathcal{E}_0$ is fixed in the isogeny class, along with bases (P_A, Q_A) of $\mathcal{E}_0[\ell_A^{n_A}]$ and (P_B, Q_B) of $\mathcal{E}_0[\ell_B^{n_B}]$.

First, key generation. Alice samples a random a in $\mathbb{Z}/\ell_A^{n_A}\mathbb{Z}$ as her private key; the point $P_A + [a]Q_A$ has exact order $\ell_A^{n_A}$, and generates the kernel of an $\ell_A^{n_A}$-isogeny $\phi_A : \mathcal{E}_0 \to \mathcal{E}_A \cong \mathcal{E}_0/\langle P_A + [a]Q_A \rangle$, which she computes as a series of ℓ_A-isogenies. Her public key is $(\mathcal{E}_A, \phi_A(P_B), \phi_A(Q_B))$. Bob samples a private key b in $\mathbb{Z}/\ell_B^{n_B}\mathbb{Z}$ and computes the $\ell_B^{n_B}$-isogeny $\phi_B : \mathcal{E}_0 \to \mathcal{E}_B \cong \mathcal{E}_0/\langle P_B + [b]Q_B \rangle$ as a series of ℓ_B-isogenies; his public key is $(\mathcal{E}_B, \phi_B(P_A), \phi_B(Q_A))$. There is plenty of redundant information in these public keys, and they can be compressed following the suggestions in [40].

To complete the key exchange, Alice computes the $\ell_A^{n_A}$-isogeny $\phi_A' : \mathcal{E}_B \to \mathcal{E}_{BA} = \mathcal{E}_B/\langle \phi_B(P_A) + [a]\phi_B(Q_A) \rangle$, and Bob computes the $\ell_B^{n_B}$-isogeny $\phi_B' : \mathcal{E}_A \to \mathcal{E}_{AB} = \mathcal{E}_A/\langle \phi_A(P_B) + [b]\phi_A(Q_B) \rangle$. The shared secret is the j-invariant $j(\mathcal{E}_{AB}) = j(\mathcal{E}_{BA})$ in $\mathbb{F}_{p^2}$; the curves $\mathcal{E}_{AB}$ and $\mathcal{E}_{BA}$ have the same j-invariant because both are isomorphic to $\mathcal{E}_0/\langle P_A + [a]Q_A + P_B + [b]Q_B \rangle$. The relationships between these curves is illustrated in Fig. 2.

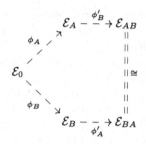

Fig. 2. Supersingular isogeny Diffie–Hellman.

Jao, De Feo, and Plût specified efficient algorithms for SIDH in [45]. The first competitive public implementation was due to Costello, Longa, and Naehrig [41]. A lot of effort has since been put into improving the algorithmic and space efficiency of SIDH [39, 40, 57], and optimizing arithmetic in its specialized finite fields [25, 26]. One particlarly nice feature of SIDH in comparison to commutative isogeny DH is that the isogenies in SIDH all have degree either ℓ_A or ℓ_B; both are fixed, and typically tiny, so computing the individual isogeny steps is much faster in the supersingular protocol, and requires much less code and precomputation.

Recovering the private key from an SIDH public key— a noncommutative analogue of vectorization— amounts to computing an isogeny between the base curve $\mathcal{E}_0$ and the target public key curve $\mathcal{E}_A$. We can do this using an algorithm due to Delfs and Galbraith [47], inspired by an algorithm for the ordinary case due to Galbraith, Hess, and Smart [66]. The algorithm walks randomly through the supersingular isogeny graph from the starting and ending curves, until curves defined over $\mathbb{F}_p$ are detected—and then finding a path between those two curves, to complete the desired isogeny, is analogous to breaking a CSIDH key (though with completely different security parameters). Alternatively, Adj, Cervantes–Vázquez, Chi–Dominguez, Menezes, and Rodríguez–Henríquez have given a useful analysis of the van Oorschot–Wiener algorithm applied to this problem [2]. The asymptotic cost of either approach is in $O(\sqrt[4]{p})$ $\mathbb{F}_{p^2}$-operations if $\ell_A^{n_A} \sim \ell_B^{n_B} \sim \sqrt{p}$. In the quantum setting, we can apply Tani's claw-finding algorithm [131] to find a curve in the intersection of the sets of curves $\ell_A^{n_A/2}$-isogenous to $\mathcal{E}_0$ and $\mathcal{E}_A$ with a query complexity of $O(\ell_A^{n_A/3})$ (or we can attack Bob's public key in $O(\ell_B^{n_B/3})$), which is $O(p^{1/6})$ when $\ell_A^{n_A} \sim \ell_B^{n_B} \sim \sqrt{p}$. The fact that the subexponential Childs–Jao–Soukharev algorithm does not apply in the noncommutative case was one of the motivations for developing SIDH.

But SIDH keys do not simply present the target curve of an unknown isogeny: they also present images of distinguished torsion bases, which may help cryptanalysis [111]. The precise nature of the cryptographic problems underlying SIDH is quite complicated, but Urbanik and Jao's survey of these problems provides useful analysis [133], while Eisentraeger, Hallgren, Lauter, Morrison, and Petit go further into the connections with the endomorphism ring [54].

Finding an isogeny between two supersingular curves over $\mathbb{F}_{p^2}$ is equivalent to determining their endomorphism rings [84,85]. This is an interesting contrast with the commutative case, where the endomorphism rings are presumed known, and in any case can be computed using Kohel's algorithm for ordinary curves [84]. As we have seen, determining the endomorphism ring in the commutative case is an important step in public key validation in commutative isogeny key exchange.

Key validation is especially problematic for SIDH. Suppose we have an algorithm which, given a prime ℓ, a positive integer n, and a curve $\mathcal{E}$, efficiently decides whether $\mathcal{E}$ is ℓ^n-isogenous to $\mathcal{E}_0$. Such an algorithm would allow us to verify whether Alice or Bob's public key was honestly generated (by calling the algorithm on $(\ell_A, n_A, \mathcal{E}_A)$ or $(\ell_B, n_B, \mathcal{E}_B)$, respectively). However, as we see in [69, Sect. 6.2] and [132], this algorithm can also be used to efficiently recover secret keys from public keys. Indeed, take Alice's public curve $\mathcal{E}_A$; there are $\ell_A + 1$ curves ℓ_A-isogenous to it. Computing each of these isogenies $\phi : \mathcal{E}_A \to \mathcal{E}_A'$, we call the algorithm on $(\ell_A, n_A - 1, \mathcal{E}_0, \mathcal{E}_A')$; if it returns true, then ϕ is the last ℓ_A-isogeny in Alice's secret key. Iterating this procedure reveals the entire key.

Problematic key validation makes defining a CCA-secure SIDH-based KEM more complicated than the equivalent in the commutative case. SIKE [9], which is the only isogeny-based candidate KEM in the NIST process, handles this by applying the Hofheinz–Hövelmanns–Kiltz a variant of the Fujisaki–Okamoto transform [64,76] to SIDH; this entails a nontrivial performance hit.

On a formal level, there are some profound differences between SIDH and classical Diffie–Hellman. The most obvious is the lack of symmetry in SIDH between Alice and Bob, whose roles are no longer interchangeable. This is reflected by their distinct and incompatible key spaces, which are in turn distinct from the shared secret space and the space the base curve lives in. Alice's private key encodes a sequence of ℓ_A-isogenies of length n_A, while Bob's encodes a sequence of ℓ_B-isogenies of length n_B. Alice's public key belongs to the space of (isomorphism classes of) elliptic curves equipped with a distinguished $\ell_B^{n_B}$-torsion basis, while Bob's is equipped with an $\ell_A^{n_A}$-torsion basis instead. The base curve $\mathcal{E}_0$ is drawn from yet another space: it is equipped with an $\ell_A^{n_A} \ell_B^{n_B}$-torsion basis.

This asymmetry might seem like a curious but minor inconvenience: the participants just need to decide who is Alice and who is Bob before each key exchange. More importantly, though, this asymmetry is incompatible with most of the theoretical machinery that we use to reason about Diffie–Hellman and its hardness. We have already seen how group Diffie–Hellman oracles create black-box field structures on prime-order groups, while HHS Diffie–Hellman oracles create a black-box group structures. In SIDH, however, a Diffie–Hellman oracle defines no binary operation on any set, let alone an interesting algebraic structure. This plurality of spaces makes it hard to adapt hidden-number-problem-style arguments [5, 22] for hardcore bits to the SIDH context in a natural way, though a valiant effort has been made by Galbraith, Petit, Shani, and Ti [67].

Remark 3. At first glance, the fact that SIKE is the only isogeny-based KEM submitted to the NIST post-quantum process, competing with 58 others mostly based on codes, lattices, and polynomial systems, might suggest that it is a strange outlier. However, this uniqueness is not so much an indicator of lack of support, so much as a sign of rare convergence and consensus in the elliptic-curve cryptography community—convergence that did not occur to the same extent in the communities working on other post-quantum paradigms. The fact that there was only one isogeny-based submission reflects the general agreement that this was the right way to do isogeny-based key agreement at that point in time. The more flexible CSIDH scheme was not developed until later, when the NIST process was already underway, and so it was not part of the conversation.

Acknowledgements. I am grateful to Luca De Feo, Florian Hess, Jean Kieffer, and Antonin Leroux for the many hours they spent discussing these cryptosystems with me; and the organisers, chairs, and community of WAIFI 2018.

References

1. Abdalla, M., Bellare, M., Rogaway, P.: DHAES: an encryption scheme based on the Diffie–Hellman problem. IACR Cryptology ePrint Archive 1999:7 (1999)
2. Adj, G., Cervantes-Vázquez, D., Chi-Domínguez, J., Menezes, A., Rodríguez-Henríquez, F.: On the cost of computing isogenies between supersingular elliptic curves. IACR Cryptology ePrint Archive 2018:313 (2018)
3. Agashe, A., Lauter, K.E., Venkatesan, R.: Constructing elliptic curves with a known number of points over a prime field. In: High Primes and Misdemeanors: Lectures in Honour of the 60th Birthday of Hugh Cowie Williams [30], pp. 1–17
4. Aguilar, C., Gaborit, P., Lacharme, P., Schrek, J., Zémor, G.: Noisy Diffie–Hellman protocols (2010). Slides presented at the recent results session of PQC 2010. https://pqc2010.cased.de/rr/03.pdf
5. Akavia, A.: Solving hidden number problem with one bit oracle and advice. In: Halevi, S. (ed.) CRYPTO 2009. LNCS, vol. 5677, pp. 337–354. Springer, Heidelberg (2009). https://doi.org/10.1007/978-3-642-03356-8_20

6. Alkim, E., Ducas, L., Pöppelmann, T., Schwabe, P.: Post-quantum key exchange - a new hope. In: Holz, T., Savage, S. (eds.) 25th USENIX Security Symposium, USENIX Security 16, Austin, TX, USA, 10–12 August 2016, pp. 327–343. USENIX Association (2016)

7. Antipa, A., Brown, D., Menezes, A., Struik, R., Vanstone, S.: Validation of elliptic curve public keys. In: Desmedt, Y.G. (ed.) PKC 2003. LNCS, vol. 2567, pp. 211–223. Springer, Heidelberg (2003). https://doi.org/10.1007/3-540-36288-6_16

8. Armknecht, F., Gagliardoni, T., Katzenbeisser, S., Peter, A.: General impossibility of group homomorphic encryption in the quantum world. In: Krawczyk, H. (ed.) PKC 2014. LNCS, vol. 8383, pp. 556–573. Springer, Heidelberg (2014). https://doi.org/10.1007/978-3-642-54631-0_32

9. Azarderakhsh, R., et al.: Supersingular isogeny key encapsulation (2017)

10. Balasubramanian, R., Koblitz, N.: The improbability that an elliptic curve has subexponential discrete log problem under the Menezes–Okamoto–Vanstone algorithm. J. Cryptol. **11**(2), 141–145 (1998)

11. Barbulescu, R., Gaudry, P., Joux, A., Thomé, E.: A heuristic quasi-polynomial algorithm for discrete logarithm in finite fields of small characteristic. In: Nguyen, P.Q., Oswald, E. (eds.) EUROCRYPT 2014. LNCS, vol. 8441, pp. 1–16. Springer, Heidelberg (2014). https://doi.org/10.1007/978-3-642-55220-5_1

12. Benaloh, J.: Simple verifiable elections. In: Wallach, D.S., Rivest, R.L. (eds.) 2006 USENIX/ACCURATE Electronic Voting Technology Workshop, EVT 2006, Vancouver, BC, Canada, 1 August 2006. USENIX Association (2006)

13. Bentahar, K.: The equivalence between the DHP and DLP for elliptic curves used in practical applications, revisited. In: Smart, N.P. (ed.) Cryptography and Coding 2005. LNCS, vol. 3796, pp. 376–391. Springer, Heidelberg (2005). https://doi.org/10.1007/11586821_25

14. Bernstein, D.J.: Curve25519: new Diffie–Hellman speed records. In: Yung, M., Dodis, Y., Kiayias, A., Malkin, T. (eds.) PKC 2006. LNCS, vol. 3958, pp. 207–228. Springer, Heidelberg (2006). https://doi.org/10.1007/11745853_14

15. Bernstein, D.J.: Differential addition chains. Preprint (2006)

16. Bernstein, D.J., Chuengsatiansup, C., Lange, T., Schwabe, P.: Kummer strikes back: new DH speed records. In: Sarkar, P., Iwata, T. (eds.) ASIACRYPT 2014. LNCS, vol. 8873, pp. 317–337. Springer, Heidelberg (2014). https://doi.org/10.1007/978-3-662-45611-8_17

17. Bernstein, D.J., et al.: Faster discrete logarithms on FPGAs. IACR Cryptology ePrint Archive 2016:382. Document ID: 01ac92080664fb3a778a430e028e55c8 (2016)

18. Bernstein, D.J., Lange, T., Schwabe, P.: On the correct use of the negation map in the Pollard rho method. In: Catalano, D., Fazio, N., Gennaro, R., Nicolosi, A. (eds.) PKC 2011. LNCS, vol. 6571, pp. 128–146. Springer, Heidelberg (2011). https://doi.org/10.1007/978-3-642-19379-8_8

19. Biasse, J., Iezzi, A., Jacobson Jr., M.: A note on the security of CSIDH. CoRR, abs/1806.03656 (2018)

20. Boneh, D.: The decision Diffie–Hellman problem. In: Buhler, J.P. (ed.) ANTS 1998. LNCS, vol. 1423, pp. 48–63. Springer, Heidelberg (1998). https://doi.org/10.1007/BFb0054851

21. Boneh, D., Lipton, R.J.: Algorithms for black-box fields and their application to cryptography (extended abstract). In: Koblitz [83], pp. 283–297

22. Boneh, D., Venkatesan, R.: Hardness of computing the most significant bits of secret keys in Diffie–Hellman and related schemes. In: Koblitz [83], pp. 129–142

23. Bonnetain, X., Schrottenloher, A.: Quantum security analysis of CSIDH and ordinary isogeny-based schemes. IACR Cryptology ePrint Archive 2018:537 (2018)
24. Bos, J.W., Costello, C., Naehrig, M., Stebila, D.: Post-quantum key exchange for the TLS protocol from the ring learning with errors problem. In: 2015 IEEE Symposium on Security and Privacy, SP 2015, San Jose, CA, USA, 17–21 May 2015, pp. 553–570. IEEE Computer Society (2015)
25. Bos, J.W., Friedberger, S.: Fast arithmetic modulo $2^x p^y \pm 1$. In: Burgess, N., Bruguera, J.D., de Dinechin, F. (eds.) IEEE Symposium on Computer Arithmetic - ARITH 2017, pp. 148–155. IEEE Computer Society (2017)
26. Bos, J.W., Friedberger, S.: Arithmetic considerations for isogeny based cryptography. IACR Cryptology ePrint Archive 2018:376 (2018)
27. Brassard, G. (ed.): CRYPTO 1989. LNCS, vol. 435. Springer, New York (1990). https://doi.org/10.1007/0-387-34805-0
28. Bröker, R., Lauter, K.E., Sutherland, A.V.: Modular polynomials via isogeny volcanoes. Math. Comput. **81**(278), 1201–1231 (2012)
29. Buchmann, J., Scheidler, R., Williams, H.C.: A key-exchange protocol using real quadratic fields. J. Cryptol. **7**, 171–199 (1994)
30. van der Poorten, A., Stein, A. (eds.): High Primes and Misdemeanors: Lectures in Honour of the 60th Birthday of Hugh Cowie Williams. Fields Institute Communications Series, vol. 42. American Mathematical Society
31. Buchmann, J., Takagi, T., Vollmer, U.: Number field cryptography. In: van der Poorten, A., Stein, A. (eds.) [30]. High Primes and Misdemeanors: Lectures in Honour of the 60th Birthday of Hugh Cowie Williams, pp. 111–125
32. Buchmann, J.A., Williams, H.C.: A key exchange system based on real quadratic fields. In: Brassard [27], pp. 335–343
33. Canetti, R., Krawczyk, H.: Analysis of key-exchange protocols and their use for building secure channels. In: Pfitzmann, B. (ed.) EUROCRYPT 2001. LNCS, vol. 2045, pp. 453–474. Springer, Heidelberg (2001). https://doi.org/10.1007/3-540-44987-6_28
34. Cassels, J.W.S.: Lectures on Elliptic Curves. London Mathematical Society Student Texts, vol. 24 Cambridge University Press (1991)
35. Castryck, W., Lange, T., Martindale, C., Panny, L., Renes, J.: CSIDH: an efficient post-quantum commutative group action. IACR Cryptology ePrint Archive 2018:383 (2018)
36. Cheon, J.H.: Security analysis of the strong Diffie–Hellman problem. In: Vaudenay, S. (ed.) EUROCRYPT 2006. LNCS, vol. 4004, pp. 1–11. Springer, Heidelberg (2006). https://doi.org/10.1007/11761679_1
37. Childs, A., Jao, D., Soukharev, V.: Constructing elliptic curve isogenies in quantum subexponential time. J. Math. Cryptol. **8**(1), 1–29 (2014)
38. Coron, J.-S., Nielsen, J.B. (eds.): EUROCRYPT 2017. LNCS, vol. 10210. Springer, Cham (2017). https://doi.org/10.1007/978-3-319-56620-7
39. Costello, C., Hisil, H.: A simple and compact algorithm for SIDH with arbitrary degree isogenies. In: Takagi and Peyrin [130], pp. 303–329
40. Costello, C., Jao, D., Longa, P., Naehrig, M., Renes, J., Urbanik, D.: Efficient compression of SIDH public keys. In: Coron and Nielsen [38], pp. 679–706
41. Costello, C., Longa, P., Naehrig, M.: Efficient algorithms for supersingular isogeny Diffie–Hellman. In: Robshaw, M., Katz, J. (eds.) CRYPTO 2016. LNCS, vol. 9814, pp. 572–601. Springer, Heidelberg (2016). https://doi.org/10.1007/978-3-662-53018-4_21
42. Costello, C., Smith, B.: Montgomery curves and their arithmetic. J. Cryptogr. Eng. **8**, 227–240 (2017)

43. Couveignes, J.M.: Hard homogeneous spaces. IACR Cryptology ePrint Archive 2006:291 (2006)
44. Cramer, R., Shoup, V.: Design and analysis of practical public-key encryption schemes secure against adaptive chosen ciphertext attack. SIAM J. Comput. **33**(1), 167–226 (2003)
45. De Feo, L., Jao, D., Plût, J.: Towards quantum-resistant cryptosystems from supersingular elliptic curve isogenies. J. Math. Cryptol. **8**(3), 209–247 (2014)
46. Déchène, I.: On the security of generalized Jacobian cryptosystems. Adv. Math. Commun. **1**(4), 413–426 (2007)
47. Delfs, C., Galbraith, S.D.: Computing isogenies between supersingular elliptic curves over $\mathbb{F}_p$. Des. Codes Cryptogr. **78**(2), 425–440 (2016)
48. den Boer, B.: Diffie–Hellman is as strong as discrete log for certain primes. In: Goldwasser, S. (ed.) CRYPTO 1988. LNCS, vol. 403, pp. 530–539. Springer, New York (1990). https://doi.org/10.1007/0-387-34799-2_38
49. Deneuville, J.-C., Gaborit, P., Zémor, G.: Ouroboros: a simple, secure and efficient key exchange protocol based on coding theory. In: Lange, T., Takagi, T. (eds.) PQCrypto 2017. LNCS, vol. 10346, pp. 18–34. Springer, Cham (2017). https://doi.org/10.1007/978-3-319-59879-6_2
50. Diem, C., Thomé, E.: Index calculus in class groups of non-hyperelliptic curves of genus three. J. Cryptol. **21**(4), 593–611 (2008)
51. Diffie, W., Hellman, M.E.: New directions in cryptography. IEEE Trans. Inf. Theory **22**(6), 644–654 (1976)
52. Ding, J.: New cryptographic constructions using generalized learning with errors problem. IACR Cryptology ePrint Archive 2012:387 (2012)
53. Ding, J., Xie, X., Lin, X.: A simple provably secure key exchange scheme based on the learning with errors problem. IACR Cryptology ePrint Archive, 2012:688 (2012)
54. Eisenträger, K., Hallgren, S., Lauter, K., Morrison, T., Petit, C.: Supersingular isogeny graphs and endomorphism rings: reductions and solutions. In: Nielsen, J.B., Rijmen, V. (eds.) EUROCRYPT 2018. LNCS, vol. 10822, pp. 329–368. Springer, Cham (2018). https://doi.org/10.1007/978-3-319-78372-7_11
55. ElGamal, T.: A public key cryptosystem and a signature scheme based on discrete logarithms. IEEE Trans. Inf. Theory **31**(4), 469–472 (1985)
56. Enge, A., Gaudry, P., Thomé, E.: An L(1/3) discrete logarithm algorithm for low degree curves. J. Cryptol. **24**(1), 24–41 (2011)
57. Faz-Hernández, A., López, J., Ochoa-Jiménez, E., Rodríguez-Henríquez, F.: A faster software implementation of the supersingular isogeny Diffie–Hellman key exchange protocol. IEEE Trans. Comput. **PP**(99), 1 (2017)
58. De Feo, L.: Mathematics of isogeny based cryptography. CoRR, abs/1711.04062 (2017)
59. De Feo, L., Kieffer, J., Smith, B.: Towards practical key exchange from ordinary isogeny graphs. IACR Cryptology ePrint Archive 2018:485 (2018)
60. Fouquet, M., Morain, F.: Isogeny volcanoes and the SEA algorithm. In: Fieker, C., Kohel, D.R. (eds.) ANTS 2002. LNCS, vol. 2369, pp. 276–291. Springer, Heidelberg (2002). https://doi.org/10.1007/3-540-45455-1_23
61. Freire, E.S.V., Hofheinz, D., Kiltz, E., Paterson, K.G.: Non-interactive key exchange. In: Kurosawa, K., Hanaoka, G. (eds.) PKC 2013. LNCS, vol. 7778, pp. 254–271. Springer, Heidelberg (2013). https://doi.org/10.1007/978-3-642-36362-7_17
62. Frey, G., Müller, M., Rück, H.: The tate pairing and the discrete logarithm applied to elliptic curve cryptosystems. IEEE Trans. Inf. Theory **45**(5), 1717–1719 (1999)

63. Fried, J., Gaudry, P., Heninger, N., Thomé, E.: A kilobit hidden SNFS discrete logarithm computation. In: Coron and Nielsen [38], pp. 202–231
64. Fujisaki, E., Okamoto, T.: Secure integration of asymmetric and symmetric encryption schemes. In: Wiener, M. (ed.) CRYPTO 1999. LNCS, vol. 1666, pp. 537–554. Springer, Heidelberg (1999). https://doi.org/10.1007/3-540-48405-1_34
65. Galbraith, S.D.: Constructing isogenies between elliptic curves over finite fields. LMS J. Comput. Math. **2**, 118–138 (1999)
66. Galbraith, S.D., Hess, F., Smart, N.P.: Extending the GHS Weil descent attack. In: Knudsen, L.R. (ed.) EUROCRYPT 2002. LNCS, vol. 2332, pp. 29–44. Springer, Heidelberg (2002). https://doi.org/10.1007/3-540-46035-7_3
67. Galbraith, S.D., Petit, C., Shani, B., Ti, Y.B.: On the security of supersingular isogeny cryptosystems. In: Cheon, J.H., Takagi, T. (eds.) ASIACRYPT 2016. LNCS, vol. 10031, pp. 63–91. Springer, Heidelberg (2016). https://doi.org/10. 1007/978-3-662-53887-6_3
68. Galbraith, S.D., Smith, B.: Discrete logarithms in generalized Jacobians. IACR Cryptology ePrint Archive 2006:333 (2006)
69. Galbraith, S.D., Vercauteren, F.: Computational problems in supersingular elliptic curve isogenies. Quantum Inf. Process. **17**, 265 (2017)
70. Gaudry, P.: Fast genus 2 arithmetic based on Theta functions. J. Math. Cryptol. **1**(3), 243–265 (2007). https://eprint.iacr.org/2005/314/
71. Gaudry, P.: Index calculus for abelian varieties of small dimension and the elliptic curve discrete logarithm problem. J. Symb. Comput. **44**(12), 1690–1702 (2009)
72. Gaudry, P., Hess, F., Smart, N.P.: Constructive and destructive facets of Weil descent on elliptic curves. J. Cryptol. **15**(1), 19–46 (2002)
73. Gaudry, P., Thomé, E., Thériault, N., Diem, C.: A double large prime variation for small genus hyperelliptic index calculus. Math. Comput. **76**(257), 475–492 (2007)
74. Grémy, L., Guillevic, A.: DiscreteLogDB, a database of computations of discrete logarithms (2017). https://gitlab.inria.fr/dldb/discretelogdb
75. Guillevic, A., Morain, F.: Discrete logarithms. In: El Mrabet and Joye [103], Chap. 9
76. Hofheinz, D., Hövelmanns, K., Kiltz, E.: A modular analysis of the Fujisaki–Okamoto transformation. In: Kalai, Y., Reyzin, L. (eds.) TCC 2017. LNCS, vol. 10677, pp. 341–371. Springer, Cham (2017). https://doi.org/10.1007/978-3-319-70500-2_12
77. Hofheinz, D., Kiltz, E.: Secure hybrid encryption from weakened key encapsulation. In: Menezes, A. (ed.) CRYPTO 2007. LNCS, vol. 4622, pp. 553–571. Springer, Heidelberg (2007). https://doi.org/10.1007/978-3-540-74143-5_31
78. Jao, D., De Feo, L.: Towards quantum-resistant cryptosystems from supersingular elliptic curve isogenies. In: Yang, B.-Y. (ed.) PQCrypto 2011. LNCS, vol. 7071, pp. 19–34. Springer, Heidelberg (2011). https://doi.org/10.1007/978-3-642-25405-5_2
79. Jao, D., Miller, S.D., Venkatesan, R.: Expander graphs based on GRH with an application to elliptic curve cryptography. J. Number Theory **129**(6), 1491–1504 (2009)
80. Kleinjung, T., Diem, C., Lenstra, A.K., Priplata, C., Stahlke, C.: Computation of a 768-bit prime field discrete logarithm. In: Coron and Nielsen [38], pp. 185–201
81. Koblitz, N.: Elliptic curve cryptosystems. Math. Comput. **48**, 203–209 (1987)
82. Koblitz, N.: Hyperelliptic cryptosystems. J. Cryptol. **1**(3), 139–150 (1989)
83. Koblitz, N. (ed.): CRYPTO 1996. LNCS, vol. 1109. Springer, Heidelberg (1996). https://doi.org/10.1007/3-540-68697-5

84. Kohel, D.R.: Endomorphism rings of elliptic curves over finite fields. Ph.D. thesis, University of California at Berkley (1996)
85. Kohel, D.R., Lauter, K., Petit, C., Tignol, J.-P.: On the quaternion ℓ-isogeny path problem. LMS J. Comput. Math. **17**(A), 418–432 (2014)
86. Kuperberg, G.: A subexponential-time quantum algorithm for the dihedral hidden subgroup problem. SIAM J. Comput. **35**(1), 170–188 (2005)
87. Kuperberg, G.: Another subexponential-time quantum algorithm for the dihedral hidden subgroup problem. In: Severini, S., Brandao, F. (eds.) 8th Conference on the Theory of Quantum Computation. Communication and Cryptography (TQC 2013). Leibniz International Proceedings in Informatics (LIPIcs), vol. 22, pp. 20–34. Schloss Dagstuhl-Leibniz-Zentrum fuer Informatik, Dagstuhl, Germany (2013)
88. Langley, A., Hamburg, M., Turner, S.: Elliptic curves for security. RFC, 7748, pp. 1–22 (2016)
89. Lenstra, A.K., Lenstra, H.W. (eds.): The Development of the Number field Sieve. LNM, vol. 1554. Springer, Heidelberg (1993). https://doi.org/10.1007/BFb0091534
90. Lenstra, A.K., Verheul, E.R.: The XTR public key system. In: Bellare, M. (ed.) CRYPTO 2000. LNCS, vol. 1880, pp. 1–19. Springer, Heidelberg (2000). https://doi.org/10.1007/3-540-44598-6_1
91. Lim, C.H., Lee, P.J.: A key recovery attack on discrete log-based schemes using a prime order subgroup. In: Kaliski, B.S. (ed.) CRYPTO 1997. LNCS, vol. 1294, pp. 249–263. Springer, Heidelberg (1997). https://doi.org/10.1007/BFb0052240
92. Lochter, M., Merkle, J.: Elliptic curve cryptography (ECC) brainpool standard curves and curve generation. RFC, 5639, pp. 1–27 (2010)
93. Marlinspike, M., Perrin, T.: The X3DH key agreement protocol (2016)
94. Martin-Lopez, E., Laing, A., Lawson, T., Alvarez, R., Zhou, X.-Q., O'Brien, J.L.: Experimental realization of Shor's quantum factoring algorithm using qubit recycling. Nat. Photon. **6**(11), 773–776, 11 (2012)
95. Maurer, U.M.: Towards the equivalence of breaking the Diffie–Hellman protocol and computing discrete logarithms. In: Desmedt, Y.G. (ed.) CRYPTO 1994. LNCS, vol. 839, pp. 271–281. Springer, Heidelberg (1994). https://doi.org/10.1007/3-540-48658-5_26
96. Maurer, U.M., Wolf, S.: The relationship between breaking the Diffie–Hellman protocol and computing discrete logarithms. SIAM J. Comput. **28**(5), 1689–1721 (1999)
97. Maze, G., Monico, C., Rosenthal, J.: Public key cryptography based on semigroup actions. Adv. Math. Commun. **1**(4), 489–507 (2007)
98. Menezes, A., Okamoto, T., Vanstone, S.A.: Reducing elliptic curve logarithms to logarithms in a finite field. IEEE Trans. Inf. Theory **39**(5), 1639–1646 (1993)
99. Mestre, J.: La méthode des graphes. Exemples et applications. In: Proceedings of the International Conference on Class Numbers and Fundamental Units of Algebraic Number Fields (Katata), pp. 217–242 (1986)
100. Miller, V.S.: Use of elliptic curves in cryptography. In: Williams, H.C. (ed.) CRYPTO 1985. LNCS, vol. 218, pp. 417–426. Springer, Heidelberg (1986). https://doi.org/10.1007/3-540-39799-X_31
101. Montgomery, P.L.: Speeding the Pollard and elliptic curve methods of factorization. Math. Comput. **48**(177), 243–264 (1987)
102. Mireles Morales, D.J.: An analysis of the infrastructure in real function fields. IACR Cryptology ePrint Archive 2008:299 (2008)
103. El Mrabet, N., Joye, M. (eds.): Guide to Pairing-Based Cryptography. Chapman and Hall/CRC, New York (2016)

104. Murty, V.K.: Abelian varieties and cryptography. In: Maitra, S., Veni Madhavan, C.E., Venkatesan, R. (eds.) INDOCRYPT 2005. LNCS, vol. 3797, pp. 1–12. Springer, Heidelberg (2005). https://doi.org/10.1007/11596219_1
105. Muzereau, A., Smart, N.P., Vercauteren, F.: The equivalence between the DHP and DLP for elliptic curves used in practical applications. LMS J. Comput. Math. **7**, 50–72 (2004)
106. National Institute of Standards and Technology (NIST). SP 800–56A recommendations for pair-wise key-establishment schemes using discrete logarithm cryptography
107. NIST. Post-quantum cryptography standardization
108. Ochoa-Jiménez, E., Rodríguez-Henríquez, F., Tibouchi, M.: Discrete logarithms. In: El Mrabet and Joye [103], Chap. 8
109. Peikert, C.: Lattice cryptography for the internet. In: Mosca, M. (ed.) PQCrypto 2014. LNCS, vol. 8772, pp. 197–219. Springer, Cham (2014). https://doi.org/10. 1007/978-3-319-11659-4_12
110. Perrin, T., Marlinspike, M.: The double ratchet algorithm (2016)
111. Petit, C.: Faster algorithms for isogeny problems using torsion point images. In: Takagi and Peyrin [130], pp. 330–353
112. Pohlig, S.C., Hellman, M.E.: An improved algorithm for computing logarithms over GF(p) and its cryptographic significance (corresp). IEEE Trans. Inf. Theory **24**(1), 106–110 (1978)
113. Pollard, J.M.: Monte Carlo methods for index computation (mod p). Math. Comput. **32**(143), 918–924 (1978)
114. Regev, O.: A subexponential time algorithm for the dihedral hidden subgroup problem with polynomial space, June 2004. arXiv:quant-ph/0406151
115. Renes, J., Schwabe, P., Smith, B., Batina, L.: μKummer: efficient hyperelliptic signatures and key exchange on microcontrollers. In: Gierlichs, B., Poschmann, A.Y. (eds.) CHES 2016. LNCS, vol. 9813, pp. 301–320. Springer, Heidelberg (2016). https://doi.org/10.1007/978-3-662-53140-2_15
116. Rescorla, E.: The transport layer security (TLS) protocol version 1.3. RFC, 8446, pp. 1–160 (2018)
117. Robert, D.: Theta functions and cryptographic applications. Ph.D. thesis, Université Henri Poincaré - Nancy I, July 2010
118. Roetteler, M., Naehrig, M., Svore, K.M., Lauter, K.E.: Quantum resource estimates for computing elliptic curve discrete logarithms. In: Takagi and Peyrin [130], pp. 241–270
119. Rostovtsev, A., Stolbunov, A.: Public-key cryptosystem based on isogenies. IACR Cryptology ePrint Archive 2006:145 (2006)
120. Rubin, K., Silverberg, A.: Torus-based cryptography. In: Boneh, D. (ed.) CRYPTO 2003. LNCS, vol. 2729, pp. 349–365. Springer, Heidelberg (2003). https://doi.org/10.1007/978-3-540-45146-4_21
121. Schnorr, C.-P.: Efficient identification and signatures for smart cards. In: Brassard [27], pp. 239–252
122. Shanks, D.: Class number, a theory of factorization and genera. Proc. Symp. PureMath. **20**, 415–440 (1971)
123. Shor, P.W.: Algorithms for quantum computation: discrete logarithms and factoring. In: Proceedings of the 35th Annual Symposium on Foundations of Computer Science, pp. 124–134. IEEE (1994)
124. Shoup, V.: Lower bounds for discrete logarithms and related problems. In: Fumy, W. (ed.) EUROCRYPT 1997. LNCS, vol. 1233, pp. 256–266. Springer, Heidelberg (1997). https://doi.org/10.1007/3-540-69053-0_18

125. Silverman, J.H.: The Arithmetic of Elliptic Curves. Graduate Texts in Mathematics, vol. 106. Springer, New York (1992)
126. Smart, N.P.: The discrete logarithm problem on elliptic curves of trace one. J. Cryptol. **12**(3), 193–196 (1999)
127. Smith, B.: Isogenies and the discrete logarithm problem in jacobians of genus 3 hyperelliptic curves. J. Cryptol. **22**(4), 505–529 (2009)
128. Stolbunov, A.: Constructing public-key cryptographic schemes based on class group action on a set of isogenous elliptic curves. Adv. Math. Commun. **4**(2), 215–235 (2010)
129. Sutherland, A.V.: Accelerating the CM method. LMS J. Comput. Math. **15**, 172–204 (2012)
130. Takagi, T., Peyrin, T. (eds.): ASIACRYPT 2017. LNCS, vol. 10625. Springer, Cham (2017). https://doi.org/10.1007/978-3-319-70697-9
131. Tani, S.: Claw finding algorithms using quantum walk. Theor. Comput. Sci. **410**(50), 5285–5297 (2009)
132. Thormarker, E.: Post-quantum cryptography: supersingular isogeny Diffie–Hellman key exchange. Ph.D. thesis, Stockholm University (2017)
133. Urbanik, D., Jao, D.: SoK: the problem landscape of SIDH. In: Proceedings of the 5th ACM on ASIA Public-Key Cryptography Workshop, APKC 2018, pp. 53–60. ACM, New York (2018)
134. van Dam, W., Hallgren, S., Ip, L.: Quantum algorithms for some hidden shift problems. SIAM J. Comput. **36**(3), 763–778 (2006)
135. Vélu, J.: Isogénies entre courbes elliptiques. C. R. Acad. Sci. Paris Sér. A-B **273**, A238–A241 (1971)
136. Wenger, E., Wolfger, P.: Harder, better, faster, stronger: elliptic curve discrete logarithm computations on FPGAs. J. Cryptogr. Eng. **6**(4), 287–297 (2016)

Elliptic Curves

A New Family of Pairing-Friendly Elliptic Curves

Michael Scott[1]($\boxtimes$) and Aurore Guillevic[2]

[1] MIRACL.com, Trim, Ireland
mike.scott@miracl.com
[2] Université de Lorraine, CNRS, Inria, LORIA, Nancy, France

Abstract. There have been recent advances in solving the finite extension field discrete logarithm problem as it arises in the context of pairing-friendly elliptic curves. This has lead to the abandonment of approaches based on supersingular curves of small characteristic, and to the reconsideration of the field sizes required for implementation based on non-supersingular curves of large characteristic. This has resulted in a revision of recommendations for suitable curves, particularly at a higher level of security. Indeed for a security level of 256 bits, the BLS48 curves have been suggested, and demonstrated to be superior to other candidates. These curves have an embedding degree of 48. The well known taxonomy of Freeman, Scott and Teske only considered curves with embedding degrees up to 50. Given some uncertainty around the constants that apply to the best discrete logarithm algorithm, it would seem to be prudent to push a little beyond 50. In this note we announce the discovery of a new family of pairing friendly elliptic curves which includes a new construction for a curve with an embedding degree of 54.

Keywords: Elliptic curves · Pairing-based cryptography
Aurifeuillean factorization

1 Introduction

One of great break-throughs in pairing-based cryptography was the discovery of the BN curves [3]. A group size of 256 bits (to match the 128-bit security level) can be supported by an elliptic curve over a field also of 256 bits, and since the embedding degree is 12, the size of the discrete logarithm (DL) problem over the extension field is 3072 bits. Which was a serendipitous direct hit on the size of DL problem believed to correspond to 128-bit security level. The fit was perfect.

Recall that protocols based on bilinear pairings typically consist of operations on three groups, denoted as $\mathbb{G}_1$, $\mathbb{G}_2$ and $\mathbb{G}_T$, and the calculation of the pairing itself, usually denoted as $w = e(P, Q)$, where the pairing takes two elliptic curve point parameters $P \in \mathbb{G}_1$ and $Q \in \mathbb{G}_2$ respectively, and evaluates to an element in the finite extension field $w \in \mathbb{G}_T$. Here $\mathbb{G}_1$ is contained in the elliptic curve $E(\mathbb{F}_p)$, $\mathbb{G}_2$ is contained in $E'(\mathbb{F}_{p^{k/d}})$, and $\mathbb{G}_T$ is contained in the finite extension

© Springer Nature Switzerland AG 2018
L. Budaghyan and F. Rodríguez-Henríquez (Eds.): WAIFI 2018, LNCS 11321, pp. 43–57, 2018.
https://doi.org/10.1007/978-3-030-05153-2_2

field $\mathbb{F}_{p^k}$, where k is the embedding degree associated with the pairing-friendly curve, and d is a divisor of k corresponding to a supported twisted curve E'. Note that $\mathbb{G}_2$ points can be manipulated on the smaller twisted curve, and transformed to a point on $E(\mathbb{F}_{p^k})$ only when needed.

The pairing calculation consists of two parts, a Miller loop followed by a final exponentiation. In real-world protocols much of the action takes place in the smallest group $\mathbb{G}_1$, although implementors have tended to concentrate more on the pairing itself. In more complex protocols products of pairings are required, and here a single final exponentiation can be applied to an amalgamation of Miller loops [20].

A BN curve is an example of a parameterised pairing-friendly curve, that is fixed polynomial formulae exist for the prime modulus p and the group order r in terms of an integer parameter u, which is chosen such that both p and r are prime. Such parameterised curves have become very popular for many reasons. First the ratio between the group and field size can be as low as one, and secondly multiple optimizations become possible. The most significant of these would be the development of the optimal ate pairing [23], for which the number of iterations of the Miller loop is reduced from the number of bits in r (as required for the original Tate pairing) to the number of bits in u. Since the degrees of the defining polynomial formulae increase with the embedding degree, rather paradoxically this implies that the number of iterations required in the Miller loop actually tends to decrease as the security level increases. Also a simpler form of final exponentiation applies [21].

However almost immediately after BN curves were introduced, Schirokauer [19] in a paper introducing the Number Field Sieve (NFS), warned us that: "Without discussing the evident difficulty of implementing the NFS for degree 12 fields, we observe that the special form of p may reduce the difficulty of computing logarithms in $\mathbb{F}_{p^{12}}$". In the absence of any concrete evidence to support this concern, the BN curve was nonetheless widely adopted. However Schirokauer's warning has proven to be prescient and the field of pairing-based cryptography has been disrupted by the recent, but not entirely surprising, discovery of a faster algorithm for solving the discrete logarithm problem in the finite extension field that arises when using these types of pairings, see [1,12,14].

A pairing-friendly curve can be characterised by the defining triplet $\{\rho, k, d\}$, where ρ is the ratio between the number of bits in p and the number of bits in r. Given a group size of g bits, the field size of $\mathbb{G}_1$ is $f_1 = \rho g$, the extension field size of $\mathbb{G}_T$ is $f_T = \rho k g$, and the field size of $\mathbb{G}_2$ is $f_2 = f_T/d$, where d is from the set of possible twists $\{1, 2, 3, 4, 6\}$, and is usually the maximum from this set that divides k.

When choosing a suitable curve, the starting point is the security level in bits, typically 128, 192 or 256. The group size should ideally be exactly twice this, and the other field sizes are then immediately fixed as shown above by the defining triplet.

For example for 128 bits of security of a BN curve, the defining triplet is $1, 12, 6$, and given $g = 256$, then $f_1 = 256$, $f_T = 3072$, and $f_2 = 512$. For a

BLS12 curve (by which we mean a BLS curve with embedding degree of 12, see below), the triplet is $3/2, 12, 6$, and given $g = 256$, then $f_1 = 384$, $f_T = 4608$, and $f_2 = 768$.

The main problem is to satisfy the security requirement for $\mathbb{G}_T$, so that it matches that for $\mathbb{G}_1$. See Table 1. Note that these numbers are rather imprecise as exact analysis is difficult. We have mainly followed the analysis of Barbulescu and Duquesne [1], extrapolating in places, rather than the less conservative estimates of Menezes, Sarkar and Singh [16]. However the estimates they provide depend on certain constants, in which one can have diminishing confidence as the security level increases.

Table 1. Recommended extension field sizes

DL Algorithm complexity	2^{128}	2^{192}	2^{256}
NFS ($L_{p^k}[1/3, 1.923]$)	3072	7680	15360
(ex)T$_{ower}$NFS medium ($L_{p^k}[1/3, 1.747]$)	3618	9241	18480
S$_{pecial}$(ex)T$_{ower}$NFS medium ($L_{p^k}[1/3, 1.526]$)	5004	12871	27410

Basically, according to current knowledge, only the NFS and TNFS estimates apply to finite fields of prime extension degree. The exTNFS estimates apply to composite order extensions, and the SexTNFS estimates to parameterised prime, composite order extensions, like the BN curves. It is now clear that BN curves are not quite as perfect as originally thought. As Barbulescu and Duquesne put it: "Variants of NFS where p is parameterized are considered to be the dream situation for an attacker", although they do go on to offer some reassurance that they do not expect any further improvements in the SexTNFS algorithm.

We should say a word about $\mathbb{G}_2$. Since this is of a size an integer multiple of $\mathbb{G}_1$, we can be confident that if $\mathbb{G}_1$ is secure then so is $\mathbb{G}_2$. However in the optimal ate pairing [23], each iteration of the Miller loop typically involves at least a point doubling in $\mathbb{G}_2$. Therefore we would like $\mathbb{G}_2$ to be as small as possible, and therefore the twist d to be as large as possible. The maximum attainable on an elliptic curve is $d = 6$, and this can only happen if $6|k$. But inevitably as the embedding degree k increases, so must $\mathbb{G}_2$. Ideally we would not want $\mathbb{G}_2$ growing too large, as elliptic curve cryptography over large extension fields will be very slow (and probably best implemented using affine coordinates).

From an implementation point of view the ideal solution is one that keeps f_1 as small as possible, while meeting all of the security constraints. This assumes that the value of ρ is small, that the embedding degree is such that we serendipitously hit the appropriate target in Table 1, and that a sextic twist applies and so the embedding degree is a multiple of 6.

An alternative response might be to revert to the Cocks-Pinch construction [17], avoiding parameterized curves, while continuing to use composite order extensions. It is not difficult to generate such curves for $k = 0 \mod 6$ such that sextic twists can be supported, although only for $\rho \geq 2$. The idea would be to revert to the original Tate pairing and adopt the lower exTNFS estimates. However this is unlikely to prove competitive in practice.

2 BLS and KSS Curves

BLS curves are the original small discriminant parameterised family of families of pairing-friendly elliptic curves [2]. For any positive embedding degree $k = 0 \bmod 6$ (unless $18|k$), they provide simple formulae from which can be derived pairing friendly curves which support the maximal twist of $d = 6$, and have a relatively small ρ value given by $\rho = (2+k/3)/\varphi(k)$ [9]. Observe that the value of ρ decreases with increasing values of k. Having a range of embedding degrees to choose from makes it easier to hit the optimal values in Table 1 for any security level.

For example Barbulescu and Duquesne [1] have demonstrated that the BLS12 curve is a good fit for the 128-bit level of security, and the BLS24 curve is the best choice for 192 bits of security. In another recent paper Kiyomura et al. [15], reacting to the new understanding, demonstrated that a BLS48 curve is also the best choice of pairing-friendly curve to meet the new estimates for the 256-bit level of security. In this case the security requirement could be met with a group size of 512 bits, a modulus of 576 bits (as $\rho = 9/8$), and a finite field extension size of $48 \cdot 576 = 27648$.

However the BLS curves do not exist for $18|k$, as in these cases the polynomial formula for p is not irreducible, and therefore cannot generate primes [9]. Serendipitously for the cases of $k = 18$ and $k = 36$ there do exist the alternative KSS curves [13], which, as luck would have it, provide curves with the same ρ values as determined by the above formula for the missing BLS curves. However since the taxonomy of Freeman, Scott and Teske does not explore beyond $k = 50$, the situation for $k = 54$ is currently unknown. But if a BLS curve did exist for $k = 54$, then from the formula given above, it would have a ρ value of $10/9$.

3 The New Discovery

The new curve was discovered using the KSS method as described in [13]. It was immediately observed that the new curve found with embedding degree 54 is not of the form of a typical KSS curve, where integer solutions exist only in a restricted set of residue classes. Recall that the KSS method also rediscovers the BN curves [3]. It appeared possible that the new family of curves was, like the BN curves, a "sporadic", and not related to any existing family. On the other hand it has a certain symmetry, which suggested that it might be a member of an as-yet undiscovered family of families.

We found that $-\zeta_{54} - \zeta_{54}^{10}$ as a suitable element $\in \mathbb{Q}(\zeta_{54})$, and following the KSS method [13] from there we obtained the solution

$$p = 1 + 3u + 3u^2 + 3^5 u^9 + 3^5 u^{10} + 3^6 u^{10} + 3^6 u^{11} + 3^9 u^{18} + 3^{10} u^{19} + 3^{10} u^{20}$$
$$r = 1 + 3^5 u^9 + 3^9 u^{18}$$
$$t = 1 + 3^5 u^{10}$$
$$c = 1 + 3u + 3u^2$$

$$(1)$$

where p is the prime modulus, r is the prime order of the pairing-friendly group, t is the trace of the Frobenius, and c is a cofactor. It can be verified that the Complex Multiplication (CM) discriminant is $D = 3$ because $4p - t^2 = 3f^2$, for some polynomial f. This implies that the curve has twists of degree 6 which, as with the BN and BLS curves, facilitates an important optimization. Observe that the prime p can be any of 1, 3, 5 or 7 mod 8 depending on the choice of u. The total number of points on the curve will be $\#E = cr$.

Recall that the embedding degree is the smallest value of k such that $r|(p^k - 1)$ [17]. In this case it is easily confirmed that $k = 54$. The value of $\rho = \deg(p)/\deg(r) = 10/9$, which is close to the ideal value of 1.

4   Aurifeuillean Construction of Pairing-Friendly Curves

The method of discovery does not explain the particular form of the new curve, or whether or not it is a member of a larger "family of families". To answer these questions we take a different approach.

The Aurifeuillean factorization of cyclotomic polynomials is a well-known tool used in the Cunningham project[1] [7] to factor large integers of the form $b^n \pm 1$, where b is a square-free basis, $b \in \{2, 3, 5, 6, 7, 10, 11, 12\}$. Some of the factors can be obtained with the algebraic factorisation: since $u^n - 1 = \prod_{d|n} \Phi_d(u)$, where Φ_d is the d-th cyclotomic polynomial, then $b^n - 1 = \prod_{d|n} \Phi_d(b)$. For some combinations of bases and powers, more factors can be obtained with the Aurifeuillean factorization of the cyclotomic polynomials [5, 18, 22]:

Lemma 4.1. *Let $k > 1$ be an integer and let $\Phi_k(u)$ denote the k-th cyclotomic polynomial. Let a be a square-free integer and s an integer. Then $\Phi_k(as^2)$ will factor if*

- *$a \equiv 1 \pmod{4}$ and $a \equiv k \pmod{2a}$*
- *or $a \equiv 2, 3 \pmod{4}$ and $2a \equiv k \pmod{4a}$.*

This fact is already known for pairing-friendly curves: it was used to compute the order of the supersingular elliptic curves in characteristic 3 of embedding degree 6 (see for instance [8, Table 1]). Let $E/\mathbb{F}_{3^\ell}$ be a supersingular elliptic curve defined over $\mathbb{F}_{3^\ell}$ for an odd ℓ. If the curve has embedding degree 6, then the order

---

[1] http://www.cerias.purdue.edu/homes/ssw/cun/index.html.

of $E(\mathbb{F}_{3^\ell})$ is a divisor of $\Phi_6(3^\ell)$, where $\Phi_6(u) = u^2 - u + 1$. Assuming that $\ell = 2m+1$, the Aurifeuillean factorization is $\Phi_6(3u^2) = (3u^2 + 3u + 1)(3u^2 - 3u + 1)$. Replacing u by 3^m so that $3u^2 = 3 \cdot 3^{2m} = 3^\ell$, we get $\Phi_6(3^{2m+1}) = (3^{2m+1} + 3^{m+1} + 1)(3^{2m+1} - 3^{m+1} + 1)$, and $\#E(\mathbb{F}_{3^\ell}) = 3^\ell - 3^{(\ell+1)/2} + 1$. We will apply this tool to find new families of pairing-friendly ordinary curves in large characteristic for $k = 3^j$ and $k = 2 \cdot 3^j$. Moreover this factorization pattern provides a larger framework for pairing-friendly curve construction, and a general point of view for the construction of MNT curves, BN curves and Freeman's curves.

4.1 Aurifeuillean Construction

We combine the Brezing-Weng method [6], providing *cyclotomic families*, with Lemma 4.1, to obtain Algorithm 1. The idea is to look for an integer a where $-2k \leq a \leq 2k$, and satisfying Lemma 4.1 so that $\Phi_k(au^2) = r(u)r(-u)$, and we continue as for the Brezing-Weng method, with $r(u)$ a factor of $\Phi_k(au^2)$, and $\zeta_k = au^2$. The number field $K = \mathbb{Q}[u]/(r(u))$ contains a square root of a. When a is positive, we can choose $D = a$, or a multiple of a.

Algorithm 1. Aurifeuillean construction of pairing-friendly curves

---

Input: embedding degree k, small discriminant D
Output: family (r, t, y, p) of a pairing-friendly elliptic curve of embedding
degree k, or $\perp$

1 $\rho_{\min} \leftarrow 2$
2 **for** $a \in \{-2k, \ldots, -3, -2, 2, 3, \ldots, 2k\}$ **do**
3   **if** *IsSquareFree(a)* and $((a = 1 \bmod 4$ and $k = a \bmod 2a)$ or $(a = 2, 3 \bmod 4$
        and $k = 2a \bmod 4a))$ **then**
4          $r(u) =$ an irreducible factor of $\Phi_k(au^2)$, s.t. $\Phi_k(au^2) = r(u)r(-u)$
5          $K = \mathbb{Q}(\omega) = \mathbb{Q}[u]/(r(u))$
6   **if** $-D$ *is a square in* K **then**
7              $S = 1/\sqrt{-D} \in K$
8   write S as a polynomial $s(u)$ s.t. $S = s(\omega)$ in K
9   **for** $e = 1 \ldots k - 1$, $\gcd(e, k) = 1$ **do**
10                 $t(u) = (au^2)^e + 1 \bmod r(u)$
11                 $y(u) = (t(u) - 2)s(u) \bmod r(u)$
12                 $p(u) = (t^2(u) + Dy^2(u))/4$
13   **if** $p(u)$ *represents primes (cf. [9, Def. 2.5]) and*
                   $\deg p / \deg r < \rho_{\min}$ **then**
14                     $\rho_{\min} \leftarrow \deg p / \deg r$
15                     $F_{\min} \leftarrow (r, t, y, p)$
16 **if** $\rho_{\min} = 2$ **then**
17   **return** $\perp$
18 **else**
19   **return** $F_{\min}$

We ran Algorithm 1 for $7 \leq k \leq 100$, and small D ($1 \leq D \leq 100$). We checked that the polynomials $r(x)$ and $p(x)$ can give prime integers (there exist several x_0 such that $p(x_0)$ and $r(x_0)$ are prime at the same time). For $k = 7$ and $k = 35$, $p(x)$ does not represent primes (p is always even for $k = 7$ and $21 \mid p$ for $k = 35$). We obtained new families of pairing-friendly curves of ρ value as good as [9] for $k \in \{9, 15, 21, 30, 33, 39, 42, 45, 51, 54, 57, 66, 69, 75, 78, 81, 87, 90, 93\}$. Each time the best ρ was obtained for $a = 3$ or $a = -3$, and $D = 3$. For $k = 12$ and $a = \pm 6$, the output is the BN curve family. As an example, we give the parameters obtained for $k = 15$. The family for $k = 9$ (Example 4.5) falls in our new construction that we present in the next section.

Example 4.2. Aurifeuillean construction for $k = 15$. We obtain a family with $D = 3$, $\deg r(u) = 8$ and $\rho = 4/3$ as in [9]. $\Phi_{15}(-3u^2) = r(u)r(-u)$ where $r(u) = 81u^8 + 81u^7 + 54u^6 + 27u^5 + 9u^4 + 9u^3 + 6u^2 + 3u + 1$. There are two choices for $t(u)$, producing two families with $\rho = 4/3$:

$$
\begin{aligned}
t_1(u) &= 54u^6 + 3u + 1 = (-3u^2)^8 + 1 \bmod r(u) \\
y_1(u) &= -18u^5 - u - 1 \\
p_1(u) &= 729u^{12} + 243u^{10} + 81u^7 + 54u^6 + 27u^5 + 3u^2 + 3u + 1 \\
t_2(u) &= -27u^6 - 3u + 1 = (-3u^2)^{13} + 1 \bmod r(u) \\
y_2(u) &= -27u^6 - 18u^5 - u - 1 \\
p_2(u) &= 729u^{12} + 729u^{11} + 243u^{10} + 81u^7 + 54u^6 + 27u^5 + 3u^2 + 1
\end{aligned}
$$

4.2 Aurifeuillean Family for $k = 3^j$

We now explain the generalization of the $k = 54$ family to any $k = 3^j$ using an Aurifeuillean factorization of $\Phi_{3^j}(u)$. We start by the general expression:

$$\Phi_{3^j}(u) = \Phi_3(u^{3^{j-1}}) = u^m + u^{m/2} + 1, \text{ where } m = \varphi(3^j) = 2 \cdot 3^{j-1}. \quad (2)$$

If $k = 3^j$, then to obtain an Aurifeuillean factorization, we take $a = -3 \equiv 1 \bmod 4$, we need $k = a \bmod 2a \Leftrightarrow 3^j = 3 \bmod 6$, and indeed this is always the case. The degree of $\Phi_k(u)$ is $m = 2 \cdot 3^{j-1}$, in particular m is even and $m/2$ is odd. We obtain the Aurifeuillean factorization:

$$
\begin{aligned}
\Phi_{3^j}(-3u^2) &= 3^m u^{2m} - 3^{m/2} u^m + 1 \\
&= (3^{m/2} u^m + 3^{(m+2)/4} u^{m/2} + 1)(3^{m/2} u^m - 3^{(m+2)/4} u^{m/2} + 1) \\
&= r(u)r(-u)
\end{aligned}
$$

where

$$r(u) = 3^{m/2} u^m + 3^{(m+2)/4} u^{m/2} + 1.$$

We choose $D = 3$, we compute $\sqrt{-3}$ and its inverse modulo $r(u)$:

$$
\begin{aligned}
\sqrt{-3} &= -2 \cdot 3^{(m+2)/4} u^{m/2} - 3 \\
1/\sqrt{-3} &= -\sqrt{-3}/3 = 2 \cdot 3^{(m-2)/4} u^{m/2} + 1.
\end{aligned}
$$

We know that $-3u^2$ is a primitive k-th root of unity in $K = \mathbb{Q}(u)/(r(u))$. All the $(-3u^2)^e$ for $3 \nmid e$ are primitive k-th roots of unity modulo $r(u)$, i.e. $e \neq 0 \bmod 3$. We have

$$t(u) = (-3u^2)^e + 1 \bmod r(u)$$
$$y(u) = (t(u) - 2)/\sqrt{-D} \bmod r(u)$$
$$= ((-3u^2)^e - 1)(2 \cdot 3^{(m-2)/4}u^{m/2} + 1) \bmod r(u).$$

To get $\rho = \deg p/\deg r$ as small as possible, we want to minimize

$$\max(\deg t(u), \deg y(u)).$$

The possible degrees e to get the smallest possible degree of $p(u)$ are listed in Table 2, with the values of $t(u), y(u) \bmod r(u)$, and $p(u)$.

4.3   Aurifeuillean Family for $k = 2 \cdot 3^j$

We proceed the same way as in Sect. 4.2. The general expression for $\Phi_{2 \cdot 3^j}(u)$ is

$$\Phi_{2 \cdot 3^j}(u) = \Phi_{3^j}(-u) = u^m - u^{m/2} + 1, \text{ where } m = \varphi(2 \cdot 3^j) = 2 \cdot 3^{j-1}. \quad (3)$$

To obtain an Aurifeuillean factorization of $\Phi_{2 \cdot 3^j}(u)$, we choose $a = 3$ (so $a = 3 \bmod 4$) and the condition $k = 2a \pmod{4a} \Leftrightarrow k = 2 \cdot 3^j = 6 \bmod 12$ is always satisfied since $6 \mid k$ but $4 \nmid k$. We obtain

$$\Phi_{2 \cdot 3^j}(3u^2) = \Phi_{3^j}(-3u^2) = 3^m u^{2m} - 3^{m/2}u^m + 1$$
$$= r(u)r(-u)$$

where again

$$r(u) = 3^{m/2}u^m + 3^{(m+2)/4}u^{m/2} + 1.$$

We take $D = 3$, and we know that $3u^2$ is a primitive k-th root (a primitive $2 \cdot 3^j$-th root) of unity in $K = \mathbb{Q}(\omega)$ where ω is a root of $r(u)$. In the same way as previously we obtain Table 3. The polynomial $r(u)$ is the same but the trace differs and the embedding degree of the family is doubled.

As speculated in Sect. 3, the $k = 54$ family found with the KSS method is indeed a member of a larger "family of families".

Construction 4.3. For $n = 3^j$ and $m = \varphi(n)$, then a pairing-friendly curve with embedding degree of $k = 2n$ if j is odd, and $k = n$ if j is even, with discriminant $D = 3$, and with a ρ value of $(m+2)/m$, can be found as

$$r(u) = 1 + 3^{(m+2)/4}u^{m/2} + 3^{m/2}u^m$$
$$t(u) = 1 - 3u - 2 \cdot 3^{(m+2)/4}u^{1+m/2}$$
$$c(u) = 1 + 3u^2 \qquad\qquad (4)$$
$$p(u) = c(u) \cdot r(u) + t(u) - 1$$

Table 2. For $k = 3^j$, values of e such that $\gcd(e, k) = 1$, $t(u) = (-3u^2)^e + 1 \bmod r(u)$, $p(u)$ is irreducible, and $\deg p(u)$ is minimal. For even j, $\rho = (m+2)/m$. For odd j, $p(u)$ is not irreducible for the first three values $e = (m+2)/4, 3(m-2)/4+2, m+(m+2)/4$, only the last three ones with $\rho = (m+4)/m$ are possible.

even j			
e	$(m+2)/4$	deg	ρ
$t(u) \bmod r(u)$	$3^{(m+2)/4}u^{m/2+1} + 1$	$m/2+1$	
$y(u) \bmod r(u)$	$-3^{(m+2)/4}u^{m/2+1} - 2 \cdot 3^{(m-2)/4}u^{m/2} - 2u - 1$	$m/2+1$	
$p(u)$	$(3u^2 + 3u + 1)r(u) + t(u) - 1$	$m+2$	$(m+2)/m$
e	$(m+2)/4 + m/2$		
$t(u) \bmod r(u)$	$-2 \cdot 3^{(m+2)/4}u^{m/2+1} - 3u + 1$	$m/2+1$	
$y(u) \bmod r(u)$	$-2 \cdot 3^{(m-2)/4}u^{m/2} + u - 1$	$m/2$	
$p(u)$	$(3u^2 + 1)r(u) + t(u) - 1$	$m+2$	$(m+2)/m$
e	$(m+2)/4 + m$		
$t(u) \bmod r(u)$	$3^{(m+2)/4}u^{m/2+1} + 3u + 1$	$m/2+1$	
$y(u) \bmod r(u)$	$3^{(m+2)/4}u^{m/2+1} - 2 \cdot 3^{(m-2)/4}u^{m/2} + u - 1$	$m/2+1$	
$p(u)$	$(3u^2 - 3u + 1)r(u) + t(u) - 1$	$m+2$	$(m+2/m)$
even and odd j			
e	1		
$t(u) \bmod r(u)$	$-3u^2 + 1$	2	
$y(u) \bmod r(u)$	$-2 \cdot 3^{(m+2)/4}u^{m/2+2} - 2 \cdot 3^{(m-2)/4}u^{m/2} - 3u^2 - 1$	$m/2+2$	
$p(u)$	$(9u^4 + 6u^2 + 1)r(u) + t(u) - 1$	$m+4$	$(m+4)/m$
e	$1 + m/2$		
$t(u) \bmod r(u)$	$-3^{(m+6)/4}u^{m/2+2} - 3u^2 + 1$	$m/2+2$	
$y(u) \bmod r(u)$	$3^{(m+2)/4}u^{m/2+2} - 2 \cdot 3^{(m-2)/4}u^{m/2} + 3u^2 - 1$	$m/2+2$	
$p(u)$	$(9u^4 - 3u^2 + 1)r(u) + t(u) - 1$	$m+4$	$(m+4)/m$
e	$1 + m$		
$t(u) \bmod r(u)$	$3^{(m+6)/4}u^{m/2+2} + 2 \cdot 3u^2 + 1$	$m/2+2$	
$y(u) \bmod r(u)$	$3^{(m+2)/4}u^{m/2+2} - 2 \cdot 3^{(m-2)/4}u^{m/2} - 1$	$m/2+2$	
$p(u)$	$(9u^4 - 3u^2 + 1)r(u) + t(u) - 1$	$m+4$	$(m+4)/m$

Construction 4.4. *For $n = 3^j$ and $m = \varphi(n)$, then a pairing-friendly curve with embedding degree of $k = n$ or $k = 2n$, with discriminant $D = 3$, and with a ρ value of $(m+4)/m$, can be found as*

$$r(u) = 1 + 3^{(m+2)/4}u^{m/2} + 3^{m/2}u^m$$
$$t(u) = 1 + 3\,\epsilon\,u^2$$
$$c(u) = 9u^4 - 6\,\epsilon\,u^2 + 1 \tag{5}$$
$$p(u) = c(u) \cdot r(u) + t(u) - 1$$
$$\epsilon = (-1)^{k \bmod 2}$$

This hypothesis has been tested for all applicable embedding degrees less than 1000. However it is not particularly useful for cases other than $k = 54$. For

Table 3. For $k = 2 \cdot 3^j$, values of e such that $\gcd(e, k) = 1$, $t(u) = (3u^2)^e + 1 \bmod r(u)$, $p(u)$ is irreducible, and $\deg p(u)$ is minimal. For odd j, $\rho = (m+2)/m$. For even j, $p(u)$ is not irreducible for the first three values $e = (m+2)/4, m+(m+2)/4, 2m+(m+2)/4$, only the last three ones with $\rho = (m+4)/m$ are possible.

j odd		deg	ρ
e	$(m+2)/4$, $e = 5 \bmod 6$		
$t(u) \bmod r(u)$	$3^{(m+2)/4}u^{m/2+1} + 1$	$m/2 + 1$	
$y(u) \bmod r(u)$	$-3^{(m+2)/4}u^{m/2+1} - 2 \cdot 3^{(m-2)/4}u^{m/2} - 2u - 1$	$m/2 + 1$	
$p(u)$	$(3u^2 + 3u + 1)r(u) + t(u) - 1$	$m + 2$	$(m+2)/m$
e	$m + (m+2)/4$, $e = 5 \bmod 6$		
$t(u) \bmod r(u)$	$3^{(m+2)/4}u^{m/2+1} + 3u + 1$	$m/2 + 1$	
$y(u) \bmod r(u)$	$3^{(m+2)/4}u^{m/2+1} - 2 \cdot 3^{(m-2)/4}u^{m/2} + u - 1$	$m/2 + 1$	
$p(u)$	$(3u^2 - 3u + 1)r(u) + t(u) - 1$	$m + 2$	$(m+2)/m$
e	$2m + (m+2)/4$, $e = 5 \bmod 6$		
$t(u) \bmod r(u)$	$-2 \cdot 3^{(m+2)/4}u^{m/2+1} - 3u + 1$	$m/2 + 1$	
$y(u) \bmod r(u)$	$-2 \cdot 3^{(m-2)/4}u^{m/2} + u - 1$	$m/2$	
$p(u)$	$(3u^2 + 1)r(u) + t(u) - 1$	$m + 2$	$(m+2)/m$
j odd and even			
e	1		
$t(u) \bmod r(u)$	$3u^2 + 1$	2	
$y(u) \bmod r(u)$	$2 \cdot 3^{(m+2)/4}u^{m/2+2} - 2 \cdot 3^{(m-2)/4}u^{m/2} + 3u^2 - 1$	$m/2 + 2$	
$p(u)$	$(9u^4 - 6u^2 + 1)r(u) + t(u) - 1$	$m + 4$	$(m+4)/m$
e	$m + 1$, $e = 1 \bmod 6$		
$t(u) \bmod r(u)$	$-3^{(m+6)/4}u^{m/2+2} - 2 \cdot 3u^2 + 1$	$m/2 + 2$	
$y(u) \bmod r(u)$	$-3^{(m+2)/4}u^{m/2+2} - 2 \cdot 3^{(m-2)/4}u^{m/2} - 1$	$m/2 + 2$	
$p(u)$	$(9u^4 + 3u^2 + 1)r(u) + t(u) - 1$	$m + 4$	$(m+4)/m$
e	$2m + 1$, $e = 1 \bmod 6$		
$t(u) \bmod r(u)$	$3^{(m+6)/4}u^{m/2+2} + 3u^2 + 1$	$m/2 + 2$	
$y(u) \bmod r(u)$	$-3^{(m+2)/4}u^{m/2+2} - 2 \cdot 3^{(m-2)/4}u^{m/2} - 3u^2 - 1$	$m/2 + 2$	
$p(u)$	$(9u^4 + 3u^2 + 1)r(u) + t(u) - 1$	$m + 4$	$(m+4)/m$

the embedding degrees that arise from these formulae which are less than 54 (6 and 9), there already exist curves with the same or better ρ value. The higher values of embedding degree (81, 486) are probably not useful in practice.

4.4 Applications

Our Aurifeuillean Constructions 4.3 and 4.4 for $k = 3^j$ and $k = 2 \cdot 3^j$ can be applied when the Brezing–Weng construction and the Construction 6.6 of [9] do not provide a satisfying result. The Brezing-Weng fails for $k = 54$ ($p(u)$ is never irreducible) and $k = 90$: p is not irreducible, or does not generate primes. The construction 6.6 of [9] fails for $18 \mid k$ ($k = 18, 36, 54, 72, 90$). We can alternatively use the Aurifeuillean construction when $18 \mid k$ and $8 \nmid k$, that is $k \in \{18, 54, 90\}$.

Unfortunately for $k = 18$ the Aurifeuillean construction gives $\rho = 5/3$, larger than $\rho = 4/3$ achieved by [13] (referenced as Construction 6.12 in [9]). The construction provides a new family for $k = 54$ with $\rho = 10/9$. For $k = 90$ however, the coefficients of $p(u)$ are very large and such a large embedding degree is unlikely to be used in pairing-based cryptography.

Our family also covers $k = 9$ and as a conclusion we provide our alternative choice for $k = 9$ and $D = 3$.

Example 4.5. Aurifeuillean construction for $k = 9$. $\Phi_9(-3u^2) = r(u)r(-u)$ where $r(u) = 27u^6 + 9u^3 + 1$. Three are three choices for the trace $t(u)$. With $D = 3$, we obtain $\rho = 4/3$.

$$
\begin{aligned}
t_1(u) &= -18u^4 - 3u + 1 = (-3u^2)^5 + 1 \bmod r(u) \\
y_1(u) &= -6u^3 + u - 1 \\
p_1(u) &= 81u^8 + 27u^6 + 27u^5 - 18u^4 + 9u^3 + 3u^2 - 3u + 1 \\[4pt]
\hline
t_2(u) &= 9u^4 + 3u + 1 = (-3u^2)^8 + 1 \bmod r(u) \\
y_2(u) &= 9u^4 - 6u^3 + s - 1 \\
p_2(u) &= 81u^8 - 81u^7 + 27u^6 + 27u^5 - 18u^4 + 9u^3 + 3u^2 + 1 \\[4pt]
\hline
t_3(u) &= 9u^4 + 1 = (-3u^2)^2 + 1 \bmod r(u) \\
y_3(u) &= -9u^4 - 6u^3 - 2u - 1 \\
p_3(u) &= 81u^8 + 81u^7 + 27u^6 + 27u^5 + 36u^4 + 9u^3 + 3u^2 + 3u + 1
\end{aligned}
$$

MNT Curves as Aurifeuillean Curves for $k = 3$. The MNT construction provides three families of curves of embedding degree 3, 4 and 6 respectively. The curve for $k = 3$ can be obtained with the Aurifeuillean factorization of $\Phi_3(u)$, and the two curves for $k = 4, 6$ with the cyclotomic construction. We start with $\Phi_3(-3u^2) = (3u^2 + 3u + 1)(3u^2 - 3u + 1) = r(u)r(-u)$. The two choices for the trace are $-3u^2 + 1 \bmod r(u) = -3u + 2 = t_1$ and $(-3u^2)^2 + 1 \bmod r(u) = 3u - 1 = t_2$. Since $t_1 = t_2(-u + 1)$, we continue with $t = 3u - 1$, and compute $p(u) = r(u) + t(u) - 1 = 3u^2 - 1$. We obtain the first MNT curve (see Table 4), with the change of variable $l = 2u$ (indeed, $p = 3u^2 - 1$ is always even for odd u, so the MNT family takes $l = 2u$). The CM equation is $4p - t^2 = 3u^2 + 6u - 5$, and requires to solve a Pell equation as in the original paper. The MNT curve families for $k = 4$ and $k = 6$ do not correspond to the Aurifeuillean construction, but to a cyclotomic construction ($r(u) = \Phi_k(u)$). The two Aurifeuillean constructions (without choosing $-D$ as a square in $\mathbb{Q}(\zeta_k)$) for $k = 4, 6$ produce supersingular curves of characteristic 2 and 3 respectively. We summarise this in Table 4.

Galbraith, McKee and Valença Factorisation Patterns. Galbraith, McKee and Valença already investigated the strategy of finding $q(l)$ such that $\Phi_k(q(l))$ splits into two quadratic factors for $k = 3, 4, 6$ or two quartic factors for $k = 5, 8, 10, 12$ in [10]. They obtained Aurifeuillean factorisation patterns for $k = 3, 4, 5, 6, 10, 12$. Their work allowed Freeman to obtain $k = 10$ curves and Barreto and Naehrig to obtain $k = 12$ curves, both with $\rho = 1$.

Table 4. Correspondence between MNT families, cyclotomic construction and Aurifeuillean factorization

k	MNT		Cyclotomic	Aurifeuillean
3	$t(u)$	$-1 \pm 6l$	$u+1, -u \ (\zeta_3 = u, -u-1)$	$-3u+2, 3u-1$
	$r(u)$	$12l^2 \mp 6l + 1$	$\Phi_3(u) = u^2 + u + 1$	$3u^2 - 3u + 1, \ \Phi_3(-3u^2) = r(u)r(-u)$
	$p(u)$	$12l^2 - 1$	$(u+1)^2, u^2$	$3u^2 - 6u + 2, \ 3u^2 - 1$
	Dy^2	$12l^2 \pm 12l - 5$	$3(u+1)^2, 3u^2$	$3u^2 - 12u + 4, \ 3u^2 + 6u - 5$
			supersingular, $q = p^2$	MNT with $l = 2u$
4	$t(u)$	$-l, \ l+1$	$\pm u + 1 \ (\zeta_4 = \pm u)$	$2u, -2(u-1)$
	$r(u)$	$l^2 + 2l + 2, \ l^2 + 1$	$\Phi_4(u) = u^2 + 1$	$2u^2 - 2u + 1, \ \Phi_4(\pm 2u^2) = r(u)r(-u)$
	$p(u)$	$l^2 + l + 1$	$u^2 \pm u + 1$	$2u^2, 2(u-1)^2$
	Dy^2	$3l^2 + 4l + 4$	$3u^2 \pm 2u + 3$	$4u^2, 4(u-1)^2$
			MNT with $l = u, u-1$	supersingular, $q = 2^\ell$
6	$t(u)$	$1 \pm 2l$	$u+1, -u+2 (\zeta_6 = u, -u+1)$	$3u, -3(u-1)$
	$r(u)$	$4l^2 \mp 2l + 1$	$\Phi_6(u) = u^2 - u + 1$	$3u^2 - 3u + 1, \ \Phi_6(3u^2) = r(u)r(-u)$
	$p(u)$	$4l^2 + 1$	$u^2 + 1, u^2 - 2u + 2$	$3u^2, 3(u-1)^2$
	Dy^2	$12l^2 - 4l + 3$	$3u^2 - 2u + 3, 3u^2 - 4u + 4$	$3u^2, 3(u-1)^2$
			MNT with $l = 2u$	supersingular, $q = 3^\ell$

4.5 Further Investigations

Granville and Pleasants [11] investigated the possibility that there are other identities still to be discovered. Wagstaff [24] used the Cunningham tables to try unsuccessfully to discover new identities. His results tend to confirm the theoretical results of [11] that under reasonable definitions, Schinzel found the last Aurifeuillean-like factorizations.

It seems very unlikely that a new mysterious pairing-friendly family as the Barreto-Naehrig curves with $\rho = 1$ will be discovered with similar techniques: we ran Algorithm 1 for $k \leq 100$ without success.

New discoveries are still possible, but will more probably arise with large computer search for factorization of $\Phi_k(g(u))$ for polynomials $g(u)$ of degree strictly larger than 2, as for $k = 8$.

5 An Example Construction

An actual curve can be generated using the seed value $u = C404042_{16}$, which has a low Hamming weight of 6. Then the curve

$$y^2 = x^3 + 12$$

is a pairing-friendly elliptic curve with a group order r of 512 bits, and a modulus p of 569 bits. Given the embedding degree of 54, the finite extension field is of size 30726 bits, comfortably, but not excessively, above the size recommended for an overall security equivalent to 256 bits (from Table 1). The embedding degree $k = 54$ is obviously of the desirable form $k = 2^i 3^j$, which simplifies implementation [15].

To derive an optimal pairing, following [23] we find the shortest vector in a lattice and observe that $u + up^9 + p^{10} = 0 \bmod r$. Then an optimal ate pairing is defined as

$$t(Q,P) = (f_{u,Q}^{p^9+1}(P).l_{p^9uQ+p^{10}Q,uQ}(P).l_{p^{10}Q,p^9uQ}(P))^{(p^{54}-1)/r}$$

The Miller loop (of just 28 iterations, given the bit length of u) provides $f_{u,Q}(P)$ and the value of uQ, after which two line function evaluations, some cheap applications of the Frobenius operator, and a final exponentiation complete the calculation.

Implementation will require the construction of a tower of extensions. Since $\mathbb{G}_2$ is over $E'(\mathbb{F}_{p^9})$ it would make sense to use a 1-3-9-18-54 towering, similar to that recommended for the $k = 18$ case [4]. It is apparent from the defining equation that $p = 1 \bmod 3$. It is also clear that u must be even to generate primes. So we make the substitution $u = 2v$ and from there it is straightforward if tedious to coerce Fermat's identity $p = a^2 + 3b^2$ where

$$
\begin{aligned}
a &= 1 + 3v + 2^8 3^5 v^9 + 2^{10} 3^5 v^{10} \\
b &= v + 2^8 3^4 v^9
\end{aligned}
\tag{6}
$$

As demonstrated by Benger and Scott [4] this implies by Euler's conjecture that $x^{54} - 2$ is irreducible over $\mathbb{F}_p$ as long as $3 \nmid b$, which is equivalent to the simple condition that $3 \nmid u$.

6 Conclusion

We present a new family of pairing friendly curves with an embedding degree of $k = 54$, which fills a gap that might be useful in the event of a deeper understanding emerging of the true difficulty of the discrete logarithm problem as it applies to high-security pairing-based cryptography. Motivated by this discovery we place it into a wider context, and identify it as just one member of a larger family of curves. The $k = 54$ solution may have been previously overlooked as the limit of practical interest was at one time conservatively estimated as being 50 [9]. We also strived to find a solution for the next "missing" case of $k = 72$ (by which we mean an embedding degree which is a multiple of six, and which is not covered by the BLS construction) but failed despite an extensive computer search. Nevertheless clearly the KSS method is a powerful tool for discovering families of pairing-friendly curves.

References

1. Barbulescu, R., Duquesne, S.: Updating key size estimations for pairings. J. Cryptol. (2018). https://doi.org/10.1007/s00145-018-9280-5. http://eprint.iacr.org/2017/334

2. Barreto, P.S.L.M., Lynn, B., Scott, M.: Constructing elliptic curves with prescribed embedding degrees. In: Cimato, S., Persiano, G., Galdi, C. (eds.) SCN 2002. LNCS, vol. 2576, pp. 257–267. Springer, Heidelberg (2003). https://doi.org/10.1007/3-540-36413-7_19

3. Barreto, P.S.L.M., Naehrig, M.: Pairing-friendly elliptic curves of prime order. In: Preneel, B., Tavares, S. (eds.) SAC 2005. LNCS, vol. 3897, pp. 319–331. Springer, Heidelberg (2006). https://doi.org/10.1007/11693383_22

4. Benger, N., Scott, M.: Constructing tower extensions of finite fields for implementation of pairing-based cryptography. In: Hasan, M.A., Helleseth, T. (eds.) WAIFI 2010. LNCS, vol. 6087, pp. 180–195. Springer, Heidelberg (2010). https://doi.org/10.1007/978-3-642-13797-6_13

5. Brent, R.P.: On computing factors of cyclotomic polynomials. Math. Comp. **61**(203), 131–149 (1993). https://doi.org/10.2307/2152941

6. Brezing, F., Weng, A.: Elliptic curves suitable for pairing based cryptography. Des. Codes Cryptogr. **37**(1), 133–141 (2005). https://eprint.iacr.org/2003/143

7. Brillhart, J., Lehmer, D.H., Selfridge, J.L., Tuckerman, B., Wagstaff Jr., S.S.: Factorizations of $b^n \pm 1$, $b = 2, 3, 5, 6, 7, 10, 11, 12$ up to High Powers. Contemporary Mathematics, 2nd edn, vol. 22. American Mathematical Society, Providence (1988). https://homes.cerias.purdue.edu/ssw/cun/

8. Estibals, N.: Compact hardware for computing the tate pairing over 128-bit-security supersingular curves. In: Joye, M., Miyaji, A., Otsuka, A. (eds.) Pairing 2010. LNCS, vol. 6487, pp. 397–416. Springer, Heidelberg (2010). https://doi.org/10.1007/978-3-642-17455-1_25

9. Freeman, D., Scott, M., Teske, E.: A taxonomy of pairing-friendly elliptic curves. J. Cryptol. **23**(2), 224–280 (2010). http://eprint.iacr.org/2006/372

10. Galbraith, S.D., McKee, J.F., Valença, P.C.: Ordinary Abelian varieties having small embedding degree. Finite Fields Appl. **13**(4), 800–814 (2007). https://eprint.iacr.org/2004/365

11. Granville, A., Pleasants, P.: Aurifeuillian factorization. Math. Comp. **75**(253), 497–508 (2006). https://doi.org/10.1090/S0025-5718-05-01766-7

12. Joux, A., Pierrot, C.: The special number field sieve in $\mathbb{F}_{p^n}$ - application to pairing-friendly constructions. In: Cao, Z., Zhang, F. (eds.) Pairing 2013. LNCS, vol. 8365, pp. 45–61. Springer, Cham (2014). https://doi.org/10.1007/978-3-319-04873-4_3

13. Kachisa, E.J., Schaefer, E.F., Scott, M.: Constructing brezing-weng pairing-friendly elliptic curves using elements in the cyclotomic field. In: Galbraith, S.D., Paterson, K.G. (eds.) Pairing 2008. LNCS, vol. 5209, pp. 126–135. Springer, Heidelberg (2008). https://doi.org/10.1007/978-3-540-85538-5_9

14. Kim, T., Barbulescu, R.: Extended tower number field sieve: a new complexity for the medium prime case. In: Robshaw, M., Katz, J. (eds.) CRYPTO 2016. LNCS, vol. 9814, pp. 543–571. Springer, Heidelberg (2016). https://doi.org/10.1007/978-3-662-53018-4_20

15. Kiyomura, Y., Inoue, A., Kawahara, Y., Yasuda, M., Takagi, T., Kobayashi, T.: Secure and efficient pairing at 256-bit security level. In: Gollmann, D., Miyaji, A., Kikuchi, H. (eds.) ACNS 2017. LNCS, vol. 10355, pp. 59–79. Springer, Cham (2017). https://doi.org/10.1007/978-3-319-61204-1_4

16. Menezes, A., Sarkar, P., Singh, S.: Challenges with assessing the impact of NFS advances on the security of pairing-based cryptography. In: Phan, R.C.-W., Yung, M. (eds.) Mycrypt 2016. LNCS, vol. 10311, pp. 83–108. Springer, Cham (2017). https://doi.org/10.1007/978-3-319-61273-7_5

17. El Mrabet, N., Joye, M. (eds.): Guide to Pairing-Based Cryptography. Chapman and Hall/CRC, Boca Raton (2016). https://www.crcpress.com/Guide-to-Pairing-Based-Cryptography/El-Mrabet-Joye/p/book/9781498729505

18. Schinzel, A.: On primitive prime factors of $a^n - b^n$. Proc. Cambridge Philos. Soc. **58**(4), 555–562 (1962). https://doi.org/10.1017/S0305004100040561

19. Schirokauer, O.: The number field sieve for integers of low weight. Math. Comput. **79**(269), 583–602 (2010). https://doi.org/10.1090/S0025-5718-09-02198-X. http://eprint.iacr.org/2006/107

20. Scott, M.: On the efficient implementation of pairing-based protocols. In: Chen, L. (ed.) IMACC 2011. LNCS, vol. 7089, pp. 296–308. Springer, Heidelberg (2011). https://doi.org/10.1007/978-3-642-25516-8_18

21. Scott, M., Benger, N., Charlemagne, M., Dominguez Perez, L.J., Kachisa, E.J.: On the final exponentiation for calculating pairings on ordinary elliptic curves. In: Shacham, H., Waters, B. (eds.) Pairing 2009. LNCS, vol. 5671, pp. 78–88. Springer, Heidelberg (2009). https://doi.org/10.1007/978-3-642-03298-1_6

22. Stevenhagen, P.: On Aurifeuillian factorizations. Nederl. Akad. Wetensch. Indag. Math. **49**(4), 451–468 (1987). https://doi.org/10.1016/1385-7258(87)90009-6

23. Vercauteren, F.: Optimal pairings. IEEE Trans. Inf. Theory **56**, 455–461 (2009). https://eprint.iacr.org/2008/096

24. Wagstaff Jr., S.S.: The search for Aurifeuillian-like factorizations. J. Integers **12A**(6), 1449–1461 (2012). https://homes.cerias.purdue.edu/~ssw/cun/mine.pdf

Superspecial Hyperelliptic Curves of Genus 4 over Small Finite Fields

Momonari Kudo[1]([✉]) and Shushi Harashita[2]

[1] Kobe City College of Technology, 8-3, Gakuen-Higashimachi,
Nishi-ku, Kobe 651-2194, Japan
m-kudo@math.kyushu-u.ac.jp
[2] Graduate School of Environment and Information Sciences,
Yokohama National University, 79-7, Tokiwadai, Hodogaya-ku,
Yokohama 240-8501, Japan
harasita@ynu.ac.jp

Abstract. In this paper, we enumerate *superspecial* hyperelliptic curves of genus 4 over finite fields $\mathbb{F}_q$ for small q. This complements our preceding results in the non-hyperelliptic case. We give a feasible algorithm to enumerate superspecial hyperelliptic curves of genus g over $\mathbb{F}_q$ in the case that q and $2g+2$ are coprime and $q > 2g+1$. We executed the algorithm for $(g, q) = (4, 11^2), (4, 13^2), (4, 17^2)$ and $(4, 19)$ with our implementation on a computer algebra system Magma. Moreover, we found many *maximal* hyperelliptic curves and some *minimal* hyperelliptic curves over $\mathbb{F}_q$ from among enumerated superspecial curves.

Keywords: Hyperelliptic curves · Superspecial curves
Maximal curves

1 Introduction

Let $\mathbb{F}_q$ denote the finite field with q elements. A curve C of genus g over $\mathbb{F}_q$ is called *maximal* (resp. *minimal*) if the number of $\mathbb{F}_q$-rational points on C attains the Hasse-Weil upper bound $q + 1 + 2g\sqrt{q}$ (resp. the Hasse-Weil lower bound $q+1-2g\sqrt{q}$). These curves are interesting objects in their own right, and also are useful in applications such as coding theory (e.g., [11, 21]). Specifically, algebraic geometric codes produced from curves with many rational points have both high information rate and high error-correcting rate; for such curves the sum of these two quantities is large. Thus, it has been a central problem to find maximal curves. However, the number of maximal curves over $\mathbb{F}_q$ of genus g for a fixed pair (g, q) is very small, compared with the whole set of curves over $\mathbb{F}_q$ of genus g, and thus it is not easy at all to find maximal curves. The notion of *superspecial* curves helps us to find such curves. In general, a curve over a field K of positive characteristic is said to be *superspecial* if its Jacobian is isomorphic to a product of supersingular elliptic curves over the algebraic closure $\overline{K}$ of K. It is known that any maximal or minimal curve C over $\mathbb{F}_{p^2}$ is superspecial, where p is a prime. Conversely any superspecial curve over an algebraically closed field descends to

© Springer Nature Switzerland AG 2018
L. Budaghyan and F. Rodríguez-Henríquez (Eds.): WAIFI 2018, LNCS 11321, pp. 58–73, 2018.
https://doi.org/10.1007/978-3-030-05153-2_3

a maximal *or* minimal curve over $\mathbb{F}_{p^2}$, see the proof of [4, Theorem 1.1]. In the hyperelliptic case, we can say *more*: The existence of a superspecial hyperelliptic curve of genus g in characteristic p implies that there exists a maximal curve of genus g over $\mathbb{F}_{p^2}$ *and also* a minimal curve of genus g over $\mathbb{F}_{p^2}$. We will review this fact in Sect. 2.2. This work focuses on enumerating superspecial curves to find *all* maximal hyperelliptic curves among them.

In the literature, there are many works on the enumeration of superspecial curves of genus g:

- If $g \leq 3$, some theoretical approaches to enumerate superspecial/maximal curves are available, which are based on Torelli's theorem (cf. [3], [25, Proposition 4.4] for $g = 1$, [10,13,22] for $g = 2$, and [9,12] for $g = 3$). In particular, there exists a maximal curve of genus g over $\mathbb{F}_{p^{2e}}$ if $g = 2$ and $p^{2e} \neq 4, 9$ (cf. [22, Theorem 3]) and if $g = 3$, $p \geq 3$ and e is odd (cf. [12, Theorem 1]).
- If $g \geq 4$, however, these approaches are not so effective; different from the case of $g \leq 3$, the dimension of the moduli space of curves of genus g is strictly less than that of the moduli space of principally polarized abelian varieties of dimension g. Thus the case of $g = 4$ is the next target. For $p = 5$, Fuhrmann-Garcia-Torres [5] found a maximal curve C_0 of genus 4 over $K = \mathbb{F}_{5^2}$, and proved that it gives a unique isomorphism class over $\overline{K}$. In recent years, enumerations of superspecial curves of genus 4 in some small characteristic have been completed in [14,16,17]. Specifically, the isomorphism classes of superspecial *non-hyperelliptic* curves of genus 4 over $\mathbb{F}_q$ are determined for $q = 5^{2e-1}, 5^{2e}, 7^{2e-1}, 7^{2e}$ and 11^{2e-1}, where e is a natural number. Note that all the maximal curves over $K = \mathbb{F}_{5^2}$ enumerated in [14] are included in the unique isomorphism class of C_0 over $\overline{K}$. In the hyperelliptic case, the existence of superspecial curves is known for some q (e.g., [23,24]). Note that this study is motivated to enumerate *all* superspecial curves, while the papers [23] and [24] characterize *specific* superspecial curves. In particular, the paper [24] uses Serre's covering result in order to study the maximality of a specific curve.

In this work, we enumerate superspecial *hyperelliptic* curves of genus 4 in characteristic $p \leq 19$. Note that we do not use Serre's covering result, but apply techniques in computer algebra such as Gröbner bases. This work also complements our preceding results in [14,16,17] for non-hyperelliptic curves. Thanks to Ekedahl [4, Theorem 1.1], there is no superspecial hyperelliptic curve of genus 4 if $p \leq 7$. Our results are the theorems (Theorems 1–3) below, see Table 1 for a summary of results in $g = 4$. Note that the number of isomorphism classes of superspecial curves over $\mathbb{F}_{p^a}$ depends on the parity of a (cf. [16, Proposition 2.3.1]).

Theorem 1. *There is no superspecial hyperelliptic curve of genus 4 in characteristic 11 and 13.*

Theorem 2. *There exist precisely 5 (resp. 25) superspecial hyperelliptic curves of genus 4 over $\mathbb{F}_{17}$ (resp. $\mathbb{F}_{17^2}$) up to isomorphism over $\mathbb{F}_{17}$ (resp. $\mathbb{F}_{17^2}$). Moreover, there exist precisely two isomorphism classes of superspecial hyperelliptic curves of genus 4 over an algebraically closed field in characteristic 17.*

Theorem 3. *There exist precisely* 12 *superspecial hyperelliptic curves of genus* 4 *over* $\mathbb{F}_{19}$ *up to isomorphism over* $\mathbb{F}_{19}$. *Moreover, there exist precisely* 2 *superspecial hyperelliptic curves of genus* 4 *over* $\mathbb{F}_{19}$ *up to isomorphism over the algebraic closure.*

Table 1. Main references to enumerations of isomorphism classes of superspecial curves of genus $g = 4$ over $\mathbb{F}_q$, where q is a power of a prime p.

q	Non-hyperelliptic	Hyperelliptic	q	Non-hyperelliptic	Hyperelliptic
$p \leq 3$	Non-existence		13^{2e-1}		Non-existence
	by Ekedahl [4]		13^{2e}		by Theorem 1
5^{2e-1}	[16, Theorem A]		17^{2e-1}	Not yet	Theorem 2
5^{2e}	[14, Theorem A]	Non-existence	17^{2e}		Theorem 2
7^{2e-1}	Non-existence	by Ekedahl [4]	19^{2e-1}		Theorem 3
7^{2e}	by [14, Theorem B]		19^{2e}		Not yet
11^{2e-1}	[16, Theorem B]	Non-existence	$p \geq 23$	Not yet	
11^{2e}	Not yet	by Theorem 1		(Existences for some p, cf. [6,23])	

Note that we have explicit defining equations of the superspecial hyperelliptic curves in Theorems 2 and 3 (but omit them in the statements). Such explicit equations also define maximal or minimal curves over $\mathbb{F}_{p^2}$. For example, we found the following superspecial curve over $\mathbb{F}_{17^2}$; for each $a \in \mathbb{F}_{17}^\times$ consider the hyperelliptic curve $C_a : H_a(x, y) = y^2 - (x^{10} + ax^7 + (13a^2)x^4 + (12a^3)x) = 0$ over $\mathbb{F}_{17^2}$, which is included in one of the 25 isomorphism classes of the superspecial hyperelliptic curves over $\mathbb{F}_{17^2}$. Then $C_a : H_a(x, y) = 0$ is a maximal curve over $\mathbb{F}_{17^2}$. Indeed, the number of its $\mathbb{F}_{17^2}$-rational points is 426, which coincides with the Hasse-Weil upper bound $q + 1 + 2g\sqrt{q}$ for $q = 17^2$. Moreover, each C_a is not $\mathbb{F}_{17^2}$-isomorphic to $y^2 = x^{10} + x$, which is a maximal curve of known type (cf. [23, 24]). This means that we obtain a maximal curve of new type. For the other equations, see a table of the web page of the first author [27].

We prove Theorems 1–3 with help of computational results. Our computational methods are (A) Algorithm to enumerate superspecial hyperelliptic curves, (B) Reduction of defining equations of hyperelliptic curves, and (C) Isomorphism testing. Our enumeration method (A) is based on the computation of Cartier-Manin matrices (cf. [7,19], [26, Sect. 2]), and reduces our enumeration problem into solving multivariate systems over finite fields. The method (A) is also viewed as a hyperelliptic curve-version of algorithms for non-hyperelliptic curves given in [14,16,17]. The method (B) reduces parameters of defining equations as much as possible. Namely, it reduces the number of variables in multivariate systems to be solved, which clearly makes our algorithm (A) efficient. The method (C) gives an algorithm to classify isomorphism classes of arbitrary hyperelliptic curves of given genus over a finite field. Note that in this paper we do not mention the asymptotic complexity but the practicality of our enumeration algorithm only.

Notation. For a field K, we denote by $K^\times$ its multiplicative group. The general linear group of degree n over K is denoted by $\mathrm{GL}_n(K)$.

2   Preliminaries

Let K be a field of odd characteristic $p > 0$. In this section, we review some basic facts on hyperelliptic curves over K and their superspeciality.

2.1   Hyperelliptic Curves

Let C be a hyperelliptic curve over K. By definition, there exists a morphism $\pi : C \to \mathbf{P}^1$ of degree 2 over $\overline{K}$, where $\mathbf{P}^1$ denotes the projective line. It is known (cf. [8, Proposition 5.3]) that the pencil π is unique up to automorphisms of $\mathbf{P}^1$, whence we may assume that π is defined over K. Let g be the genus of C. It is known that π is ramified over distinct $2g + 2$ points. Write $\mathbf{P}^1 := \operatorname{Proj}(K[X, Z])$. Let $K(C)$ (resp. $K(x)$) be the field of rational functions on C (resp. in $x = X/Z$). The π induces a quadratic extension $K(C)/K(x)$. Let $y \in K(C)$ be a generator of the different ideal of $K(C)/K(x)$ with respect to $K[x]$, the ring of integers of $K(x)$. As y^2 belongs to $K[x]$, we see that C is realized as the desingularization of the homogenization of

$$y^2 = f(x), \tag{1}$$

where $f(x)$ is a polynomial over K with non-zero discriminant. Assume that the cardinality of K is greater than $2g + 1$. If necessarily, by an automorphism of $\mathbf{P}^1$ over K we translate the ramified points of π outside $\infty := (1 : 0) \in \mathbf{P}^1$. Then $f(x)$ is of degree $2g + 2$.

Remark 1. If $f(x) = 0$ has a root α in K, by the transformation $x' = 1/(x - \alpha)$ and $y' = y/(x - \alpha)^g$ we have another realization of the curve:

$$y'^2 = \phi(x'), \tag{2}$$

where $\phi(x')$ is a polynomial over K of degree $2g + 1$. However as $f(x)$ does not always have a rational root, we can not use the model of odd degree.

The next lemma describes the set of isomorphisms between two hyperelliptic curves C_1 and C_2. This in particular gives a criterion for whether C_1 and C_2 are isomorphic to each other or not.

Lemma 1. *Let $f_1(x)$ and $f_2(x)$ be elements of $K[x]$ of degree $2g + 2$. Let C_1 and C_2 be the hyperelliptic curves over K defined by $y^2 = f_1(x)$ and $y^2 = f_2(x)$ respectively. Set $F_i(X, Z) = Z^{2g+2} f_i(X/Z) \in K[X, Z]$. Let k be a field containing K. The set of k-isomorphisms from C_1 to C_2 is bijective to*

$$\left(\{h \in \mathrm{GL}_2(k) \mid F_1((X, Z) \cdot {}^t h) = \lambda^2 F_2(X, Z) \text{ for some } \lambda \in k^\times\}/\sim\right) \times \{\pm 1\},$$

where we say $h_1 \sim h_2$ if $h_1 = \mu h_2$ for some $\mu \in k^\times$.

Proof. Let φ be a k-isomorphism from C_1 to C_2. Let π_i be morphisms from C_i to $\mathbf{P}^1$ of degree 2 (chosen over K as above). The composition $\pi_2 \circ \varphi$ is a morphism

C_1 to $\mathbf{P}^1$ of degree 2 over k. The uniqueness of such morphisms implies that there exists a k-automorphism ψ of $\mathbf{P}^1$ commuting the following diagram:

$$
\begin{array}{ccc}
C_1 & \xrightarrow{\ \varphi\ } & C_2 \\
{\scriptstyle \pi_1}\downarrow & & \downarrow{\scriptstyle \pi_2} \\
\mathbf{P}^1 & \xrightarrow{\ \psi\ } & \mathbf{P}^1
\end{array}
$$

The automorphism ψ is represented by an element $h \in \mathrm{GL}_2(k)$ up to scalar multiplications. Clearly φ sends the different ideal $\mathfrak{d}_1$ of $k(C_1)/k(\mathbf{P}^1)$ to the different ideal $\mathfrak{d}_2$ of $k(C_2)/k(\mathbf{P}^1)$, whence the generator y of $\mathfrak{d}_1$ of the equation $y^2 = f_1(x)$ defining C_1 is sent to that of $\mathfrak{d}_2$ up to a scalar multiplication. Thus, for some scalar $\lambda \in k^\times$ we have the equation

$$
F_1((X, Z) \cdot {}^t h) = \lambda^2 F_2(X, Z). \tag{3}
$$

Conversely for $h = \begin{pmatrix} a & b \\ c & d \end{pmatrix} \in \mathrm{GL}_2(k)$ satisfying (3) for some $\lambda \in k^\times$, we have the two isomorphisms $C_1 \to C_2$ defined by $(x, y) \mapsto \left(\frac{ax+b}{cx+d}, \pm\lambda \frac{y}{(cx+d)^{g+1}} \right)$.   □

2.2   Superspecial Curves and Maximal Curves

Let C be a nonsingular projective curve over a perfect field K. We say that C is *superspecial* if its Jacobian $\mathrm{Jac}(C)$ is the product of some supersingular elliptic curves over the algebraic closure $\overline{K}$. It is well-known that C is superspecial if and only if the Cartier operator on the cohomology group $H^0(C, \Omega^1_C)$ is zero (cf. [20]). The Cartier operator on $H^0(C, \Omega^1_C)$ with respect to a fixed basis of $H^0(C, \Omega^1_C)$ is called a Cartier-Manin matrix of C, which for hyperelliptic curves will be reviewed in the next subsection.

As mentioned in the introduction, it is known that any superspecial hyperelliptic curve over an algebraically closed field can descend to a maximal curve over $\mathbb{F}_{p^2}$ *and also* to a minimal curve over $\mathbb{F}_{p^2}$. This is deduced from the following facts.

1. If C is hyperelliptic, then the automorphism group of C is isomorphic to the automorphism group of the Jacobian variety $\mathrm{Jac}(C)$ with the principal polarization of C. This fact implies that giving a descent datum of C is equivalent to giving that of its Jacobian with the principal polarization.
2. For superspecial C, by definition we have $\mathrm{Jac}(C) \simeq E^g$ for a supersingular elliptic curve E, where g is the genus of C. It is known that E descends to an elliptic curve E_0 over $\mathbb{F}_{p^2}$ over which the Frobenius is the multiplication by p and $-p$ respectively.
3. A polarization on E^g is, by definition, a homomorphism from E^g to its dual (which is isomorphic to E^g), but such a homomorphism is always defined over $\mathbb{F}_{p^2}$, i.e., is induced from an $\mathbb{F}_{p^2}$-homomorphism from E_0^g to itself.

See the proof of [4, Theorem 1.1] and [18, Sect. 1.2] for the second and third facts. The second and third facts imply that $\mathrm{Jac}(C)$ with the principal polarization descend to a principally polarized abelian variety over $\mathbb{F}_{p^2}$ on which the Frobenius is p and $-p$ respectively. By the first fact, if C is hyperelliptic, then C descends to a curve C_0 over $\mathbb{F}_{p^2}$ such that the Frobenius on the first étale cohomology group $H^1_{\text{ét}}(C_0, \mathbb{Z}_l)$ for a prime $l \neq p$ is the multiplication by p and $-p$ respectively. The curve C_0 is minimal in the former case and is maximal in the latter case.

Thus, we conclude that the existence of a superspecial hyperelliptic curve of genus g in characteristic p implies the existence of a maximal curve of genus g over $\mathbb{F}_{p^2}$ and that of a minimal curve of genus g over $\mathbb{F}_{p^2}$.

2.3   Cartier-Manin Matrices of Hyperelliptic Curves

As in the previous subsection, let K be a perfect field of odd characteristic $p > 0$. The *Cartier-Manin matrix* of a curve C over K is defined as the matrix representing the Cartier operator on $H^0(C, \Omega^1_C)$, see [26, Sect. 2]. Here $H^0(C, \Omega^1_C)$ is the space of holomorphic differentials of C. In the following proposition, we introduce a well-known method (cf. [7,19], [26, Sect. 2]) to obtain the Cartier-Manin matrix of a hyperelliptic curve.

Proposition 1. *Let C be a hyperelliptic curve $y^2 = f(x)$ of genus g over K, where $d = \deg(f)$ is either $2g + 1$ or $2g + 2$. Then the Cartier-Manin matrix of the hyperelliptic curve C is the $g \times g$ matrix whose (i, j)-entry is the coefficient of x^{pi-j} in $f^{(p-1)/2}$ for $1 \leq i, j \leq g$.*

As we mentioned in Sect. 2.2, a non-singular curve is superspecial if and only if its Cartier-Manin matrix is the zero matrix. From Proposition 1, we have the corollary below. By this corollary, one can decide whether a given hyperelliptic curve is superspecial or not by computing its Cartier-Manin matrix.

Corollary 1. *Let C be a hyperelliptic curve $y^2 = f(x)$ of genus g over K. Then C is superspecial if and only if the coefficients of x^{pi-j} in $f^{(p-1)/2}$ are equal to 0 for all positive integers $1 \leq i, j \leq g$.*

3   Enumeration of Superspecial Hyperelliptic Curves

Let $K = \mathbb{F}_q$ be a finite field of odd characteristic p, where q is a power of p. In this section, we give algorithms to enumerate superspecial hyperelliptic curves and to determine their isomorphism classes. As we mentioned in Sects. 1 and 2.2, a curve over K is superspecial if and only if its Cartier-Manin matrix is zero. Recall from Proposition 1 of Sect. 2.3 that the Cartier-Manin matrix of a hyperelliptic curve $y^2 = f(x)$ of genus g with $\deg(f) = 2g + 1$ or $2g + 2$ is determined from certain coefficients in the multiple $f^{(p-1)/2}$.

In this section, we give three computational techniques for determining the isomorphism classes of superspecial hyperelliptic curves of genus g over finite fields: (A) Algorithm to enumerate superspecial hyperelliptic curves, (B) Reduction of defining equations of hyperelliptic curves, and (C) Isomorphism testing. Specifically, based on Corollary 1, we shall reduce our enumeration problem into a computational problem that we solve multivariate systems over finite fields.

3.1   (A): Algorithm to Enumerate Superspecial Hyperelliptic Curves

Recall from Sect. 2.1 that a hyperelliptic curve of genus g over K is given by the equation $y^2 = f(x)$ for some polynomial f of degree $2g + 2$ with non-zero discriminant. Write $f(x) = a_d x^d + a_{d-1} x^{d-1} + a_{d-2} x^{d-2} + \cdots + a_1 x + a_0$ with $a_k \in K$ for $0 \le k \le d$. Let $\mathcal{S}$ be the set of the coefficients of the g^2 monomials in $f(x)^{(p-1)/2}$ given in Proposition 1. Based on Proposition 1 together with Corollary 1, we give a strategy to enumerate superspecial hyperelliptic curves of genus g over K:

- Enumerate $(a_0, \ldots, a_d) \in K^{d+1}$ such that all elements of $\mathcal{S}$ are zero and such that the discriminant of $f(x)$ is not zero.

In other words, by regarding all elements of $\mathcal{S}$ as algebraic relations on a_i's, it suffices to compute all roots $(a_0, \ldots, a_d) \in K^{d+1}$ of the multivariate system $P = 0$ for all $P \in \mathcal{S} \subset K[a_0, \ldots, a_d]$ such that f has no double root in the algebraic closure $\overline{K}$. Here, we show a concrete method (*Enumeration Method* below) for the enumeration, and write down its pseudocode in Algorithm 1. This method is viewed as a hyperelliptic curve-version of algorithms given in [14,16,17] for non-hyperelliptic curves.

Enumeration Method. With notation as above, we conduct the following:

0. Regard some unknown coefficients in $f(x)$ as indeterminates. Choose an integer $0 \le s_1 \le d + 1$. For simplicity, let $a_0, \ldots, a_{s_1-1}$ be indeterminates here. The remaining part $(a_{s_1}, \ldots, a_d)$ runs through $K^{\oplus d+1-s_1}$.

For each element $(c_{s_1}, \ldots, c_d) \in K^{\oplus d+1-s_1}$, proceed with the following four steps:

1. Put

$$f(x) := c_d x^d + c_{d-1} x^{d-1} + \cdots + c_{s_1} x^{s_1} + a_{s_1-1} x^{s_1-1} + \cdots + a_1 x + a_0.$$

   Compute $h := f^{p-1}$ over $K[a_0, \ldots, a_{s_1-1}][x]$.
2. Let $\mathcal{S}$ be the set of the coefficients of the g^2 monomials in $h = f^{p-1}$, given in Proposition 1. Note that $\mathcal{S} \subset K[a_0, \ldots, a_{s_1-1}]$.
3. Regard some unknown coefficients among $a_0, \ldots, a_{s_1-1}$ as indeterminates. For simplicity, let $a_0, \ldots, a_{s_2-1}$ with $s_2 \le s_1$ be indeterminates here. The remaining part $(a_{s_2}, \ldots, a_{s_1-1})$ runs through $K^{\oplus s_1-s_2}$.
4. For each $(c_{s_2}, \ldots, c_{s_1-1}) \in K^{\oplus s_1-s_2}$, proceed with the following three steps 4a – 4c:

Algorithm 1. Algorithm to enumerate superspecial hyperelliptic curves

---

Input: An integer g, a prime number p, and $q = p^s$ for some $s \geq 1$
Output: The set $\mathcal{F}$ of polynomials $f(x)$ over $K = \mathbb{F}_q$ of degree $2g + 2$ such that the
    curves $y^2 = f(x)$ are superspecial hyperelliptic curves of genus g
1: $\mathcal{F} \leftarrow \emptyset$
2: $d \leftarrow 2g + 2$
3: Choose an integer $0 \leq s_1 \leq d + 1$ and let $a_0, \ldots, a_{s_1-1}$ be indeterminates
4: **for** $(c_{s_1}, \ldots, c_d) \in K^{\oplus d+1-s_1}$ **do**
5:     $f(x) \leftarrow c_d x^d + c_{d-1} x^{d-1} + \cdots + c_{s_1} x^{s_1} + a_{s_1-1} x^{s_1-1} + \cdots + a_1 x + a_0$
6:     Compute $h := f^{(p-1)/2}$ over $K[a_0, \ldots, a_{s_1-1}][x]$
7:     $\mathcal{S} \leftarrow$ the set of the coefficients of the g^2 monomials in h, given in Proposition 1.
8:     Choose an integer $0 \leq s_2 \leq s_1$
9:     **for** $(c_{s_2}, \ldots, c_{s_1-1}) \in K^{\oplus s_1-s_2}$ **do**
10:       $\mathcal{S}' \leftarrow \{P(a_0, \ldots, a_{s_2-1}, c_{s_2}, \ldots, c_{s_1-1}) : P \in \mathcal{S}\}$
11:       Compute the roots of the system (4) over K (with Gröbner basis algorithms)
12:       $V \leftarrow \{(c_0, \ldots, c_{s_2-1}) \in K^{\oplus s_2} : P'(c_0, \ldots, c_{s_2-1}) = 0$ for all $P' \in \mathcal{S}'\}$
13:       **for** $(c_0, \ldots, c_{s_2-1}) \in V$ **do**
14:         $f_{\text{sol}} \leftarrow c_d x^d + c_{d-1} x^{d-1} + \cdots + c_{s_2} x^{s_2} + c_{s_2-1} x^{s_2-1} + \cdots + c_1 x + c_0$
15:         Decide whether f_{sol} has no double root in $\overline{K}$ or not (this can be done by
            constructing the minimal splitting field of f_{sol})
16:         **if** f_{sol} has no double root in $\overline{K}$ **then**
17:           $\mathcal{F} \leftarrow \mathcal{F} \cup \{f_{\text{sol}}\}$
18:         **end if**
19:       **end for**
20:     **end for**
21: **end for**
22: **return** $\mathcal{F}$

4a. For each $P \in \mathcal{S}$, substitute respectively $c_{s_2}, \ldots, c_{s_1-1}$ into $a_{s_2}, \ldots, a_{s_1-1}$
    of the coefficients in P. Put

$$\mathcal{S}' := \{P(a_0, \ldots, a_{s_2-1}, c_{s_2}, \ldots, c_{s_1-1}) : P \in \mathcal{S}\}.$$

    Note that $\mathcal{S}' \subset K[a_0, \ldots, a_{s_2-1}]$.
4b. Compute the roots of the multivariate system

$$P'(a_0, \ldots, a_{s_2-1}) = 0 \text{ for all } P' \in \mathcal{S}' \tag{4}$$

    over K with Gröbner basis algorithms.
4c. For each root of the above system, substitute it into unknown coefficients
    in f, and decide whether f has no double root in $\overline{K}$ or not (this can be
    done by constructing the minimal splitting field of f). If f has no double
    root in $\overline{K}$, store f.

Remark 2. Our enumeration method adopts the *hybrid approach* by Bettale,
Faugère and Perret [1] to solve multivariate systems over finite fields. In their
approach, exhaustive search and Gröbner basis algorithms are mixed for efficiency, but there is a trade-off between them. Note that an optimal choice of

coefficients to be regarded as indeterminants is not unique, and deeply depends on the situation. In our case, such a choice is heuristically decided from experimental computations for each situation (Propositions 2–6 in the next section).

3.2 (B): Reduction of Defining Equations of Hyperelliptic Curves

In this subsection, we give a reduction of defining equations of hyperelliptic curves. Let C be a hyperelliptic curve over K. Let $y^2 = f(x)$ be a defining equation of C. Remark that a good method of reduction over an algebraically closed field is to translate three ramified points of the corresponding morphism $C \to \mathbf{P}^1$ of degree 2 to $\{0, 1, \infty\}$, but we can not adopt this method, because the ramified points are not necessarily K-rational points. In this paper we use an elementary reduction:

Lemma 2. *Assume that p and $2g + 2$ are coprime. Let $\epsilon \in K^\times \smallsetminus (K^\times)^2$. Any hyperelliptic curve C of genus g over K is the desingularization of the homogenization of*

$$cy^2 = x^{2g+2} + bx^{2g} + a_{2g-1}x^{2g-1} + \cdots + a_1 x + a_0$$

for $a_i \in K$ for $i = 0, 1, \ldots, 2g - 1$ where $b = 0, 1, \epsilon$ and $c = 1, \epsilon$.

Proof. As in Sect. 2.1, a hyperelliptic curve C over K is realized as $y^2 = f(x)$ for a polynomial $f(x)$ of degree $2g + 2$ over K. This can be expressed as $cy^2 = h(x)$ for $c \in K^\times$ and for a monic polynomial $h(x)$ of degree $2g+2$ over K. Considering the transformation $(x, y) \mapsto (x, \alpha y)$ for some $\alpha \in K^\times$, one may assume $c = 1$ or ϵ. Considering $x \to x + a$, we can transform $h(x)$ to a polynomial with no x^{2g+1}-term, i.e., we may assume C is defined by an equation of the form

$$cy^2 = x^{2g+2} + a_{2g}x^{2g} + a_{2g-1}x^{2g-1} + \cdots + a_1 x + a_0.$$

The transformation $(x, y) \mapsto (\beta x, \beta^{g+1} y)$ for some $\beta \in K^\times$ and the multiplication by $\beta^{-(2g+2)}$ to the whole, we may assume that $a_{2g} = 0, 1$ or ϵ.

Remark 3. Let $h(x)$ be a monic polynomial over K with non-zero discriminant. Let $\epsilon \in K^\times \smallsetminus (K^\times)^2$. Let C_1 and C_2 be the hyperelliptic curves defined by $y^2 = h(x)$ and $\epsilon y^2 = h(x)$ respectively. The transformation $(x, y) \mapsto (x, \sqrt{\epsilon}y)$ gives an isomorphism from C_1 to C_2 over $K[\sqrt{\epsilon}]$. In particular, C_1 is superspecial if and only if C_2 is superspecial.

3.3 (C): Isomorphism Testing

We suppose that p and $2g+2$ are coprime. Let C_1 and C_2 be hyperelliptic curves of genus g over K. As we showed in Sect. 2.1 and in Lemma 2, defining equations of C_1 and C_2 are given by $H_1(x, y) = c_1 y^2 - f_1(x)$ and $H_2(x, y) = c_2 y^2 - f_2(x)$ respectively for some c_1 and c_2 in $K^\times$ and for some polynomials $f_1(x)$ and $f_2(x)$ in $K[x]$ of degree $2g + 2$. Let F_i denote the homogenization of f_i with respect to an extra variable z for each $1 \leq i \leq 2$. By Lemma 1, we have a fact that C_1 and C_2 are isomorphic over K if and only if there exists $h \in \mathrm{GL}_2(K)$ such that $h \cdot F_1 = \lambda^2 F_2$ for some $\lambda \in K^{\times}$. In other words, there exist $h \in \mathrm{GL}_2(K)$ and $\lambda \in K^{\times}$ such that all the coefficients in $F := h \cdot F_1 - \lambda F_2$ are zero, where $h \cdot F_1(x, z) := F_1((x, z) \cdot {}^t h)$. Based on this fact, we write down a method (*Isomorphism Testing Method* below) to determine whether C_1 and C_2 are isomorphic over K (or over $\overline{K}$) for $K = \mathbb{F}_q$. Here q is a power of the characteristic p of K. The correctness of this computational method is straightforward from its construction.

Isomorphism Testing Method. For the inputs $H_1(x, y) = c_1 y^2 - f_1(x)$, $H_2(x, y) = c_2 y^2 - f_2(x)$, and q as above, the following 5 steps decide whether $C_1 : H_1(x, y) = 0$ and $C_2 : H_2(x, y) = 0$ are isomorphic over K or not (resp. $\overline{K}$ or not):

1. Let $b_1, b_2, b_3, b_4, b_5, \lambda$ and μ be indeterminates, and set

$$F_i(x, z) := c_i^{-1} z^{2g+2} f_i(x/z) \text{ for } i = 1, 2, \quad \text{and} \quad h := \begin{pmatrix} b_1 & b_2 \\ b_3 & b_4 \end{pmatrix}$$

   where h is a square matrix whose entries are indeterminates.
2. Compute $F(x, z) := F_1((x, z) \cdot {}^t h) - \lambda^2 F_2(x, z)$ over the polynomial ring $K[b_1, b_2, b_3, b_4, b_5, \lambda, \mu][x, z]$ whose coefficient ring is also a polynomial ring.
3. Let $\mathcal{C}_F$ be the set of the coefficients of the non-zero terms in $F(x, z)$. We put

$$\mathcal{C} := \mathcal{C}_F \cup \{(b_1 b_4 - b_2 b_3) b_5 - 1\} \cup \{\lambda \mu - 1\},$$

   and $\mathcal{C}' := \mathcal{C} \cup \{b_i^q - b_i : 1 \leq i \leq 4\} \cup \{\lambda^q - \lambda\}$ (resp. $\mathcal{C}' := \mathcal{C}$). Note that $b_1 b_4 - b_2 b_3 = \det(h)$.
4. Test whether the multivariate system defined by the ideal $\langle \mathcal{C}' \rangle$ has a root over K (resp. $\overline{K}$) or not. One can do this by computing the reduced Gröbner basis in $K[b_1, b_2, b_3, b_4, b_5, \lambda, \mu]$ with respect to some term order.
5. If the system in Step 4 has a root over K (resp. $\overline{K}$), then $C_1 : H_1(x, y) = 0$ and $C_2 : H_2(x, y) = 0$ are isomorphic over K (resp. $\overline{K}$). Otherwise C_1 and C_2 are not isomorphic over K (resp. $\overline{K}$).

4   Main Results

In this section, we prove Theorems 1–3 stated in Sect. 1. As an application of the theorems, we also found hyperelliptic curves of genus 4 over $\mathbb{F}_q$ such that they are maximal as curves over $\mathbb{F}_{p^2}$ for $q = 17, 17^2$ and 19. We choose a primitive element ζ of $\mathbb{F}_q$ for each $17, 17^2$ and 19. Specifically, we take $\zeta = 3$ for $q = 17$, $\zeta = -8 + \sqrt{61}$ for $q = 17^2$, and $\zeta = 2$ for $q = 19$.

4.1   Proofs of and Corollaries of the Main Theorems

In the following proofs, we use computational results, which shall be given in the next subsection (Sect. 4.2).

Proofs of Theorems 1–3. We here prove the case of $q = 11^2$ only since the other cases are proved by a similar idea together with Propositions 3–6. Let C be a hyperelliptic curve of genus $g = 4$ over $K = \mathbb{F}_q$. Since $p = 11$ is coprime to $2g + 2 = 2 \cdot 4 + 2 = 10$, it follows from Lemma 2 that C is given by $y^2 - f(x)$ or $\epsilon y^2 - f(x)$ for $\epsilon \in K^\times \setminus (K^\times)^2$. Here $f(x)$ is a polynomial of the form

$$f(x) = x^{10} + a_8 x^8 + a_7 x^7 + \cdots + a_1 x + a_0$$

for some $a_i \in K$ with $0 \le i \le 8$ such that it has no double root over the algebraic closure $\overline{K}$. By Proposition 2 in the next subsection (Sect. 4.2), there does not exist such an $f(x)$ that $C : y^2 = f(x)$ is superspecial. It follows from Remark 3 that there is no such an $f(x)$ that $C : \epsilon y^2 = f(x)$ is superspecial. □

Some Corollaries. By Theorem 1, we relax the restriction on non-hyperelliptic curves in [16, Theorem B] (or [17, Main Theorem]).

Corollary 2. *There exist precisely 30 (resp. 9) superspecial curves of genus 4 over $\mathbb{F}_{11}$ up to isomorphism over $\mathbb{F}_{11}$ (resp. the algebraic closure of $\mathbb{F}_{11}$).*

Since a maximal or minimal (hyperelliptic) curve over $\mathbb{F}_{p^2}$ is superspecial, we have the following corollaries (Corollaries 3 and 4 below) from Theorem 1:

Corollary 3. *There does not exist any maximal (resp. minimal) hyperelliptic curve of genus 4 over $\mathbb{F}_{121}$.*

Corollary 4. *There does not exist any maximal (resp. minimal) hyperelliptic curve of genus 4 over $\mathbb{F}_{169}$.*

In contrast to the cases of $p \le 13$, it is shown in Theorems 2 and 3 that there exist superspecial hyperelliptic curves of genus 4 over $\mathbb{F}_p$ and $\mathbb{F}_{p^2}$ in the cases of $p = 17$ and 19. Computing the number of rational points of the enumerated curves by a computer, we have the following corollaries.

Corollary 5. *There exist precisely 2 (resp. 2) maximal (resp. minimal) hyperelliptic curves of genus 4 over $\mathbb{F}_{17^2}$ up to isomorphism over $\mathbb{F}_{17^2}$.*

Corollary 6. *There exist maximal hyperelliptic curves of genus 4 over $\mathbb{F}_{19^2}$. There also exists a minimal hyperelliptic curve of genus 4 over $\mathbb{F}_{19^2}$.*

Maximal curves and minimal curves in Corollaries 5 and 6 will be introduced in Sect. 4.3.

4.2   Computational Parts of Our Proofs of the Main Theorems

We show computational results for the proofs of the main theorems. The computational results are proved by executing *Enumeration Method* in Sect. 3.1 and *Isomorphism Testing Method* in Sect. 3.3. All our computations were conducted on a computer with ubuntu 16.04 LTS OS at 3.40 GHz CPU (Intel Core i7-6700) and 15.6 GB memory We implemented and executed the computations over Magma V2.22-7 [2] in its 64-bit version.

In Propositions 2, 3, 4, 5 and 6, below, we show our computational results for $q = 11^2$, 13^2, 17, 17^2 and 19, respectively. We give a computational proof of Proposition 5 only, and omit those of the other propositions since our computational settings in all the proofs are almost the same (our proofs of Propositions 2, 3, 4 and 6 will be given in a separated pdf [15]). We also note that the enumeration for $q = 17^2$ is more expensive than those for $q = 11^2$, 13^2, 17 and 19 due to the largest cardinality of $\mathbb{F}_q$.

Proposition 2. *Consider the polynomial $f(x) \in \mathbb{F}_{121}[x]$ of the form*

$$f(x) = x^{10} + a_8 x^8 + a_7 x^7 + \cdots + a_1 x + a_0 \tag{5}$$

for $a_i \in \mathbb{F}_{121}$ with $0 \le i \le 8$. Then there does not exist $(a_0, \ldots, a_8) \in (\mathbb{F}_{121})^{\oplus 9}$ such that $C : y^2 = f(x)$ is a superspecial hyperelliptic curve of genus 4 over $\mathbb{F}_{121}$.

Proposition 3. *Consider the polynomial $f(x) \in \mathbb{F}_{169}[x]$ of the form*

$$f(x) = x^{10} + a_8 x^8 + a_7 x^7 + \cdots + a_1 x + a_0 \tag{6}$$

for $a_i \in \mathbb{F}_{169}$ with $0 \le i \le 8$. Then there does not exist $(a_0, \ldots, a_8) \in (\mathbb{F}_{169})^{\oplus 9}$ such that $C : y^2 = f(x)$ is a superspecial hyperelliptic curve of genus 4 over $\mathbb{F}_{169}$.

Proposition 4. *Consider the polynomial $f(x) \in \mathbb{F}_{17}[x]$ of the form*

$$f(x) = x^{10} + a_8 x^8 + a_7 x^7 + \cdots + a_1 x + a_0, \tag{7}$$

for $a_i \in \mathbb{F}_{17}$ with $0 \le i \le 8$. Then there exist precisely 5 (resp. 2) superspecial hyperelliptic curves $C : cy^2 = f(x)$ with $c = 1$ or ϵ, up to isomorphism over $\mathbb{F}_{17}$ (resp. the algebraic closure of $\mathbb{F}_{17}$), such that $f(x)$ are of the form (7). Here ϵ is an element of $\mathbb{F}_{17}^{\times} \setminus (\mathbb{F}_{17}^{\times})^2$. Representatives of the 5 isomorphisms classes over $\mathbb{F}_{17}$ are given by

1. $y^2 = x^{10} + x$,
2. $y^2 = x^{10} + x^7 + 13x^4 + 12x$,
3. $y^2 = x^{10} + x^7 + 14x^6 + 6x^5 + 12x^3 + 5x^2 + 7x + 6$,
4. $y^2 = x^{10} + x^8 + x^7 + 15x^6 + 4x^5 + 12x^4 + 15x^3 + 11x^2 + 9x + 4$, *and*
5. $y^2 = x^{10} + x^8 + 2x^7 + 9x^5 + x^4 + 10x^3 + 8x^2 + 11x + 5$,

and those of the 2 isomorphism classes over the algebraic closure are given by

1. $y^2 = x^{10} + x$, *and*
2. $y^2 = x^{10} + x^7 + 13x^4 + 12x$.

Proposition 5. *Consider the polynomial $f(x) \in \mathbb{F}_{17^2}[x]$ of the form*

$$f(x) = x^{10} + a_8 x^8 + a_7 x^7 + \cdots + a_1 x + a_0, \tag{8}$$

for $a_i \in K = \mathbb{F}_{17^2}$ with $0 \le i \le 8$. Then there exist precisely 25 (resp. 2) superspecial hyperelliptic curves $C : cy^2 = f(x)$ with $c = 1$ or ϵ, up to isomorphism over $\mathbb{F}_{17^2}$ (resp. the algebraic closure of $\mathbb{F}_{17^2}$), such that $f(x)$ are of the form (8).

Proof. We prove the assertion by executing *Enumeration Method* (Algorithm 1 for its pseudocode) and *Isomorphism Testing Method* given in Sects. 3.1 and 3.3 respectively. We first conduct *Enumeration Method*. In the following, we describe details of our computation, and also show our choices of coefficients to be regarded as indeterminates and a term ordering in the algorithm.

0. We regard the $s_1 := 8$ coefficients a_i for $0 \leq i \leq 7$ as indeterminates. For the Gröbner basis computation in $\mathbb{F}_{17^2}[a_0, a_1, a_2, a_3, a_4, a_5, a_6, a_7]$ below, we adopt the graded reverse lexicographic (grevlex) order with $a_7 \prec a_6 \prec a_5 \prec a_4 \prec a_3 \prec a_2 \prec a_1 \prec a_0$.

For each $c_8 \in \{0, 1, \zeta\}$, we proceed with the following four steps:

1. Put $f(x) := x^{10} + c_8 x^8 + a_7 x^7 + \cdots + a_1 x + a_0$, and compute $h := f^{p-1}$ over $\mathbb{F}_{17^2}[a_0, \ldots, a_7][x]$.
2. Let $\mathcal{S}$ be the set of the coefficients of the g^2 monomials in $h = f^{p-1}$, given in Proposition 1. Note that $\mathcal{S} \subset \mathbb{F}_{17^2}[a_0, \ldots, a_7]$.
3. We regard 6 unknown coefficients in f as indeterminates. Specifically, we keep $a_0, \ldots, a_5$ being indeterminates, whereas we substitute some elements of $\mathbb{F}_{17^2}$ into a_6 and a_7 in the next step.
4. For each $(c_6, c_7) \in (\mathbb{F}_{17^2})^{\oplus 2}$, proceed with the following three steps 4a–4c:
   4a. For each $P \in \mathcal{S}$, substitute (c_6, c_7) into (a_6, a_7) of the coefficients in P. Put $\mathcal{S}' := \{P(a_0, \ldots, a_5, c_6, c_7) : P \in \mathcal{S}\}$. Note that $\mathcal{S}' \subset \mathbb{F}_{17^2}[a_0, \ldots, a_5]$.
   4b. Compute the roots of the multivariate system $P'(a_0, \ldots, a_5) = 0$ for all $P' \in \mathcal{S}'$ over $\mathbb{F}_{17^2}$ with Gröbner basis algorithms.
   4c. For each root $(c_0, \ldots, c_5)$ of the system constructed in Step 4b, substitute it into unknown coefficients in f. Namely, we set $f_{\text{sol}} := x^{10} + c_8 x^8 + c_7 x^7 + \cdots + c_1 x + c_0$. Decide whether f_{sol} has no double root in the algebraic closure or not (this can be done by constructing the minimal splitting field of f_{sol}). If f_{sol} has no double root in the algebraic closure, store f_{sol}.

As a computational result, we obtain the set $\mathcal{F}$ of all the polynomials $f(x)$ of the form (8) such that $C : y^2 = f(x)$ are superspecial hyperelliptic curves of genus 4 over $\mathbb{F}_{17^2}$. Put $\mathcal{H}_0 := \{cy^2 - f(x) : c = 1, \epsilon \text{ and } f(x) \in \mathcal{F}\}$. For each pair (H_1, H_2) of elements in $\mathcal{H}_0$ with $H_1 \neq H_2$, we execute *Isomorphism Testing Method* given in Sect. 3.3. We obtain a subset $\mathcal{H} \subset \mathcal{H}_0$ such that for each pair (H_1, H_2) of elements in $\mathcal{H}$ with $H_1 \neq H_2$, the two hyperelliptic curves $C_1 : H_1(x, y) = 0$ and $C_2 : H_2(x, y) = 0$ are not isomorphic over $\mathbb{F}_{17^2}$. The obtained set $\mathcal{H}$ consists of 25 elements. This shows the assertion over $\mathbb{F}_{17^2}$. Similarly, by executing *Isomorphism Testing Method* again for pairs of elements in $\mathcal{H}$, we obtain representatives of the isomorphism classes over the algebraic closure of $\mathbb{F}_{17^2}$. The resulting set consists of 2 elements. □

Proposition 6. *Consider the polynomial $f(x) \in \mathbb{F}_{19}[x]$ of the form*

$$f(x) = x^{10} + a_8 x^8 + a_7 x^7 + \cdots + a_1 x + a_0 \qquad (9)$$

for $a_i \in \mathbb{F}_{19}$ with $0 \leq i \leq 8$. Then there exist precisely 12 (resp. 2) superspecial hyperelliptic curves $C : cy^2 = f(x)$ with $c = 1$ or ϵ, up to isomorphism over

$\mathbb{F}_{19}$ (*resp. the algebraic closure of* $\mathbb{F}_{19}$), *such that* $f(x)$ *are of the form* (9). *Representatives of the 12 isomorphisms classes over* $\mathbb{F}_{19}$ *are given by*

1. $y^2 = x^{10} + 1$,
2. $y^2 = x^{10} + 2$,
3. $y^2 = x^{10} + x^7 + 4x^6 + 15x^5 + 6x^4 + 8x^3 + 5x^2 + 12x + 1$,
4. $y^2 = x^{10} + x^8 + 7x^6 + x^4 + x^2 + 7$,
5. $y^2 = x^{10} + x^8 + x^7 + 12x^6 + x^5 + 10x^4 + 9x^3 + 8x^2 + 9x + 3$,
6. $y^2 = x^{10} + x^8 + x^7 + 13x^6 + 9x^5 + 14x^4 + 4x^3 + 11x^2 + 3x + 8$,
7. $y^2 = x^{10} + x^8 + 2x^7 + 6x^6 + 18x^5 + 4x^4 + 13x^3 + 18x^2 + 10x + 14$,
8. $y^2 = x^{10} + x^8 + 2x^7 + 12x^6 + 18x^4 + 5x^3 + x^2 + 7$,
9. $y^2 = x^{10} + x^8 + 4x^7 + 8x^6 + 8x^5 + 3x^4 + 11x^3 + 8x^2 + 8x + 4$,
10. $y^2 = x^{10} + 2x^8 + 9x^6 + 8x^4 + 16x^2 + 15$,
11. $y^2 = x^{10} + 2x^8 + x^7 + 12x^6 + 9x^5 + 2x^3 + 4x^2 + 7x + 4$, *and*
12. $y^2 = x^{10} + 2x^8 + 3x^7 + 17x^6 + 9x^5 + 2x^3 + 12x^2 + 2x + 4$,

and those of the 2 isomorphism classes over the algebraic closure are given by

1. $y^2 = x^{10} + 1$, *and*
2. $y^2 = x^{10} + x^7 + 4x^6 + 15x^5 + 6x^4 + 8x^3 + 5x^2 + 12x + 1$.

Remark 4. 1. We have explicit defining equations of the 25 superspecial curves in Proposition 5 but omit them in the statement. See a table at [27] for the equations.
2. The source codes and the log files together with detailed information on timing are available at [27].
3. In our implementations, we used the Magma function Variety to solve multivariate systems over finite fields. We also used FactorisationOverSplittingField in order to decide whether a univariate polynomial over a finite field has no double root or not.

4.3   Application to Finding Maximal Curves and Minimal Curves

We found maximal curves and minimal curves over $\mathbb{F}_{p^2}$ for $p = 17$ and 19 among enumerated superspecial hyperelliptic curves, see also a table on the web page of the first author [27]. We here introduce several explicit equations (cf. the example C_a given in Sect. 1 is included in one of the $\mathbb{F}_{17}$-isomorphism classes of superspecial hyperelliptic curves over $\mathbb{F}_{17}$ enumerated in Proposition 4).

1. Recall from Corollary 5 that there exist precisely 2 (resp. 2) maximal (resp. minimal) hyperelliptic curves over $\mathbb{F}_{17^2}$ up to isomorphism over $\mathbb{F}_{17^2}$. Specifically, the two maximal curves are given by $y^2 = x^{10} + x$ and $y^2 = x^{10} + x^7 + 13x^4 + 12x$, respectively. The two minimal curves are given by

$$y^2 = x^{10} + x^8 + \zeta^{16}x^7 + \zeta^{83}x^6 + \zeta^{276}x^5 + \zeta^{164}x^4 + \zeta^{102}x^3 + \zeta^{111}x^2 + \zeta^2 x + \zeta^{152},$$
$$y^2 = x^{10} + x^8 + \zeta^{22}x^7 + \zeta^{250}x^6 + \zeta^{89}x^5 + \zeta^{182}x^4 + \zeta^9 x^3 + \zeta^{225}x^2 + \zeta^{282}x + \zeta^{113}$$

respectively, where we take $\zeta = -8 + \sqrt{61} \in \mathbb{F}_{17^2}$. The above four equations are obtained in Proposition 5 as representatives of the $\mathbb{F}_{17^2}$-isomorphism classes of the superspecial hyperelliptic curves of genus 4 over $\mathbb{F}_{17^2}$.

2. There exist maximal curves and minimal curves over $\mathbb{F}_{19^2}$, see Corollary 6. Specifically, the following hyperelliptic curves (1) – (5) are maximal as curves over $\mathbb{F}_{19^2}$: (1) $y^2 = x^{10} + 1$, (2) $y^2 = x^{10} + 2$, (3) $y^2 = x^{10} + x^7 + 4x^6 + 15x^5 + 6x^4 + 8x^3 + 5x^2 + 12x + 1$, (4) $y^2 = x^{10} + x^8 + 7x^6 + x^4 + x^2 + 7$, and (5) $y^2 = x^{10} + 2x^8 + 9x^6 + 8x^4 + 16x^2 + 15$. On the other hand, the following curve is minimal: (6) $y^2 = x^{10} + x^8 + 2x^7 + 12x^6 + 18x^4 + 5x^3 + x^2 + 7$. The above six equations are listed in Proposition 6 as representatives of the $\mathbb{F}_{19}$-isomorphism classes of the superspecial hyperelliptic curves of genus 4 over $\mathbb{F}_{19}$.

Remark 5. 1. We have that 3 of the defining equations listed in Proposition 4 define maximal curves over $\mathbb{F}_{17^2}$, but omit them here (cf. [27]).
2. Note that the maximal hyperelliptic curve $y^2 = x^{10} + x$ (resp. $y^2 = x^{10} + 1$) over $\mathbb{F}_{17^2}$ (resp. $\mathbb{F}_{19^2}$) is of known type, see e.g., [23] for more general results on the existence of such a kind of maximal hyperelliptic curves.

5   Concluding Remark

We enumerated the isomorphism classes of superspecial hyperelliptic curves of genus 4 over finite fields $\mathbb{F}_q$ for $q = 11$, 11^2, 13, 13^2, 17, 17^2 and 19. Specifically, the enumerations were theoretically reduced into computational problems. To solve the problems in real time, we proposed three computational methods. With our methods, we have succeeded in finishing all required computations within a day in total. Our computational results show the non-existence of a superspecial hyperelliptic curve in characteristic $p = 11$ and 13. They also provide explicit defining equations for the enumerated superspecial hyperelliptic curves in characteristic $p = 17$ and 19. Many of them are maximal curves over $\mathbb{F}_{17^2}$ and $\mathbb{F}_{19^2}$, respectively. Indeed, we found that 3 (resp. 2) among the 5 (resp. 25) superspecial curves over $\mathbb{F}_{17}$ (resp. $\mathbb{F}_{17^2}$) are maximal curves over $\mathbb{F}_{17^2}$, and that 5 among the 12 superspecial curves over $\mathbb{F}_{19}$ are those over $\mathbb{F}_{19^2}$.

References

1. Bettale, L., Faugère, J.-C., Perret, L.: Hybrid approach for solving multivariate systems over finite fields. J. Math. Crypt. **3**, 177–197 (2009)
2. Bosma, W., Cannon, J., Playoust, C.: The Magma algebra system. I. The user language. J. Symb. Comput. **24**, 235–265 (1997)
3. Deuring, M.: Die Typen der Multiplikatorenringe elliptischer Funktionenkörper. Abh. Math. Sem. Univ. Hamburg **14**(1), 197–272 (1941)
4. Ekedahl, T.: On supersingular curves and abelian varieties. Math. Scand. **60**, 151–178 (1987)
5. Fuhrmann, R., Garcia, A., Torres, F.: On maximal curves. J. Number Theory **67**, 29–51 (1997)
6. van der Geer, G., et al.: Tables of Curves with Many Points (2009). http://www.manypoints.org. Accessed 5 Apr 2018

7. González, J.: Hasse-Witt matrices for the Fermat curves of prime degree. Tohoku Math. J. **49**(2), 149–163 (1997). MR 1447179 (98b:11064)
8. Hartshorne, R.: Algebraic Geometry, GTM 52. Springer, Heidelberg (1977). https://doi.org/10.1007/978-1-4757-3849-0
9. Hashimoto, K.: Class numbers of positive definite ternary quaternion Hermitian forms. Proc. Japan Acad. Ser. A Math. Sci. **59**(10), 490–493 (1983)
10. Hashimoto, K., Ibukiyama, T.: On class numbers of positive definite binary quaternion Hermitian forms II. J. Fac. Sci. Univ. Tokyo Sect. IA Math. **28**(3), 695–699 (1982)
11. Hurt, N.E.: Many Rational Points: Coding Theory and Algebraic Geometry. Kluwer Academic Publishers, Dordrecht (2003)
12. Ibukiyama, T.: On rational points of curves of genus 3 over finite fields. Tohoku Math. J. **45**, 311–329 (1993)
13. Ibukiyama, T., Katsura, T.: On the field of definition of superspecial polarized abelian varieties and type numbers. Compositio Math. **91**(1), 37–46 (1994)
14. Kudo, M., Harashita, S.: Superspecial curves of genus 4 in small characteristic. Finite Fields Their Appl. **45**, 131–169 (2017)
15. Kudo, M., Harashita, S.: Enumerating superspecial hyperelliptic curves of genus 4 over small finite fields, in preparation
16. Kudo, M. and Harashita, S.: Enumerating superspecial curves of genus 4 over prime fields, arXiv: 1702.05313 [math.AG] (2017)
17. Kudo, M., Harashita, S.: Enumerating superspecial curves of genus 4 over prime fields (abstract version of [16]). In: Proceedings of The Tenth International Workshop on Coding and Cryptography 2017 (WCC 2017), 18–22 September 2017, Saint-Petersburg, Russia (2017). http://wcc2017.suai.ru/proceedings.html
18. Li, K.-Z., Oort, F.: Moduli of Supersingular Abelian Varieties. Lecture Notes in Mathematics, vol. 1680. Springer, Berlin (1998). https://doi.org/10.1007/BFb0095931
19. Manin, Y. I.: On the theory of Abelian varieties over a field of finite characteristic, AMS Transl. Ser. 2 **50**, 127–140 (1966). Translated by G. Wagner (originally published in Izv. Akad. Nauk SSSR Ser. Mat. **26**, 281–292 (1962))
20. Nygaard, N.O.: Slopes of powers of Frobenius on crystalline cohomology. Ann. Sci. École Norm. Sup. **14**(4), 369–401 (1982, 1981)
21. Özbudak, F., Saygı, Z.: Explicit maximal and minimal curves over finite fields of odd characteristics. Finite Fields Their Appl. **42**, 81–92 (2016)
22. Serre, J.-P.: Nombre des points des courbes algebrique sur $\mathbb{F}_q$. Théor. Nombres Bordeaux **83**(2), 22 (1983, 1982)
23. Tafazolian, S.: A note on certain maximal hyperelliptic curves. Finite Fields Their Appl. **18**, 1013–1016 (2012)
24. Tafazolian, S., Torres, F.: On the curve $y^n = x^m + x$ over finite fields. J. Number Theory **145**, 51–66 (2014)
25. Xue, J., Yang, T.-C., Yu, C.-F.: On superspecial abelian surfaces over finite fields. Doc. Math. **21**, 1607–1643 (2016)
26. Yui, N.: On the Jacobian varieties of hyperelliptic curves over fields of characisctic $p > 2$. J. Algebr. **52**, 378–410 (1978)
27. Data base of superspecial curves of genus 4 over finite fields and their algebraic closures. http://www2.math.kyushu-u.ac.jp/~m-kudo/Ssp-curves-genus-4.html

Fast Computation of Isomorphisms Between Finite Fields Using Elliptic Curves

Anand Kumar Narayanan[✉]

Laboratoire d'Informatique de Paris 6, Sorbonne Universite, Paris, France
anand.narayanan@lip6.fr

Abstract. We propose a randomized algorithm to compute isomorphisms between finite fields using elliptic curves. To compute an isomorphism between two fields of cardinality q^n, our algorithm takes

$$n^{1+o(1)} \log^{1+o(1)} q + \max_{\ell} \left(\ell^{n_\ell + 1 + o(1)} \log^{2+o(1)} q + O(\ell \log^5 q) \right)$$

time, where ℓ runs through primes dividing n but not $q(q-1)$ and n_ℓ denotes the highest power of ℓ dividing n. Prior to this work, the best known run time dependence on n was quadratic. Our run time dependence on n is at worst quadratic but is subquadratic if n has no large prime factor. In particular, the n for which our run time is nearly linear in n have natural density at least $3/10$. The crux of our approach is finding a point on an elliptic curve of a prescribed prime power order or equivalently finding preimages under the Lang map on elliptic curves over finite fields. We formulate this as an open problem whose resolution would solve the finite field isomorphism problem with run time nearly linear in n.

1 Introduction

1.1 Computing Isomorphisms Between Finite Fields

Every finite field has prime power cardinality, for every prime power there is a finite field of that cardinality and every two finite fields of the same cardinality are isomorphic. This now well known result due to Moore [Moo1889] poses two algorithmic problems. The first concerns field construction: given a prime power, construct a finite field of that cardinality. The second is the isomorphism problem: compute an isomorphism between two explicitly presented finite fields of the same cardinality.

Field construction is performed by constructing an irreducible polynomial of appropriate degree over the underlying prime order field, with all known efficient

Supported by NSF grant #CCF-1423544 and European Union's H2020 Programme (grant agreement #ERC-669891).

L. Budaghyan and F. Rodríguez-Henríquez (Eds.): WAIFI 2018, LNCS 11321, pp. 74–91, 2018.
https://doi.org/10.1007/978-3-030-05153-2_4

unconditional constructions requiring randomness. The fastest known construction, due to Couveignes and Lercier [CL13] uses elliptic curve isogenies. In practice, a polynomial is chosen at random and tested for irreducibility [Ben81]. Such non canonical construction of finite fields motivates the isomorphism problem in several applications. For instance in cryptography, the discrete logarithm problem over small characteristic finite fields is often posed over fields constructed using random irreducible polynomials. In cryptanalysis, the quasi-polynomial algorithm [BGJT14] for discrete logarithms works over fields constructed using irreducible polynomials of a special form. An isomorphism computation is thus required as a preprocessing step in cryptanalysis.

Zierler noted that the isomorphism problem reduces to root finding over finite fields and hence has efficient randomized algorithms [Zie74]. Remarkably, the isomorphism problem was shown to be in deterministic polynomial time by Lenstra [Len87]. Allombert [All02] proposed a linear algebraic randomized algorithm, close in spirit with Lenstra's algorithm but markedly faster. Employing the (randomized) polynomial factorization algorithm of Kaltofen and Shoup [KS99] (implemented using Kedlaya-Umans fast modular composition [KU08]) to find roots, Zierler's approach yields the fastest previously known algorithm for computing isomorphisms. Our main result is an algorithm with improved run time in most cases.

An alternate approach relying on cyclotomy instead of root finding was introduced by Pinch [Pin92] and improved upon by Rains [Rai08] to give the fastest algorithm in practice. The cyclotomic method of Pinch requires that the finite fields in question contain certain small order roots of unity. To remove this requirement, Pinch [Pin92] proposed using elliptic curves over finite fields. This way, instead of roots of unity, one seeks rational points of small order on elliptic curves. Our algorithm, although very different, relies on elliptic curves as well. We take inspiration from the aforementioned algorithm of Couveniges and Lercier [CL13]. While [CL13] used elliptic curve isogenies to solve field construction in nearly linear time, we solve the isomorphism problem. Our algorithm may also be viewed as an extension of Allombert's [All02] using elliptic curves.

A critical component of our approach is a method to reduce the isomorphism problem for arbitrary degrees to prime power degrees in nearly linear time. The reduction is mostly subtle linear algebra and similar to a theorem of Shoup [Sho95][Theorem 5][1]. We invoke elliptic curves only to solve the prime power cases.

Soon after posting a preprint version of the current paper online [Nar2016], I was informed of concurrent related work that later appeared here [BDDFS17]. Therein Brieulle, De Feo, Doliskani, Flori and Schost address the very same isomorphism problem (and more generally finite field embedding problems) using elliptic curve based techniques similar to ours. In contrast to our emphasis on establishing complexity theoretic bounds on the isomorphism problem, their goals are directed towards obtaining fast practical algorithms. In particular,

---

[1] We thank an anonymous referee for pointing out the similarity.

they present an open source implementation of their algorithm which appears to be the current state of the art in practice.

1.2   Computing Isomorphisms and Root Finding

We formally pose the isomorphism problem stating the manner in which the input fields and the output isomorphism are represented. Let q be a power of a prime p and let $\mathbb{F}_q$ denote the finite field with q elements. Fix an algebraic closure $\overline{\mathbb{F}}_q$ of $\mathbb{F}_q$ and let $\sigma : \overline{\mathbb{F}}_q \longrightarrow \overline{\mathbb{F}}_q$ denote the qth power Frobenius endomorphism. We consider two finite fields of cardinality q^n to be given through two monic irreducible degree n polynomials $f(x), g(x) \in \mathbb{F}_q[x]$. The fields are then constructed as $\mathbb{F}_q(\alpha)$ and $\mathbb{F}_q(\beta)$ where $\alpha, \beta \in \overline{\mathbb{F}}_q$ are respectively roots of $f(x), g(x)$. Without loss of generality [CL13], all our algorithms assume the base field $\mathbb{F}_q$ to be given as the quotient of the polynomial ring over $\mathbb{Z}/p\mathbb{Z}$ by a monic irreducible polynomial over $\mathbb{Z}/p\mathbb{Z}$.

An isomorphism $\phi : \mathbb{F}_q(\alpha) \longrightarrow \mathbb{F}_q(\beta)$ that fixes $\mathbb{F}_q$ is completely determined by the image $\phi(\alpha)$. We call the unique $r_\phi(x) \in \mathbb{F}_q[x]$ of degree less than n such that $\phi(\alpha) = r_\phi(\beta)$ as the polynomial representation of ϕ. We are justified in seeking the polynomial representation of ϕ since given $r_\phi(x)$, one may compute the image of an element in $\mathbb{F}_q(\alpha)$ under ϕ in time nearly linear in n using fast modular composition [KU08]. For an $r(x) \in \mathbb{F}_q[x]$ of degree less than n, $r(x)$ is the polynomial representation of an isomorphism from $\mathbb{F}_q(\alpha)$ to $\mathbb{F}_q(\beta)$ if and only if $r(\beta)$ is a root of $f(x)$. Hence the problem of computing the polynomial representation of an isomorphism that fixes $\mathbb{F}_q$ is identical to the following root finding problem.

ISOMORPHISM PROBLEM: Given monic irreducibles $f(x), g(x) \in \mathbb{F}_q[x]$ of degree n, find a root of $f(x)$ in $\mathbb{F}_q(\beta)$ where $\beta \in \overline{\mathbb{F}}_q$ is a root of $g(x)$.

There are two input size parameters, namely n and $\log q$. Prior to our work, the best known run time was quadratic in n resulting from using [KS99, KU08] to find roots in the ISOMORPHISM PROBLEM. We are primarily interested in lowering the run time exponent in n. Our run time dependence on $\log q$ will be polynomial but not optimized for. Here on, all our algorithms are Las Vegas randomized and by run time we mean expected run time.

1.3   Summary of Results

We present an algorithm for the ISOMORPHISM PROBLEM with run time

$$n^{1+o(1)} \log^{1+o(1)} q + \max_\ell \left(\ell^{n_\ell+1+o(1)} \log^{2+o(1)} q + O(\ell \log^5 q) \right)$$

where ℓ runs through primes dividing n but not $q(q-1)$ with n_ℓ the highest power of ℓ dividing n. Evidently, our run time depends on the prime factorization of n. Although at worst quadratic in n, we next argue it is subquadratic for *most* n. If n has a large (say $\Omega(n)$) prime factor not dividing $q(q-1)$, our running time exponent in n is 2. In all other cases, it is less than 2. Call n with largest prime factor at most $n^{1/c}$ as $n^{1/c}$-powersmooth. For $n^{1/c}$-powersmooth n with $1 < c \le 2$, our run time exponent in n is at most $2/c$. The natural density of $n^{1/c}$-powersmooth n tends to the Dickman-de Bruin function $\rho(c)$ and for $1 < c \le 2$, $\rho(c) = 1 - \log c$ [Gra08]. In particular, $n^{1/1.1}$-powersmooth n have density $1 - \log(1.1) > 9/10$. Hence the n with run time exponent in n at most $2/1.1 \approx 1.8$ have density at least $9/10$. Likewise, $n^{1/2}$-powersmooth n have density at least $3/10$. Hence the n with run time linear in n have density at least $3/10$.

The paper is organized as follows. In Sect. 2, the ISOMORPHISM PROBLEM is reduced in linear time to subproblems, each one corresponding to a prime power ℓ^{n_ℓ} dividing n. A key component in the reduction is a fast linear algebraic algorithm (Lemma 2.1) that takes a polynomial relation between two $\alpha, \beta \in \overline{\mathbb{F}}_q$ of the same degree and computes a root of the minimal polynomial of α in $\mathbb{F}_q(\beta)$. In Sect. 3, subproblems corresponding to prime powers ℓ^{n_ℓ} such that ℓ divides $q - 1$ are solved in linear time using Kummer theory. Likewise, in Sect. 4, subproblems corresponding to powers of the characteristic p are solved in linear time using Artin-Schreier theory. The key in both these special cases is a new recursive algorithm to evaluate the action of idempotents in the Galois group ring that appear in the proof of Hilbert's theorem 90. In Sect. 5, the generic case of a prime power ℓ^{n_ℓ} where $\ell \nmid (q - 1)p$ is handled using an elliptic curve $E/\mathbb{F}_q$ with $\mathbb{F}_q$ rational ℓ torsion. The analogue of Hilbert's theorem 90 in this context is Lang's theorem which states that the first cohomology group $H^1(\mathbb{F}_q, E)$ is trivial [Lan78]. In Subsect. 5.2, the ISOMORPHISM PROBLEM is reduced to computing discrete logarithms in the $\mathbb{F}_q$ rational ℓ torsion subgroup of E. The crux of the reduction is to compute a preimage of a non trivial $\mathbb{F}_q$ rational ℓ torsion point under the Lang map. In Subsect. 5.3, we devise a fast algorithm to compute such a preimage using ℓ isogenies and solve the ISOMORPHISM PROBLEM of degree ℓ^{n_ℓ} in $\ell^{n_\ell+1+o(1)} \log^{1+o(1)} q + O(\ell \log^5 q)$ time. In Subsect. 5.3, we pose an algorithmic Problem 5.7 concerning Lang's theorem, a solution to which would solve the ISOMORPHISM PROBLEM in subquadratic time for all n.

Fast modular composition and fast modular power projection [KU08], key ingredients in our algorithm, are considered impractical with no existing implementations. Practical implications of our algorithm are thus unclear.

We also extend our algorithm to solve the following more general root finding problem: given a polynomial over $\mathbb{F}_q$ and a positive integer n, find its roots in $\mathbb{F}_{q^n}$ (see Remark 2.4). The construction of $\mathbb{F}_{q^n}$ could be given or left to the algorithm. The former allows one to compute embeddings of one finite field in another.

2   Reduction of the Isomorphism Problem to Prime Power Degrees

For $\alpha \in \overline{\mathbb{F}}_q$, call $[\mathbb{F}_q(\alpha) : \mathbb{F}_q]$ the degree of α. For $\alpha, \beta \in \overline{\mathbb{F}}_q$, call $\alpha \sim \beta$ if and only if there is an integer j such that $\alpha = \sigma^j(\beta)$. That is, $\alpha \sim \beta$ means they have the same minimal polynomial.

Lemma 2.1. *There is an $n^{1+o(1)} \log^{1+o(1)} q$ time algorithm that given the minimal polynomial $g(x) \in \mathbb{F}_q[x]$ of an $\alpha \in \overline{\mathbb{F}}_q$ of degree n and $f_1(x), f_2(x) \in \mathbb{F}_q[x]$ of degree less than n such that $f_1(\alpha)$ is of degree n, finds an $r(x) \in \mathbb{F}_q[x]$ such that $r(\beta)$ is a root of $g(x)$ for all $\beta \in \overline{\mathbb{F}}_q$ satisfying $f_1(\alpha) \sim f_2(\beta)$.*

Proof. Since α and $f_1(\alpha)$ both have degree n, α is in $\mathbb{F}_q(f_1(\alpha))$ and there is a unique $h(x) = \sum_{i=0}^{n-1} h_i x^i \in \mathbb{F}_q[x]$ such that $h(f_1(\alpha)) = \alpha$. We next describe how to compute $h(x)$.

Pick $u \in \mathbb{F}_q^n$ uniformly at random and consider the $\mathbb{F}_q$-linear functional

$$\mathcal{U} : \mathbb{F}_q(\alpha) \longrightarrow \mathbb{F}_q, y \longmapsto u^t y$$

where $y \in \mathbb{F}_q^n$ is an element of $\mathbb{F}_q(\alpha)$ written in the standard basis $(1, \alpha, \alpha^2, \ldots, \alpha^{n-1})$.

Abusing notation, let $h = (h_0, h_1, \ldots, h_{n-1})^t$ denote the coefficient vector of $h(x)$. We will determine $h(x)$ by solving the linear system

$$\mathcal{U}(\alpha^i h(f_1(\alpha))) = \mathcal{U}(\alpha f_1(\alpha)^i), i \in \{0, 1, \ldots, 2n - 2\}$$

in its coefficients. Let A be the n by $2n - 1$ matrix whose i^{th} column consists of $f_1(\alpha)^{i-1}$ written in the standard basis. Multiplication by α is an $\mathbb{F}_q$ linear transformation on $\mathbb{F}_q(\alpha)$ with matrix representation on the standard basis being the companion matrix

$$X := \begin{bmatrix} 0 & 0 & 0 & \ldots & 0 & -g_0 \\ 1 & 0 & 0 & \ldots & 0 & -g_1 \\ 0 & 1 & 0 & \ldots & 0 & -g_2 \\ \vdots & \vdots & \vdots & \ddots & \vdots & \vdots \\ 0 & 0 & 0 & \ldots & 1 & -g_{n-1} \end{bmatrix}$$

with respect to $g(x) = \sum_{i=0}^{n-1} g_i x^i + x^n$. Let $(a_0, a_1, \ldots, a_{2n-2}) := u^t A$ and $(b_0, b_1, \ldots, b_{2n-2}) := u^t X A$. Since X has at most $2n - 1$ non zero coefficients, $u^t X$ can be computed with number of $\mathbb{F}_q$-operations bounded linearly in n. Given $u, f_1(x)$ and $g(x)$, to compute $u^t A$ is an instance of the modular power projection problem. Likewise computing $u^t X A$ given $u^t X, f_1(x)$ and $g(x)$. By [KU08], each of these modular power projection instances can be solved in $n^{1+o(1)} \log^{1+o(1)} q)$ time. The aforementioned linear system in matrix form is

$$\begin{bmatrix} a_0 & a_1 & a_2 & \ldots & a_{n-1} \\ a_1 & a_2 & a_3 & \ldots & a_n \\ a_2 & a_3 & a_4 & \ldots & a_{n+1} \\ \vdots & \vdots & \vdots & \ddots & \vdots \\ a_{n-1} & a_n & a_{n+1} & \ldots & a_{2n-2} \end{bmatrix} \begin{bmatrix} h_0 \\ h_1 \\ h_3 \\ \vdots \\ h_{n-1} \end{bmatrix} = \begin{bmatrix} b_0 \\ b_1 \\ b_2 \\ \vdots \\ b_{2n-2} \end{bmatrix} \tag{1}$$

and by [Sho99] has full rank with probability at least $1/2$ for a randomly chosen u. One of its solutions is the coefficient vector h of the $h(x)$ we seek. Being Toeplitz, in $n^{1+o(1)} \log^{1+o(1)} q)$ time, we can test if it is full rank and if so find the solution

h. Once $h(x)$ is found, using [CL13, Corollary 1] to compose polynomials, within time stated in the lemma, we output $h(f_2(x))$ as $r(x)$. The output is correct since $h(f_2(\beta)) \sim h(f_1(\alpha)) = \alpha$. □

Lemma 2.2. *There is an algorithm that given the minimal polynomial $g(x) \in \mathbb{F}_q[x]$ of an $\alpha \in \overline{\mathbb{F}}_q$ of degree m and a positive integer n dividing m, finds an element $\alpha_n \in \mathbb{F}_q(\alpha)$ of degree n and its minimal polynomial over $\mathbb{F}_q$ in time $m^{1+o(1)} \log^{2+o(1)} q$.*

Proof. Pick $\beta \in \mathbb{F}_q(\alpha)$ uniformly at random and set $\alpha_n := \sum_{i=0}^{m/n-1} \sigma^{ni}(\beta)$, the trace of β down to $\mathbb{F}_{q^n} \subseteq \mathbb{F}_q(\alpha)$. By iterated Frobenius [vzGS92, KU08], this trace computation can be performed in the time stated in the lemma. Compute the minimal polynomial $M(x) \in \mathbb{F}_q[x]$ of α_n over $\mathbb{F}_q$ using [Sho99], [KU08, Sect. 8.4], again, in time stated in the lemma. If the degree of $M(x)$ is n, output α_n and $M(x)$. Since the trace down to $\mathbb{F}_{q^n}$ maps a random element from $\mathbb{F}_q(\alpha)$ to a random element in $\mathbb{F}_{q^n}$, we succeed with probability at least $1/2$. □

We next reduce ISOMORPHISM PROBLEM to itself restricted to prime power input degree.

Lemma 2.3. *Let $n = \prod_\ell \ell^{n_\ell}$ be the factorization of n into prime powers. In $n^{1+o(1)} \log^{2+o(1)} q$ time, ISOMORPHISM PROBLEM with inputs of degree n may be reduced to identical problems; one for each prime ℓ dividing n with inputs of degree ℓ^{n_ℓ}.*

Proof. Consider an input $f(x), g(x) \in \mathbb{F}_q[x]$ to ISOMORPHISM PROBLEM. Let $\alpha, \beta \in \overline{\mathbb{F}}_q$ respectively be roots of $f(x), g(x)$. Compute the factorization $n = \prod_\ell \ell^{n_\ell}$ of n into prime powers. For each prime ℓ dividing n, using Lemma 2.2, compute $\alpha_\ell \in \mathbb{F}_q(\alpha)$ and $M_\ell(x) \in \mathbb{F}_q[x]$ such that α_ℓ has degree ℓ^{n_ℓ} and M_ℓ is the minimal polynomial of α_ℓ. Likewise compute $\beta_\ell \in \mathbb{F}_q(\beta)$ and $N_\ell(x) \in \mathbb{F}_q[x]$ such that β_ℓ has degree ℓ^{n_ℓ} and N_ℓ is the minimal polynomial of β_ℓ. Since $\mathbb{F}_{q^{\ell^{n_\ell}}}$ and $\mathbb{F}_{q^{n/\ell^{n_\ell}}}$ are linearly disjoint over $\mathbb{F}_q$, both $\sum_{\ell|n} \alpha_\ell$ and $\sum_{\ell|n} \beta_\ell$ have degree n. For each ℓ dividing n, solve ISOMORPHISM PROBLEM with input $M_\ell(x), N_\ell(x)$ and find a root β'_ℓ of $M_\ell(x)$ in $\mathbb{F}_q(\beta_\ell)$. Now for all ℓ dividing n, $\alpha_\ell \sim \beta'_\ell$. Applying Lemma 2.1 to the relation $\sum_{\ell|n} \alpha_\ell \sim \sum_{\ell|n} \beta'_\ell$, we solve ISOMORPHISM PROBLEM with input $f(x), g(x)$. □

Remark 2.4. Consider the problem of finding a root of a degree m polynomial $f(x) \in \mathbb{F}_q[x]$ in $\mathbb{F}_{q^n}$, where $\mathbb{F}_{q^n}$ is constructed as $\mathbb{F}_q[x]/(g(x))$ for a monic irreducible $g(x)$. Either $g(x)$ is given or constructed in linear time using [CL13]. We show that this problem reduces to the ISOMORPHISM PROBLEM in time linear in m and n. In fact, the reduction finds not just one but all the roots of $f(x)$ in an implicit form. The output is a set of roots of $f(x)$ whose orbit under σ is the set of all roots of $f(x)$. For $f(x)$ to have a root in $\mathbb{F}_{q^n}$, $f(x)$ has to have an irreducible factor of degree dividing n. Since the number of factors of n is at most $\log n$, using [KU08], in $m^{1+o(1)} \log^{2+o(1)} q \log^{1+o(1)} n$ time, we may enumerate all irreducible factors of $f(x)$ of degree dividing n. For each such irreducible factor $h(x)$, using Lemma 2.2, identify a subfield of $\mathbb{F}_{q^n}$ and find a root $h(x)$ in the subfield by solving the ISOMORPHISM PROBLEM.

3 Root Finding in Kummer Extensions of Finite Fields

Using Kummer theory, we solve the ISOMORPHISM PROBLEM restricted to the case when n is a power of a prime ℓ dividing $q - 1$. The novelty here is a fast recursive evaluation of the idempotent appearing in the standard proof of (cyclic) Hilbert's theorem 90.

Lemma 3.1. *There is an algorithm that given a finite extension $L/\mathbb{F}_q$, an integer $m \le [L : \mathbb{F}_q]$ and a $\zeta \in L$ such that $\zeta \in K := \{\beta \in L | \sigma^m(\beta) = \beta\}$ and $\zeta^{[L:K]} = 1$, finds an $\alpha \in L$ such that $\sigma^m(\alpha) = \zeta\alpha$ in $[L : \mathbb{F}_q]^{1+o(1)} \log^{2+o(1)} q$ time.*

Proof. Since the norm of ζ from L down to K is $\zeta^{[L:K]} = 1$, an α as claimed in the lemma exists by Hilbert's theorem 90 applied to the cyclic extension L/K. We next describe an algorithm that finds such an α in the stated time.

Define $\tau := \zeta^{-1}\sigma^m$, viewed as a K-linear endomorphism on L. By independence of characters, $\sum_{i=0}^{[L:K]-1} \tau^i$ is non zero. Pick $\theta \in L$ uniformly at random. If $\sum_{i=0}^{[L:K]-1} \tau^i(\theta) \ne 0$ (which happens with probability at least $1/2$), set $\alpha = \sum_{i=0}^{[L:K]-1} \tau^i(\theta)$. Since $\zeta^{-1} \in \mathbb{F}_q$ and $\zeta^{-[L:K]} = 1$,

$$\tau(\alpha) = \sum_{i=0}^{[L:K]-1} \zeta^{-i}\sigma^{mi}(\alpha) = \alpha \Rightarrow \tau(\alpha) = \alpha \Rightarrow \zeta^{-1}\sigma^m(\alpha) = \alpha \Rightarrow \sigma^m(\alpha) = \zeta\alpha.$$

We next demonstrate $\sum_{i=0}^{[L:K]-1} \tau^i(\theta)$ can be computed fast given $\theta \in L$. Our approach is similar to the iterated Frobenius trace computation of von zur Gathen and Shoup [vzGS92].

Let L be given as $\mathbb{F}_q(\eta)$ for some $\eta \in \overline{\mathbb{F}}_q$ with minimal polynomial $g(x) \in \mathbb{F}_q[x]$. By repeated squaring, in time $\tilde{O}([L : \mathbb{F}_q] \log^2 q)$ compute η^q. For a positive integer b, let Σ_b denote the partial sum $\sum_{i=0}^{b-1} \tau^i(\theta)$. Our goal is to compute $\Sigma_{[L:K]}$. For every positive integer b,

$$\sum_{i=0}^{2b-1} \tau^i(\theta) = \sum_{i=0}^{b-1} \tau^i(\theta) + \sum_{i=b}^{2b-1} \tau^i(\theta) = \sum_{i=0}^{b-1} \tau^i(\theta) + \tau^b \left(\sum_{i=0}^{2b-1} \tau^i(\theta) \right)$$

$$\Rightarrow \sum_{i=0}^{2b-1} \tau^i(\theta) = \sum_{i=0}^{b-1} \tau^i(\theta) + \zeta^{-b}\sigma^b \left(\sum_{i=0}^{b-1} \tau^i(\theta) \right) \Rightarrow \Sigma_{2b} = \Sigma_b + \zeta^{-b}\sigma^{bm}(\Sigma_b). \quad (2)$$

Given Σ_b and η^q, $\sigma^{bm}(\Sigma_b)$ can be computed in $[L : \mathbb{F}_q]^{1+o(1)} \log^{1+o(1)} q$ time using the Frobenius representation of [vzGS92] and fast modular composition [KU08]. Hence, given Σ_b, computing Σ_{2b} using Eq. 2 takes $[L : \mathbb{F}_q]^{1+o(1)} \log^{1+o(1)} q$ time, which evidently is independent of b and m.

Set $c = \lfloor \log_2[L : K] \rfloor$ and compute Σ_{2^c} by successively computing $\Sigma_0, \Sigma_2, \Sigma_4, \ldots, \Sigma_c$ using equation 2. Since $c \le \log_2[L : K]$, this takes $[L :$

$\mathbb{F}_q]^{1+o(1)} \log^{1+o(1)} q$ time. If $[L : K]$ is not a power of 2, we recursively compute $\Sigma_{[L:K]-c}$. With the knowledge of Σ_c and $\Sigma_{[L:K]-c}$, Σ_m may be computed in $\tilde{O}([L : \mathbb{F}_q] \log q)$ time [vzGS92,KU08] as

$$\Sigma_{[L:K]} = \Sigma_c + \zeta^{-c} \sigma^{mc}(\Sigma_{[L:K]-c}). \tag{3}$$

Since $[L : K] - c \leq [L : K]/2$, at most $\log_2[L : K]$ recursive calls are made in total. $\qquad\square$

We next state the algorithm followed by proof of correctness and implementation details.

Algorithm 1. Root Finding Through Kummer Theory:

Input: Monic irreducibles $g_1(x), g_2(x) \in \mathbb{F}_q[X]$ of degree ℓ^a where ℓ is a prime dividing
$\quad q - 1$ and a is a positive integer.
Output: A root of $g_1(x)$ in $\mathbb{F}_q(\beta_2)$ where $\beta_2 \in \bar{\mathbb{F}}_q$ is a root of $g_2(x)$.
1: Find a primitive ℓ^{th} root of unity $\zeta_\ell \in \mathbb{F}_q$.
2: Construct $\mathbb{F}_q(\beta_1) \cong \mathbb{F}_{q^{\ell^a}}$ where β_1 is a root of $g_1(x)$.
$\quad \triangleright$ Apply lemma 3.1 with $\left(L = \mathbb{F}_q(\beta_1), m = \ell^{a-1}, \zeta = \zeta_\ell\right)$ and find $\alpha_1 \in \mathbb{F}_q(\beta_1)$ such
$\quad$ that
$$\sigma^{\ell^{a-1}}(\alpha_1) = \zeta_\ell \alpha_1.$$
$\quad \triangleright$ Compute α_1^ℓ. (α_1^ℓ will have degree ℓ^{a-1}.)
3: Construct $\mathbb{F}_q(\beta_2) \cong \mathbb{F}_{q^{\ell^a}}$ where β_2 is a root of $g_2(x)$.
$\quad \triangleright$ Apply lemma 3.1 with $\left(L = \mathbb{F}_q(\beta_2), m = \ell^{a-1}, \zeta = \zeta_\ell\right)$ and find $\alpha_2 \in \mathbb{F}_q(\beta_2)$ such
$\quad$ that
$$\sigma^{\ell^{a-1}}(\alpha_2) = \zeta_\ell \alpha_2.$$
$\quad \triangleright$ Compute α_2^ℓ. (α_2^ℓ will have degree ℓ^{a-1}.)
4: If $a = 1$,
$\quad \triangleright$ Find an $e \in \mathbb{F}_q$ such that $e^\ell = \alpha_1^\ell / \alpha_2^\ell$.
$\quad \triangleright$ Apply lemma 2.1 to $\alpha_1 \sim e\alpha_2$ and find a root of $g_1(x)$ in $\mathbb{F}_q(\beta_2)$.
5: If $a \neq 1$,
$\quad \triangleright$ Find the minimal polynomials $h_1(x), h_2(x)$ over $\mathbb{F}_q$ of $\alpha_1^\ell, \alpha_2^\ell$ respectively.
$\quad \triangleright$ Recursively find a root α of $h_1(x)$ in $\mathbb{F}_q(\alpha_2^\ell) = \mathbb{F}_q[x]/(h_2(x))$. ($h_1(x)$ and $h_2(x)$
$\quad$ have degree ℓ^{a-1}.)
$\quad \triangleright$ Find a $\gamma \in \mathbb{F}_q(\alpha_2^\ell)$ such that $\gamma^\ell = \alpha/\alpha_2^\ell$.
$\quad \triangleright$ Apply lemma 2.1 to $\alpha_1 \sim \gamma\alpha_2$ and find a root of $g_1(x)$ in $\mathbb{F}_q(\beta_2)$.

We next argue that Algorithm 1 runs to completion and is correct.

Since ℓ divides $q - 1$, there is a primitive ℓ^{th} root of unity in $\mathbb{F}_q$, as required in Step 1.

In Step 2, α_1^ℓ is claimed to have degree ℓ^{a-1}. Let b be the degree of α_1^ℓ. Since

$$\sigma^{\ell^{a-1}}(\alpha_1^\ell) = \left(\sigma^{\ell^{a-1}}(\alpha_1)\right)^\ell = \zeta_\ell^\ell \alpha_1^\ell = \alpha_1^\ell,$$

b divides ℓ^{a-1}. Since $\sigma^b(\alpha_1^\ell) = \alpha_1^\ell$, $\sigma^b(\alpha_1)/\alpha_1$ is an ℓth root of unity. Thus $\sigma^{\ell b}(\alpha_1) = \alpha_1$ implying the degree of α_1 divides ℓb. Since $\zeta_\ell \neq 1$, α_1 has degree

ℓ^a. Thus ℓ^a divides $b\ell$ and we may conclude that α_1^ℓ has degree ℓ^{a-1}. Likewise, in Step 3, α_2 has degree ℓ^{a-1}.

In Step 4, since $a = 1$, $\alpha_1^\ell, \alpha_2^\ell \in \mathbb{F}_q$. Further $\alpha_1/\alpha_2 \in \mathbb{F}_q$ since $\sigma(\alpha_1/\alpha_2) = (\zeta\alpha_1)/(\zeta\alpha_2) = \alpha_1/\alpha_2$. Thus $\alpha_1^\ell/\alpha_2^\ell$ is an ℓth power in $\mathbb{F}_q$ ensuring that an $e \in \mathbb{F}_q$ such that $e^\ell = \alpha_1^\ell/\alpha_2^\ell$ exists. Hence $(\alpha_1/a\alpha_2)$ is an ℓth root of unity and there exists an integer i such that $\alpha_1 = \sigma^{i(\ell^{a-1})}(e\alpha_2)$. Further α_1 has degree ℓ^a. Hence Lemma 2.1, when applied to the relation $\alpha_1 \sim e\alpha_2$, correctly finds the desired output.

The recursive call in Step 5 yields a root $\alpha \in \mathbb{F}_q(\alpha_2^\ell)$ of $h_1(x)$. Hence $\alpha = \sigma^j(\alpha_1^\ell) = (\sigma^j(\alpha_1))^\ell$ for some integer j. Further, $\sigma^j(\alpha_1)/\alpha_2 \in \mathbb{F}_q(\alpha_2^\ell)$ since

$$\sigma^{\ell^{a-1}}(\sigma^j(\alpha_1)/\alpha_2) = \sigma^j(\zeta_\ell\alpha_1)/(\zeta_\ell\alpha_2) = \sigma^j(\alpha_1)/\alpha_2.$$

Hence $\alpha/\alpha_2^\ell = (\sigma^j(\alpha_1)/\alpha_2)^\ell$ is an ℓth power in $\mathbb{F}_q(\alpha_2^\ell)$ assuring the existence of a γ that is sought in Step 5. For such a γ, $\gamma^\ell = (\sigma^j(\alpha_1)/\alpha_2)^\ell$ implying $\sigma^j(\alpha_1)/(\gamma\alpha_2)$ is an ℓth root of unity. Hence, there exists an integer i such that

$$\sigma^j(\alpha_1) = \gamma\sigma^{i\ell^{a-1}}(\alpha_2) = \sigma^{i\ell^{a-1}}(\gamma\alpha_2).$$

Further α_1 has degree ℓ^a. Hence Lemma 2.1, when applied to the relation $\alpha_1 \sim \gamma c\alpha_2$, correctly finds the desired output.

3.1   Implementation and Running Time Analysis

To implement Step 1, pick a random $c \in \mathbb{F}_q$ and if $c^{\frac{(q-1)}{\ell}} \neq 1$, set $\zeta = c^{\frac{q-1}{m}}$. Else try again with a new independent choice $c \in \mathbb{F}_q$. We succeed in finding a ζ if the c chosen is not a ℓth power. This happens with probability at least $1 - 1/\ell$. The expected running time of Step 1 is hence $O(\log^2 q)$. Running times of Steps 2 and 3 are dominated by the $\ell^{a+o(1)} \log^{2+o(1)} q$ time their respective calls to Lemma 3.1 take.

In Step 4, find a root a of $x^\ell - (\alpha_1^\ell/\alpha_2^\ell) \in \mathbb{F}_q[x]$ in $\mathbb{F}_q$ using [vzGS92,KU08] in $\ell^{1+o(1)} \log^{2+o(1)} q$ time. The invocations to Lemma 2.1 in Steps 4 and 5 each take $\ell^{a+o(1)} \log^{2+o(1)} q$ time.

In Step 5, minimal polynomials of α_1 and α_2 can be computed in $\ell^{a+o(1)} \log^{1+o(1)} q$) time [KU08, Sect. 8.4]. To compute γ, find a root of $x^\ell - \alpha/\alpha_2^\ell$ in $\mathbb{F}_q(\alpha_2^\ell) = \mathbb{F}_q[x]/(h_2(x))$ using [vzGS92,KU08]. Since we a finding the root of a degree ℓ polynomial over a field of size $q^{\ell^{a-1}}$, the running time $\ell^{a+o(1)} \log^{1+o(1)} q)$ turns out to be nearly linear in ℓ^a.

Algorithm 1 makes at most one recursive call to an identical subproblem of size ℓ^{a-1}. Hence at most a recursive calls are made in total. In summary, we have the following theorem.

Theorem 3.2. *Algorithm 1 solves the* ISOMORPHISM PROBLEM *restricted to the special case when n is a power of a prime ℓ dividing $q - 1$ in $n^{1+o(1)} \log^{2+o(1)} q$ time.*

4 Root Finding in Artin-Schreier Extensions of Finite Fields

Using Artin-Schrier theory, we solve the ISOMORPHISM PROBLEM restricted to the special case when n is a power of the characteristic p. The novelty here is a fast recursive evaluation of the idempotent in the proof of the additive version of (cyclic) Hilbert's theorem 90.

Lemma 4.1. *There is an algorithm that given a finite extension $L/\mathbb{F}_q$ of degree $[L : \mathbb{F}_q]$ divisible by p, finds an $\alpha \in L$ such that $\sigma^{[L:\mathbb{F}_q]/p}(\alpha) = \alpha + 1$ in $[L : \mathbb{F}_q]^{1+o(1)} \log^{2+o(1)} q$ time.*

Proof. Let $m := [L : K]/p$ and $K := \{\beta \in L | \sigma^m(\beta) = \beta\}$. Since the trace of 1 from L down to K is 0, an α as claimed in the lemma exists by Hilbert's theorem 90 applied to the cyclic extension L/K. We next describe an algorithm that finds such an α in the stated time.

Let $Tr_{L/K} = \sum_{i=0}^{p-1} \sigma^{mi}$ denote the trace from L to K. Pick $\theta \in L$ uniformly at random. If $Tr_{L/K}(\theta) \neq 0$ (which happens with probability at least $1/2$), setting

$$\alpha := \frac{-1}{Tr_{L/K}(\theta)} \sum_{i=0}^{p-1} i\sigma^{mi}(\theta)$$

ensures $\sigma^m(\alpha) - \alpha = 1$. We next demonstrate that given $\theta \in L$, α can be computed fast.

Let L be given as $\mathbb{F}_q(\eta)$ for some $\eta \in \overline{\mathbb{F}}_q$ with minimal polynomial $g(x) \in \mathbb{F}_q[x]$. By repeated squaring, in time $O([L : \mathbb{F}_q] \log^2 q)$ compute η^q.

For a positive integer b, let Σ_b denote the partial sum $\sum_{i=0}^{b-1} i\sigma^{mi}(\theta)$ and let Γ_b denote the partial trace $\sum_{i=0}^{b-1} \sigma^{mi}(\theta)$. We intend to compute Σ_p and Γ_p to set $\alpha = \Sigma_p/\Gamma_p$.

For every positive integer b,

$$\sum_{i=0}^{2b-1} i\sigma^{mi}(\theta) = \sum_{i=0}^{b-1} i\sigma^{mi}(\theta) + \sum_{i=b}^{2b-1} i\sigma^{mi}(\theta) = \sum_{i=0}^{b-1} i\sigma^{mi}(\theta) + \sum_{i=0}^{b-1} (b+i)\sigma^{m(b+i)}(\theta)$$

$$= \sum_{i=0}^{b-1} i\sigma^{mi}(\theta) + b\sigma^{mb}\left(\sum_{i=0}^{b-1} \sigma^{mi}(\theta)\right) + \sigma^{mb}\left(\sum_{i=0}^{b-1} i\sigma^{mi}(\theta)\right).$$

$$\Rightarrow \Sigma_{2b} = \Sigma_b + b\sigma^{bm}(\Sigma_b) + \sigma^{bm}(\Gamma_b). \tag{4}$$

Likewise

$$\Gamma_{2b} = \Gamma_b + \sigma^{bm}\Gamma_b. \tag{5}$$

Given Σ_b, Γ_b and η^q, $\sigma^{mb}(\Sigma_b)$ and $\sigma^{mb}(\Gamma_b)$ can be computed in $[L : \mathbb{F}_q]^{1+o(1)} \log^{2+o(1)} q)$ time using the Frobenius representation of [vzGS92] and fast modular composition [KU08]. Hence, given Σ_b and Γ_b, computing Σ_{2b} and Γ_{2b} using Eqs. 4 and 5 takes $[L : \mathbb{F}_q]^{1+o(1)} \log^{2+o(1)} q$ time. This running time is independent of b and m.

Set $c = \lfloor \log_2 p \rfloor$ and successively compute $\Sigma_0, \Gamma_0, \Sigma_2, \Gamma_2, \Sigma_4, \Gamma_4, \ldots, \Sigma_{2^c}, \Gamma_{2^c}$ using Eqs. 4 and 5. Since $c \leq \log_2 p$, this takes $\widetilde{O}([L : \mathbb{F}_q] \log^2 q)$ time. If p is not a power of 2, we recursively compute Σ_{p-2^c} and Γ_{p-2^c}. With the knowledge of $\Sigma_{2^c}, \Gamma_{2^c}, \Sigma_{p-2^c}, \Gamma_{p-2^c}$, we may compute Σ_p and Γ_p in $\widetilde{O}([L : \mathbb{F}_q] \log q)$ time as

$$\Sigma_p = \Sigma_{2^c} + 2^c \sigma^{m2^c}(\Sigma_{p-2^c}) + \sigma^{m2^c}(\Gamma_{p-2^c}), \Gamma_p = \Gamma_2^c + \sigma^{m2^c}(\Gamma_{p-2^c}). \quad (6)$$

Since $p - 2^c \leq p/2$, at most $\log_2 p$ recursive calls are made in total.   □

We next state the algorithm followed by proof of correctness and implementation details.

Algorithm 2. Root Finding Through Artin-Schreier Theory:

Input: Monic irreducibles $g_1(x), g_2(x) \in \mathbb{F}_q[X]$ of degree p^a where a is a positive integer.

Output: A root of $g_1(x)$ in $\mathbb{F}_q(\beta_2)$ where $\beta_2 \in \overline{\mathbb{F}}_q$ is a root of $g_2(x)$.

1: Construct $\mathbb{F}_q(\beta_1) \cong \mathbb{F}_{q^{p^a}}$ where β_1 is a root of $g_1(x)$.
  ▷ Apply Lemma 4.1 with $L = \mathbb{F}_q(\beta_1)$ and find $\alpha_1 \in \mathbb{F}_q(\beta_1)$ such that

$$\sigma^{p^{a-1}}(\alpha_1) = \alpha_1 + 1.$$

  ▷ Compute $\alpha_1^p - \alpha_1$. ($\alpha_1^p - \alpha_1$ *will have degree* p^{a-1}.)
2: Construct $\mathbb{F}_q(\beta_2) \cong \mathbb{F}_{q^{p^a}}$ where β_2 is a root of $g_2(x)$.
  ▷ Apply Lemma 4.1 with $L = \mathbb{F}_q(\beta_2)$ and find $\alpha_2 \in \mathbb{F}_q(\beta_2)$ such that

$$\sigma^{p^{a-1}}(\alpha_2) = \alpha_2 + 1.$$

  ▷ Compute $\alpha_2^p - \alpha_2$. ($\alpha_2^p - \alpha_2$ *will have degree* p^{a-1}.)
3: If $a = 1$,
  ▷ Find an $e \in \mathbb{F}_q$ such that $e^p - e = (\alpha_1^p - \alpha_1) - (\alpha_2^p - \alpha_2)$.
  ▷ Apply Lemma 2.1 to $\alpha_1 \sim \alpha_2 + e$ and find a root of $g_1(x)$ in $\mathbb{F}_q(\beta_2)$.
4: If $a \neq 1$,
  ▷ Find the minimal polynomials $h_1(x), h_2(x)$ over $\mathbb{F}_q$ of $\alpha_1^p - \alpha_1, \alpha_2^p - \alpha_2$ respectively.
  ▷ Recursively find a root α of $h_1(x)$ in $\mathbb{F}_q(\alpha_2^p - \alpha_2) = \mathbb{F}_q[x]/(h_2(x))$. ($h_1(x)$ *and* $h_2(x)$ *have degree* ℓ^{a-1}.)
  ▷ Find a $\gamma \in \mathbb{F}_q(\alpha_2^\ell)$ such that $\gamma^p - \gamma = \alpha - (\alpha_2^p - \alpha_2)$.
  ▷ Apply Lemma 2.1 to $\alpha_1 \sim \alpha_2 + \gamma$ and find a root of $g_1(x)$ in $\mathbb{F}_q(\beta_2)$.

We next argue that Algorithm 2 runs to completion and is correct.

In Step 1, $\alpha_1^p - \alpha_1$ is claimed to have degree p^{a-1}. Let b be the degree of $\alpha_1^p - \alpha_1$. Since

$$\sigma^{p^{a-1}}(\alpha_1^p - \alpha_1) = \left(\sigma^{p^{a-1}}(\alpha_1)\right)^p - \sigma^{p^{a-1}}(\alpha_1) = \alpha_1^p + 1 - (\alpha_1 + 1) = \alpha_1^p - \alpha_1,$$

b divides p^{a-1}. Since α_1 is a root of $x^p - x - (\alpha_1^p - \alpha_1)$ and α_1 has degree p^a, $\alpha_1^p - \alpha_1$ has degree at most p^{a-1}. Thus $\alpha_1^p - \alpha_1$ has degree p^{a-1}. Likewise, in Step 2, $\alpha_2^p - \alpha_2$ has degree p^{a-1}.

In Step 3, since $a = 1$, $\alpha_1^p - \alpha_1, \alpha_2^p - \alpha_2 \in \mathbb{F}_q$. Further $\alpha_1 - \alpha_2$ is in $\mathbb{F}_q$ since $\sigma(\alpha_1 - \alpha_2) = (\alpha_1 + 1) - (\alpha_2 + 1) = \alpha_1 - \alpha_2$. Thus $\alpha_1 - \alpha_2 \in \mathbb{F}_q$ is a root of $x^p - x - ((\alpha_1^p - \alpha_1) - (\alpha_2^p - \alpha_2))$ ensuring that an $e \in \mathbb{F}_q$ such that $e^p - e = (\alpha_1^p - \alpha_1) - (\alpha_2^p - \alpha_2)$ exists. The roots of $x^p - x - ((\alpha_1^p - \alpha_1) - (\alpha_2^p - \alpha_2))$ are $\{e, e + 1, e + 2, \ldots, e + (p - 1)\}$. Hence $\alpha_1 - \alpha_2 = e$. Further α_1 has degree ℓ^a. Hence Lemma 2.1, when applied to the relation $\alpha_1 \sim \alpha_2 + e$, correctly finds the desired output.

The recursive call in Step 4 yields a root $\alpha \in \mathbb{F}_q(\alpha_2^p - \alpha_2)$ of $h_1(x)$. Hence $\alpha = \sigma^j(\alpha_1^p - \alpha_1) = (\sigma^j(\alpha_1))^p - \sigma^j(\alpha_1)$ for some integer j. Further, $\sigma^j(\alpha_1) - \alpha_2 \in \mathbb{F}_q(\alpha_2^p - \alpha_2)$ since

$$\sigma^{p^{a-1}}(\sigma^j(\alpha_1) - \alpha_2) = \sigma^j(\alpha_1 + 1) - (\alpha_2 + 1) = \sigma^j(\alpha_1) - \alpha_2.$$

Hence $\sigma^j(\alpha_1) - \alpha_2 \in \mathbb{F}_q(\alpha_2^p - \alpha_2)$ is a root of $x^p - x = \alpha - (\alpha_2^p - \alpha_2)$ assuring the existence of γ sought in Step 5. For such a γ, the roots of $x^p - x - (\alpha - (\alpha_2^p - \alpha_2))$ are $\{\gamma, \gamma + 1, \gamma + 2, \ldots, \gamma + (p - 1)\}$. Hence $\sigma^j(\alpha_1) - \alpha_2 = \gamma$. Further α_1 has degree ℓ^a. Hence Lemma 2.1, when applied to the relation $\alpha_1 \sim \alpha_2 + \gamma$, correctly finds the desired output.

4.1   Implementation and Run Time Analysis

Running times of Steps 1 and 2 are dominated by their respective calls to Lemma 3.1, each taking $p^{a+o(1)} \log^{2+o(1)} q$ time.

In Step 3, find a root e of $x^p - x - ((\alpha_1^p - \alpha_1) - (\alpha_2^p - \alpha_2)) \in \mathbb{F}_q[x]$ in $\mathbb{F}_q$ using [vzGS92,KU08] in $p^{1+o(1)} \log^{2+o(1)} q$ time. Invocations to Lemma 2.1 in Steps 3 and 4 take $p^{a+o(1)} \log^{2+o(1)} q$ time.

In Step 4, minimal polynomials of $\alpha_1^p - \alpha_1$ and $\alpha_2^p - \alpha_2$ can be computed in $p^{a+o(1)} \log^{1+o(1)} q$ time [KU08, Sect. 8.4]. To compute γ, find a root of $x^p - x - (\alpha - (\alpha_2^p - \alpha_2))$ in $\mathbb{F}_q(\alpha_2^p - \alpha_2) = \mathbb{F}_q[x]/(h_2(x))$ using [vzGS92,KU08]. Since we a finding the root of a degree p polynomial over a field of size $q^{p^{a-1}}$, the running time $p^{a+o(1)} \log^{1+o(1)} q$ turns out to be nearly linear in p^a.

Algorithm 2 makes at most one recursive call to an identical subproblem of size p^{a-1}. Hence at most a recursive calls are made in total. In summary, we have the following theorem.

Theorem 4.2. *Algorithm 2 solves the* ISOMORPHISM PROBLEM *restricted to the special case when* $n = p^a$ *in* $n^{1+o(1)} \log^{1+o(1)} q$ *time.*

5   Root Finding over Extensions of Finite Fields Using Elliptic Curves

We solve the ISOMORPHISM PROBLEM restricted to the case when n is a power ℓ^a of a prime $\ell \nmid q(q - 1)$ in $\ell^{a+1+o(1)} \log^{1+o(1)} q + O(\ell \log^5 q)$ time. Through this section, fix a prime ℓ such that $\ell \nmid q(q - 1)$, $\sqrt{q} \geq 5\ell^3$ and a positive integer a.

5.1 Elliptic Curves with $\mathbb{F}_q$-rational ℓ-torsion

Let E be an elliptic curve over $\mathbb{F}_q$ such that ℓ divides $|E(\mathbb{F}_q)|$ but ℓ^2 does not. Let $\sigma_E : E \longrightarrow E$ denote the qth power Frobenius endomorphism and $t \in \mathbb{Z}$ the trace of σ_E. The characteristic polynomial $P_E(X) := X^2 - tX + q \in \mathbb{Z}[X]$ of σ_E factors modulo ℓ as

$$X^2 - tX + q = (X - 1)(X - q) \mod \ell.$$

To see why 1 is a root of $P_E(X)$ modulo ℓ, observe $P_E(1) = |E(\mathbb{F}_q)|$ and $\ell \mid |E(\mathbb{F}_q)|$. The other root is q, since the product of the roots is q. By Hensel's lemma, there exists $\lambda, \mu \in \{0, 1, \dots, \ell^{a+1} - 1\}$ such that

$$X^2 - tX + q = (X - \lambda)(X - \mu) \mod \ell^{a+1},$$

where $\lambda = 1 \mod \ell$ and $\mu = q \mod \ell$. Hence there exists $P_\lambda, P_\mu \in E[\ell^{a+1}]$, each of order ℓ^{a+1} such that

$$E[\ell^{a+1}] = \langle P_\lambda \rangle \oplus \langle P_\mu \rangle, \sigma_E(P_\lambda) = \lambda P_\lambda \text{ and } \sigma_E(P_\mu) = \mu P_\mu.$$

Since $\lambda = 1 \mod \ell$ and $\ell^2 \nmid |E(\mathbb{F}_q)|$, $\lambda = 1 + \gamma \ell$ where $\gamma := (\lambda - 1)/\ell \in \mathbb{Z}_{\geq 0}$ and $\gcd(\gamma, \ell) = 1$.

5.2 Root Finding Through Discrete Logarithms in Elliptic Curve

In this subsection, we devise an algorithm for the ISOMORPHISM PROBLEM that involves discrete logarithm computations in elliptic curves. We begin with a few preparatory lemmata.

Lemma 5.1. $P_\lambda \in E(\mathbb{F}_{q^{\ell^a}})$ and $\mathrm{x}(P_\lambda)$ has degree ℓ^a.

Proof. Let c be the smallest positive integer such that $\sigma_E^c P_\lambda = P_\lambda$. To claim $P_\lambda \in E(\mathbb{F}_{q^{\ell^a}})$, it suffices to show $c = \ell^a$. Further, $c = \ell^a$ would also imply that $\mathrm{x}(P_\lambda)$ has degree ℓ^a, for if $\mathrm{x}(P_\lambda)$ were in a proper subfield of $\mathbb{F}_{q^{\ell^a}}$ then c has to be a proper divisor of ℓ^a. Since $\sigma_E(P_\lambda) = \lambda P_\lambda$ and P_λ has order ℓ^{a+1}, c equals the order of $\lambda \mod \ell^{a+1}$ in $(\mathbb{Z}/\ell^{a+1}\mathbb{Z})^\times$. For $\lambda^c = (1 + \gamma \ell)^c = 1 \mod \ell^{a+1}$ to hold, it is necessary and sufficient that ℓ^a divides $c\gamma$. Hence $c = \ell^a$. $\square$

Lemma 5.2. $E(\mathbb{F}_{q^{\ell^a}})[\ell^{a+1}] = \langle P_\lambda \rangle$

Proof. From Lemma 5.1, $\langle P_\lambda \rangle \subseteq E(\mathbb{F}_{q^{\ell^a}})$. Since $E[\ell^{a+1}] = \langle P_\lambda \rangle \oplus \langle P_\mu \rangle$, to claim the lemma it suffices to prove $E(\mathbb{F}_{q^{\ell^a}}) \cap \langle P_\mu \rangle = \{\mathcal{O}\}$. If $E(\mathbb{F}_{q^{\ell^a}}) \cap \langle P_\mu \rangle \neq \{\mathcal{O}\}$, then $\exists P \in E(\mathbb{F}_{q^{\ell^a}}) \cap \langle P_\mu \rangle$ of order ℓ. Since $P \in E(\mathbb{F}_{q^{\ell^a}})$, $\sigma_E^c P = P$ and since $P \in \langle P_\mu \rangle$, $\sigma_E P = \mu P$. Hence $\mu^{\ell^a} P = P$. Since P has order ℓ, $\mu^{\ell^a} - 1 = 0 \mod \ell$. Since ℓ is a prime, raising to ℓ^{th} powers modulo ℓ is the identity map implying $\mu = 1 \mod \ell$. Since $\gcd(\ell, q - 1) = 1$, this contradicts the fact that $\mu = q \mod \ell$. Thus $E(\mathbb{F}_{q^{\ell^a}}) \cap \langle P_\mu \rangle = \{\mathcal{O}\}$. $\square$

The group $\Sigma := \langle \sigma_E | \sigma_E^{\ell^a} = 1 \rangle$ acts on $E(\mathbb{F}_{q^{\ell^a}})$. For $P \in E(\mathbb{F}_{q^{\ell^a}})$, denote the orbit $\{P, \sigma_E P, \ldots, \sigma_E^{\ell^a-1} P\}$ of P under Σ as $\Sigma.P$.

Lemma 5.3. *The set $\langle P_\lambda \rangle \setminus \langle \ell P_\lambda \rangle$ is the following disjoint union of orbits*

$$\langle P_\lambda \rangle \setminus \langle \ell P_\lambda \rangle = \bigcup_{z=1}^{\ell-1} \Sigma.z P_\lambda.$$

Proof. For every $z \in \{0, 1, \ldots, \ell - 1\}$, $zP_\lambda \subseteq \langle P_\lambda \rangle \setminus \langle \ell P_\lambda \rangle$ and $|\Sigma.z P_\lambda| = \ell^a$. Further, $|\langle P_\lambda \rangle \setminus \langle \ell P_\lambda \rangle| = \ell^a(\ell-1)$. It is thus sufficient to prove for distinct $z_1, z_2 \in \{1, 2, \ldots, \ell - 1\}$ that $z_1 P_\lambda \cap z_2 P_\lambda = \emptyset$. If $z_1 P_\lambda = \sigma_E^j z_2 P_\lambda$ for some $z_1, z_2 \in \{1, 2, \ldots, \ell - 1\}$ and $j \in \{0, 1, \ldots, \ell^a - 1\}$ then,

$$z_1 P_\lambda = \lambda^j z_2 P_\lambda \Rightarrow z_1 - \lambda^j z_2 = 0 \mod \ell^a \Rightarrow z_1 - (1 + \gamma\ell)^j z_2 = 0 \mod \ell^a \Rightarrow z_1 = z_2 \mod \ell.$$

$\square$

Let $Tr_E : E(\mathbb{F}_{q^{\ell^a}}) \longrightarrow E(\mathbb{F}_{q^{\ell^a}})$ denote the trace like map that sends P to $\sum_{j=0}^{\ell^a-1} \sigma_E^j P$. The next lemma states that distinct Σ orbits of $\langle P_\lambda \rangle \setminus \langle \ell P_\lambda \rangle$ have distinct images under Tr_E.

Lemma 5.4. *For all $P_1, P_2 \in \langle P_\lambda \rangle \setminus \langle \ell P_\lambda \rangle$, $Tr_E(P_1) = Tr_E(P_2)$ if and only if $\Sigma.P_1 = \Sigma.P_2$.*

Proof. If $P_1, P_2 \in \langle P_\lambda \rangle \setminus \langle \ell P_\lambda \rangle$ and $\Sigma.P_1 = \Sigma.P_2$, then $\exists j \in \{0, 1, \ldots, \ell^a - 1\}$ such that $P_2 = \sigma_E^j P_1$. Hence, $Tr(P_2) = Tr_E(\sigma_E^j P_1) = \sigma_E^j Tr_E(P_1) = Tr_E(P_1)$. We next prove the converse, that is, the "only if" part of the lemma. For every $\alpha \in \mathbb{F}_q$ at most $q^{\ell^a} - 1$ elements in $\mathbb{F}_q^{\ell^a}$ have trace (down to $\mathbb{F}_q$) α. If $Tr_E(P_\lambda) = \mathcal{O}$, then

$$[E(\mathbb{F}_q) : Tr_E(E(\mathbb{F}_{q^{\ell^a}}))] \geq \ell \Rightarrow |E(\mathbb{F}_{q^{\ell^a}})| \leq \left(1 + \frac{2}{\sqrt{q}}\right) \frac{q^{\ell^a}}{\ell}.$$

This contradicts the Hasse-Weil bound $|E(\mathbb{F}_{q^{\ell^a}})| \geq q^{\ell^a} - 2\sqrt{q^{\ell^a}}$. Thus $Tr_E(P_\lambda) \neq \mathcal{O}$.

Let $P_1, P_2 \in \langle P_\lambda \rangle \setminus \langle \ell P_\lambda \rangle$ and $Tr_E(P_1) = Tr_E(P_2)$. By Lemma 5.3, there exists $z_1, z_2 \in \{1, 2, \ldots, \ell - 1\}$ such that $P_1 \in \Sigma.z_1 P_\lambda$ and $P_2 \in \Sigma.z_2 P_\lambda$. Hence $Tr_E(P_1) = Tr_E(z_1 P_\lambda) = z_1 Tr_E(P_\lambda)$. Likewise, $Tr_E(P_2) = z_2 Tr_E(P_\lambda)$. Since $Tr_E(P_1) = Tr_E(P_2)$, $(z_1 - z_2) Tr_E(P_\lambda) = \mathcal{O}$. Since $Tr_E(P_\lambda) \in E(\mathbb{F}_q)[\ell]$, $|E(\mathbb{F}_q)[\ell]| = \ell$ and $Tr(P_\lambda) \neq \mathcal{O}$, the order of $Tr_E(P_\lambda)$ is ℓ. Hence $z_1 - z_2 = 0 \mod \ell$ thereby implying $\Sigma.P_1 = \Sigma.P_2$. $\square$

Algorithm 3. Root Finding Through Elliptic Curve Discrete Logarithms

---

Input: Monic irreducibles $g_1(x), g_2(x) \in \mathbb{F}_q[X]$ of degree ℓ^a where $\ell \leq \sqrt{q}$ is a prime not dividing $q(q-1)$.

Output: A root of $g_1(x)$ in $\mathbb{F}_q(\alpha_2)$ where $\alpha_2 \in \overline{\mathbb{F}}_q$ is a root of $g_2(x)$.

1: Find an elliptic curve $E/\mathbb{F}_q$ with $\ell \| |E(\mathbb{F}_q)|$ and $\ell^2 \nmid |E(\mathbb{F}_q)|$.

2:
   ▷ Construct $\mathbb{F}_{q^{\ell^a}}$ as $\mathbb{F}_q(\alpha_1)$ where α_1 is a root of $g_1(x)$.
   ▷ Find a point $P_1 \in E(\mathbb{F}_{q^{\ell^a}})$ of order ℓ^{a+1}.
   ▷ $x(P_1)$ is obtained as $f_1(\alpha_1)$ for some $f_1(x) \in \mathbb{F}_q[x]$ of degree less than ℓ^a.
   ▷ Compute $Tr_E(P_1)$.

3:
   ▷ Construct $\mathbb{F}_{q^{\ell^a}}$ as $\mathbb{F}_q(\alpha_2)$ where α_2 is a root of $g_2(x)$.
   ▷ Find a point $P_2 \in E(\mathbb{F}_{q^{\ell^a}})$ of order ℓ^{a+1}.
   ▷ Compute $Tr_E(P_2)$.

4: Find the $z \in \{1, \ldots, \ell-1\}$ such that $Tr_E(P_1) = zTr_E(P_2)$ by solving a discrete logarithm problem in the order ℓ cyclic group $E(\mathbb{F}_q)[\ell]$.

5: Compute zP_2 and obtain $x(zP_2) = f_2(\alpha_2)$ for some $f_2(x) \in \mathbb{F}_q[x]$ of degree less than ℓ^a.

6: Apply Lemma 2.1 to the relation $f_1(\alpha_1) \sim f_2(\alpha_2)$ and output a root of $g_1(x)$ in $\mathbb{F}_q(\alpha_2)$.

We first argue that Algorithm 3 is correct. An elliptic curve $E/\mathbb{F}_q$ as required in Step 1 exists as $\ell^2 \leq \sqrt{q}$ implies ℓ has a multiple not divisible by ℓ^2 in the Hasse interval. As P_1 and P_2 are both in $E(\mathbb{F}_{q^{\ell^a}})$ and of order ℓ^{a+1}, by Lemma 5.2, $P_1, P_2 \in \langle P_\lambda \rangle \setminus \ell \langle P_\lambda \rangle$. Hence by Lemma 5.3, there exists $z \in \{1, 2, \ldots, \ell-1\}$ such that

$$P_1 = \Sigma.zP_2. \tag{7}$$

By Lemma 5.4, Eq. 7 holds if and only if

$$Tr_E(P_1) = zTr_E(P_2). \tag{8}$$

Hence z as desired in Step 4 exists and further for such a z, there exists an integer j such that $P_1 = \sigma_E^j(zP_2)$ implying $f_1(\alpha_1) \sim f_2(\alpha_2)$.

The bottleneck in the algorithm happens to be computing a point of order ℓ^{a+1} in Steps 2 and 3. An algorithm for this task is presented in the subsequent subsection. For now, we discuss the implementation of the other steps. In Step 1, we generate elliptic curves $E/\mathbb{F}_q$ by choosing a Weierstrass model over $\mathbb{F}_q$ uniformly at random. Then we compute $|E(\mathbb{F}_q)|$ using Schoof's point counting algorithm in $\tilde{O}(\log^5 q)$ time and check if $\ell \| |E(\mathbb{F}_q)|$ and $\ell^2 \nmid |E(\mathbb{F}_q)|$. Since $5\ell^3 \leq \sqrt{q}$, the probability that $\ell \| |E(\mathbb{F}_q)|$ and $\ell^2 \nmid |E(\mathbb{F}_q)|$ is close to $1/(\ell-1)$ [How, Thm 1.1]. Hence Step 1 can be completed in time $O(\ell \log^5 q)$. The iterated Frobenius algorithm of von zur Gathen and Shoup [vzGS92] implemented using fast modular composition [KU08] computes traces in finite field extensions in nearly linear time. With minor modifications (performing elliptic curve addition in place of finite field addition), it computes $Tr_E(P_1)$ and $Tr_E(P_2)$ in Steps 2 and 3 in $\ell^{a+o(1)} \log^{1+o(1)} q$ time. The discrete logarithm computation in Step 4 can be performed with $\mathcal{O}(\sqrt{\ell})$ $E(\mathbb{F}_q)$-additions by the baby step giant step algorithm. Since $z < \ell$, Step 5 only takes $O(\log(\ell))$ $E(\mathbb{F}_{q^{\ell^a}})$ additions. From Lemma 2.1, Step 6 runs in $\ell^{a+o(1)} \log^{1+o(1)} q$ time.

5.3 Lang's Theorem and Finding ℓ Power Torsion with ℓ Isogenies

In Sects. 3 and 4, we exploited certain idempotents in proofs of Hilbert's theorem 90 to solve the ISOMORPHISM PROBLEM restricted to the case where n is a power of a prime ℓ dividing $p(q-1)$ in linear time. The bottleneck in Algorithm 3 for the case $\ell \nmid p(q-1)$ is

Problem 5.5. *Given a monic irreducible $g(x) \in \mathbb{F}_q[x]$ of prime power ℓ^a degree (where $\ell \nmid p(q-1)$), an elliptic curve $E/\mathbb{F}_q$ (where ℓ divides $|E(\mathbb{F}_q)|$ but ℓ^2 does not) and $|E(\mathbb{F}_q)|$, find a generator of $E(\mathbb{F}_{q^{\ell^a}})[\ell^{a+1}]$ where $\mathbb{F}_{q^{\ell^a}}$ is constructed as $\mathbb{F}_q(\alpha)$ for some root α of $g(x)$.*

We next solve Problem 5.5 using elliptic curve isogenies.

Algorithm 4. Finding ℓ Power Torsion:

Input:
> Monic irreducible $g(x) \in \mathbb{F}_q[X]$ of degree ℓ^a where $\ell \leq \sqrt{q}$ is a prime not dividing $q(q-1)$.
> An elliptic curve $E/\mathbb{F}_q$ such that $\ell \| |E(\mathbb{F}_q)|$ and $\ell^2 \nmid |E(\mathbb{F}_q)|$.
> $|E(\mathbb{F}_q)|$.

Output: A point $P \in E$ of order ℓ^{a+1} with coordinates in $\mathbb{F}_q[x]/(g(x))$.
1: Construct $\mathbb{F}_{q^{\ell^a}}$ as $\mathbb{F}_q(\alpha)$ for a root α of $g(x)$.
2: Let $\iota : E \longrightarrow \widetilde{E}$ be the isogeny with kernel $\ker(\iota) = E(\mathbb{F}_q)[\ell]$.
   > Compute a Weierstrass equation for $\widetilde{E}/\mathbb{F}_q$.
   > Compute $\phi_\iota(x), \psi_\iota(x) \in \mathbb{F}_q[x]$ such that $\mathrm{x}(\iota(R)) = \psi_\iota(\mathrm{x}(R))/\phi_\iota(\mathrm{x}(R)), \forall R \in E$.
3: If $a = 1$
   > Find a point $\widetilde{T} \in \widetilde{E}(\mathbb{F}_q)$ of order ℓ.
   > Find a root $\gamma \in \mathbb{F}_q(\alpha)$ of $\phi_\iota(x) - \mathrm{x}(\widetilde{T})\psi_\iota(x) \in \mathbb{F}_q[x]$.
   > Output a point in E with x-coordinate γ.
4: If $a \neq 1$
   > Find $\widetilde{\alpha} \in \mathbb{F}_q(\alpha)$ of degree ℓ^{a-1} and its minimal polynomial $M(x) \in \mathbb{F}_q[x]$ by Lemma 2.2.
   > Recursively find a point $\widetilde{P} \in \widetilde{E}(\mathbb{F}_{q^{\ell^{a-1}}})$ of order ℓ^a by calling this very algorithm with input $(M(x), \widetilde{E}/\mathbb{F}_q, |E(\mathbb{F}_q)|)$.
   > Find a root $\eta \in \mathbb{F}_q(\alpha)$ of $\phi_\iota(x) - \mathrm{x}(\widetilde{P})\psi_\iota(x) \in \mathbb{F}_q(\widetilde{\alpha})[x]$.
   > Output a point in E with x-coordinate γ.

In Step 2, the Weierstrass equation for $\widetilde{E}$ and the polynomials $\psi_\iota(x)$ and $\phi_\iota(x)$ can all be computed in $\ell^{a+o(1)} \log^{2+o(1)} q$ time [CL13]. In Step 3, a point $\widetilde{T} \in \widetilde{E}(\mathbb{F}_q)$ of order ℓ can be found in $\widetilde{O}(\ell \log q)$ time as follows: generate $\widetilde{R} \in \widetilde{E}(\mathbb{F}_q)$ at random and output $\widetilde{T} = |\widetilde{E}(\mathbb{F}_q)|/\ell$ if its not the identity. Note we know don't have to compute $|\tilde{E}(\mathbb{F}_q)|$ since $|\tilde{E}(\mathbb{F}_q)| = |E(\mathbb{F}_q)|$. The root finding in Step 3 takes $\ell^{2+o(1)} \log^{2+o(1)} q$ time using [KS99,KU08]. By [CL13], a root γ of $\phi_\iota(x) - \mathrm{x}(\tilde{T})\psi_\iota(x) \in \mathbb{F}_q[x]$ has degree ℓ and the two points in E with x-coordinate γ both have order ℓ^2 and are in $E(\mathbb{F}_{q^\ell})$. Thus the output at the end of Step 3 is correct. Likewise, in Step 4, by [CL13], a root η of $\phi_\iota(x) - \mathrm{x}(\tilde{P})\psi_\iota(x)$ has degree ℓ^a and the two points in E with x-coordinate η both have order ℓ^{a+1} and are in $E(\mathbb{F}_{q^{\ell^a}})$. Hence the output at the end of Step 4 is correct. The root finding in Step 4 takes $\ell^{a+1+o(1)} \log^{1+o(1)} q$ time using [KS99,KU08] and is the bottleneck. The number of recursive calls is at most a which being logarithmic in ℓ^a can be ignored in the run time analysis.

Using Algorithm 4 as a subroutine, Algorithm 3 solves the ISOMORPHISM PROBLEM restricted to the special case when $n = \ell^a$ for some prime ℓ such that $\ell \nmid q(q-1)$ and $5\ell^3 \le \sqrt{q}$. The restriction $5\ell^3 \le \sqrt{q}$ may be removed without loss of generality. For if $5\ell^3 > \sqrt{q}$ in the ISOMORPHISM PROBLEM, we may pose the problem over a small degree extension $\mathbb{F}_{q^d}$ instead of $\mathbb{F}_q$ where d is the smallest positive integer such that $\ell \le \sqrt{q^d}$ and $\ell \nmid d$ (c.f. [Rai08]). In summary, we have proven

Theorem 5.6. *Algorithm 3 solves the* ISOMORPHISM PROBLEM *restricted to the special case when $n = \ell^a$ for some prime $\ell \nmid q(q-1)$ in $\ell^{a+1+o(1)} \log^{1+o(1)} q + O(\ell \log^5 q)$ time.*

The running time is subquadratic in the input degree ℓ^a if $a > 1$. If $a = 1$, that is, if the input degree is a prime ℓ, the running time is quadratic. The question if a sub quadratic algorithm for the later case exists remains open. We look to Lang's theorem, an elliptic curve analogue of Hilbert's theorem 90 in hopes of solving the bottleneck Problem 5.5 in subquadratic time. Lang's theorem states that the first cohomology group $H^1(\mathbb{F}_q, E)$ of an elliptic curve E over $\mathbb{F}_q$ is trivial. That is, the Lang map $\psi : E \longrightarrow E$ taking P to $\sigma_E(P) - P$ is surjective. Problem 5.5 is rephrased in terms of computing preimages under the Lang map as the following Problem 5.7. Problems 5.5 and 5.7 are equivalent since the preimage of $E(\mathbb{F}_q)[\ell] \setminus \{O\}$ under ψ is $E(\mathbb{F}_{q^{\ell^a}})[\ell^{a+1}]$.

Problem 5.7. *Given a monic irreducible $g(x) \in \mathbb{F}_q[x]$ of prime power ℓ^a degree (where $\ell \nmid p(q-1)$), an elliptic curve $E/\mathbb{F}_q$ (where ℓ divides $|E(\mathbb{F}_q)|$ but ℓ^2 does not) and $|E(\mathbb{F}_q)|$, find a preimage under the Lang map ψ of $E(\mathbb{F}_q)[\ell] \setminus \{O\}$ in $E(\mathbb{F}_{q^{\ell^a}})$ where $\mathbb{F}_{q^{\ell^a}}$ is constructed as $\mathbb{F}_q(\alpha)$ for some root α of $g(x)$.*

OPEN PROBLEM: *Solve Problem 5.5 or Problem 5.7 in time sub quadratic in ℓ^a.*

References

[All02] Allombert, B.: Explicit computation of isomorphisms between finite fields. Finite Fields Their Appl. **8**, 332–342 (2002)

[BGJT14] Barbulescu, R., Gaudry, P., Joux, A., Thomé, E.: A heuristic quasi-polynomial algorithm for discrete logarithm in finite fields of small characteristic. In: Nguyen, P.Q., Oswald, E. (eds.) EUROCRYPT 2014. LNCS, vol. 8441, pp. 1–16. Springer, Heidelberg (2014). https://doi.org/10.1007/978-3-642-55220-5_1

[Ben81] Ben-Or, M.: Probabilistic algorithms in finite fields. In: FOCS, pp. 394–398 (1981)

[Ber67] Berlekamp, E.R.: Factoring polynomials over finite fields. Bell Syst. Tech. J. **46**, 1853–1859 (1967)

[BDDFS17] Brieulle, L., De Feo, L., Doliskani, J., Flori, J.-P., Schost, E.: Computing isomorphisms and embeddings of finite fields. https://arxiv.org/abs/1705.01221. To appear in Mathematics of Computation. https://doi.org/10.1090/mcom/3363

[CZ81] Cantor, D.G., Zassenhaus, H.: A new algorithm for factoring polynomials over finite fields. Math. Comput. **36**, 587–592 (1981)

[CL13] Couveignes, J.-M., Lercier, R.: Fast construction of irreducible polynomials over finite fields. Israel J. Math. **194**(1), 77–105 (2013)

[vzGS92] von zur Gathen, J., Shoup, V.: Computing Frobenius maps and factoring polynomials. Comput. Complex. **2**, 187–224 (1992)

[Gra08] Granville, A.: Smooth numbers: computational number theory and beyond. Algorithmic Number Theory **44**, 267–323 (2008)

[How] Howe, E.: On the group orders of elliptic curves over finite fields. Compos. Math. **85**, 229–247 (1993)

[KS99] Kaltofen, E., Shoup, V.: Fast polynomial factorization over high algebraic extensions of finite fields. In: Proceedings of the 1997 International Symposium Symbolic Algebraic Computation, (ISSAC 1997), pp. 184–188 (1997)

[KU08] Kedlaya, K., Umans, C.: Fast modular composition in any characteristic. In: FOCS, pp. 146–155 (2008)

[Lan78] Lang, S.: Algebraic groups over finite fields. Am. J. Math. **78**, 555–563 (1956)

[Len87] Lenstra Jr., H.W.: Factoring integers with elliptic curves. Ann. Math. **126**(3), 649–673 (1987)

[Moo1889] Moore, E.H.: A doubly-infinite system of simple groups. Bull. New York Math. Soc. **3**, 73–78 (1893). Math. Papers read at the Congress of Mathematics (Chicago, 1893). Chicago 1896, pp. 208–242

[Nar2016] Narayanan, A.K.: Fast computation of isomorphisms between finite fields using elliptic curves. https://arxiv.org/abs/1604.03072

[Pin92] Pinch, R.G.E.: Recognizing elements of finite fields. Cryptography and Coding II, pp. 193–197 (1992)

[Rai08] Rains, E.: Efficient computation of isomorphisms between finite fields

[Sch95] Schoof, R.: Counting points on elliptic curves over finite fields. J. Theor. Nombres Bordeaux **7**, 219–254 (1995)

[Sho99] Shoup, V.: Efficient computation of minimal polynomials in algebraic extensions of finite fields. In: ISSAC 1999, pp. 53–58 (1999)

[Sho95] Shoup, V.: Fast construction of irreducible polynomials over finite fields. J. Symb. Comput. **17**, 371–391 (1994)

[Zie74] Zierler, N.: A conversion algorithm for logarithms on $GF(2^n)$. J. Pure Appl. Algebra **4**, 353–356 (1974)

Invited Talk 2

Construction of Some Codes Suitable for Both Side Channel and Fault Injection Attacks

Claude Carlet[1,2], Cem Güneri[3], Sihem Mesnager[4], and Ferruh Özbudak[5(✉)]

[1] LAGA and University of Paris VIII, Saint-Denis, France
claude.carlet@univ-paris8.fr
[2] University of Bergen, Bergen, Norway
[3] Sabancı University, FENS, 34956 Istanbul, Turkey
guneri@sabanciuniv.edu
[4] Department of Mathematics,
Universities of Paris VIII and XIII and Telecom ParisTech, Paris, France
smesnager@univ-paris8.fr
[5] Middle East Technical University, Dumlupınar Bulv. No. 1, 06800 Ankara, Turkey
ozbudak@metu.edu.tr

Abstract. Using algebraic curves over finite fields, we construct some codes suitable for being used in the countermeasure called Direct Sum Masking which allows, when properly implemented, to protect the whole cryptographic block cipher algorithm against side channel attacks and fault injection attacks, simultaneously. These codes address a problem which has its own interest in coding theory.

Keywords: SCA · FIA · MDS code · Algebraic geometry code

1 Introduction

Side-channel analysis (SCA) and fault injection attacks (FIA) are nowadays important domains of cryptanalysis. Most designers of secure embedded systems know that the implementation of cryptosystems should be protected against them. But all known protections are costly, all the more since they need to be implemented now, while the attacks can be performed in the future, which means it is advisable to include higher order protections. Hence, reducing the cost of counter-measures is an important challenge.

A well studied and efficient counter-measure against SCA is masking, which consists in splitting into several shares each sensitive variable (i.e. each variable whose value depends on data known from the attacker and of the secret key, a dependence with a few key bits being still more favorable to the attacker). But masking is not efficient against FIA and implementing masking plus a counter-measure against FIA is costly. *Direct Sum Masking* (DSM) has then been proposed in [2] as a counter-measure against both SCA and FIA. As masking, it

L. Budaghyan and F. Rodríguez-Henríquez (Eds.): WAIFI 2018, LNCS 11321, pp. 95–107, 2018.
https://doi.org/10.1007/978-3-030-05153-2_5

consists in combining the sensitive information x with a mask y, but in a different way, which is in fact more general: if x lives in (say) $\mathbb{F}_q^k$ and the mask y lives in (say) $\mathbb{F}_q^{n-k}$, the (masked) data which will be manipulated by the algorithm has the form

$$z = xG + yH,$$

where G is a $k \times n$ matrix that generates a code C of length n and dimension k, and H is an $(n-k) \times n$ matrix that generates a code D of length n and dimension $n - k$. To allow recovering x from z, the codes C and D need to satisfy $C \cap D = \{0\}$. We have then $C \oplus D = \mathbb{F}_q^n$, and (C, D) is called a linear complementary pair (LCP) of codes. The security parameter of an LCP of codes (C, D) is determined to be $\min\{d(C), d(D^\perp)\}$, where $d(.)$ means minimum distance. If D equals the dual $C^\perp$ of C in this setting, then C is called a linear complementary dual (LCD) code. The notion of LCD code is anterior to [2]. In [20], Yang and Massey who introduce it, provide a necessary and sufficient condition under which a cyclic code has a complementary dual. The security parameter of an LCD code C when used in so-called *Orthogonal Direct Sum Masking* (ODSM) is then simply $d(C)$. In [1], Bhasin et al. have shown how also using and implementing LCD codes and LCP of codes to strengthen encoded circuit against hardware Trojan horses, while minimizing the cost.

A lot of research has been carried in a very short period after the invention of ODSM to find characterizations and constructions of LCD codes. In [4], Carlet and Guilley revisited the best know constructions of linear codes and adapted them to build LCD codes. They also showed how an LCD code over a finite field of characteristic 2 can be transformed into an LCD binary code. This means that codes over all finite fields of characteristic 2 are interesting for applications. In [11], Ding et al. constructed several families of LCD cyclic codes over finite fields and analyzed their parameters. In [15], Li et al. studied a class of LCD BCH codes proposed in [14] and extended the results on their parameters. In [12], Güneri et al. characterized and studied quasi-cyclic (QC) codes that are LCD. In [13], Güneri et al. gave criteria for complementary duality of generalized quasi-cyclic codes (GQC) bearing on the component codes and exhibited some explicit long GQC that are LCD, but not quasi-cyclic. In [16], Mesnager et al. proposed a construction of algebraic geometry LCD codes which could be good candidates to be resistant against SCA. In [17] and [8], the authors studied a construction of LCD maximum distance separable (MDS) codes. Recently, Carlet et al. have completely determined in [10] all q-ary ($q > 3$) LCD codes for all possible parameters. Carlet et al. have presented in [7] a new concatenated type construction for LCD codes over small finite fields. Their construction generalizes a recent construction of Carlet et al. and of Güneri et al. and allows to construct LCD codes with improved parameters directly. In [9], Carlet et al. proposed a new characterization and parametrization of LCD codes in terms of their orthogonal or symplectic basis. However little is in known on LCP of codes. Very recently, Carlet et al. have studied in [6] LCP of codes. They have investigated constacyclic and quasi-cyclic LCP of codes and presented characterizations for LCP of constacyclic codes and LCP (C, D) of quasi-cyclic codes.

Their characterization in the constacyclic case extends the characterization of LCD cyclic codes given by Yang and Massey in [20].

But some issues with the use of LCD codes and LCP of codes in devices like smart cards (in which the algorithms are coded in software) was not completely solved in [2]. We refer, for example to [3] and [5] for further results. Indeed, since there is no possibility in such framework of hiding, like in hardware, sensitive computations in memory, it is necessary for ensuring security that each application to sensitive variables of functions, say $F : \mathbb{F}_q^k \mapsto \mathbb{F}_q^k$, be changed into the application of functions $F' : \mathbb{F}_q^n \mapsto \mathbb{F}_q^n$ such that the sensitive data which could be recovered from the output to F' (but is not in the implementation except at the very end, for avoiding leakage) equals the output to F. In other words, any function F must be *masked* into a function F', and the minimum number of intermediate data necessary for recovering the sensitive data should be as large as possible. Faults can be detected by verifying that masks have not been altered during the processing. But it seems non trivial to derive the mask from the shared output to an S-box after this S-box has been processed. Indeed, one such method (erroneous from a security standpoint) consists in recovering the unprotected information (which ruins the side-channel security), and subtract it from the shared output. This indeed allows to recover the mask, but at the cost of leaking the sensitive variable. To avoid such shortcoming, it must be considered that the mask is leaking too (since it is needed for the fault detection step), and the scheme now consists in leaking both the shared output and the mask, which changes the situation: x is now encoded into the pair $(xG + yH, y)$; the code C of generator matrix G is then changed into the code C' of generator matrix $[G : 0]$, where 0 is the all-zero matrix of dimensions $k \times (n - k)$ and the code D of generator matrix H is changed into the code D' of generator matrix $[H : I]$ where I is the identity matrix of dimensions $(n - k) \times (n - k)$. The codes C' and D' still have trivial intersection but while C and D were supplementary in $\mathbb{F}_q^n$, C' and D' are not supplementary in $\mathbb{F}_q^{2n-k}$. And instead of being $\min\{d(C), d(D^\perp)\}$, the security parameter is now $\min\{d(C'), d(D'^\perp)\} = \min\{d(C), d(D'^\perp)\}$. Note that $d(D'^\perp) \le d(D^\perp)$ since the dual distance of a linear code equals the minimum nonzero number of linearly dependent columns in its generator matrix. There is then a risk of depreciation of the security parameter. Codes are then needed whose depreciation is minimized. In Sect. 2, a solution to this problem is found by using algebraic geometry (AG) codes via the rational function field. The codes obtained in this case are maximum distance separable (MDS). In Sect. 3, AG codes from higher genera function fields are used to address the problem. In fact, the results in Sect. 2 are special cases of this more general construction. A concatenated construction is presented in Sect. 4 to descend from a code over an extension field to a code over a base field. This construction is more general than the similar results in [4] and [12] (see also [7]). Our results are then applied in Sect. 5 to present some explicit codes.

2   Construction of MDS Codes Suitable for both SCA and FIA

In this section we address the question of finding codes D (see Introduction) such that $d(D'^{\perp}) = d(D^{\perp})$. This question has its own interest in coding theory.

Nota Bene. Since this problem is general, we change "$D, H, n - k, 2n - k$" into "$C, G, n + k$" to make the notation more easily understood.

We denote by $\mathbb{F}_q$ the finite field with q elements throughout. Let $1 \leq k \leq n$ be positive integers and G be a full rank $k \times n$ matrix. Let C_1 and C_2 be $[n, k]$ and $[n + k, k]$ linear codes over $\mathbb{F}_q$ with the generating matrices G and $[G : I_k]$, respectively, where I_k denotes the $k \times k$ identity matrix. Note that the dual codes $C_1^{\perp}$ and $C_2^{\perp}$ are of dimension $n - k$ and n, respectively. Moreover, for any $x \in C_1^{\perp}$, the vector $(x, 0_k)$ obtained by appending 0's at the end is a codeword of $C_2^{\perp}$. Hence we have $d(C_2^{\perp}) \leq d(C_1^{\perp})$.

Construction of matrices G (and hence, codes C_1 and C_2) in a way that $d(C_2^{\perp})$ is as close to $d(C_1^{\perp})$ as possible was posed in [18] (and is explained again in introduction of the present paper) as a proposal to resist both SCA and FIA. The following result achieves equality of the minimum distances with maximum distance separable (MDS) codes.

Theorem 1. *Let n, k be positive integers such that*

$$k \leq n \quad and \quad n + k \leq q. \tag{1}$$

Let $\alpha_1, \alpha_2, \ldots, \alpha_n, \beta_1, \beta_2, \ldots, \beta_k \in \mathbb{F}_q$ be distinct elements. Let $f_0(x) = 1, f_1(x) = x, \ldots, f_{k-1}(x) = x^{k-1} \in \mathbb{F}_q[x]$ and consider the $k \times n$ and $k \times k$ matrices given by

$$H_1 = \begin{bmatrix} f_0(\alpha_1) & f_0(\alpha_2) & \cdots & f_0(\alpha_n) \\ f_1(\alpha_1) & f_1(\alpha_2) & \cdots & f_1(\alpha_n) \\ \vdots & \vdots & \ddots & \vdots \\ f_{k-1}(\alpha_1) & f_{k-1}(\alpha_2) & \cdots & f_{k-1}(\alpha_n) \end{bmatrix}$$

and

$$H_2 = \begin{bmatrix} f_0(\beta_1) & f_0(\beta_2) & \cdots & f_0(\beta_k) \\ f_1(\beta_1) & f_1(\beta_2) & \cdots & f_1(\beta_k) \\ \vdots & \vdots & \ddots & \vdots \\ f_{k-1}(\beta_1) & f_{k-1}(\beta_2) & \cdots & f_{k-1}(\beta_k) \end{bmatrix}.$$

Then H_2 is invertible. Moreover, let H be the $k \times n$ matrix given by

$$H = H_2^{-1} H_1.$$

Consider the linear code C_1 over $\mathbb{F}_q$ with the generator matrix H and the linear code C_2 over $\mathbb{F}_q$ with generator matrix $[H : I_k]$, which is the $k \times (n + k)$ matrix obtained by concatenation of H and I_k. We have

1. C_1 is an $[n, k]$ MDS code over $\mathbb{F}_q$. Its dual code $C_1^\perp$ is also an MDS code over $\mathbb{F}_q$ with the minimum distance $d(C_1^\perp) = k + 1$.
2. C_2 is an $[n + k, k]$ MDS linear code over $\mathbb{F}_q$. Its dual code $C_2^\perp$ is also an MDS code over $\mathbb{F}_q$ with the minimum distance $d(C_2^\perp) = k + 1$.

It is possible to weaken the condition in (1) in order to include the case $n + k = q + 1$ by using evaluations at "infinity" and using irreducible polynomials of degree $k - 1$ as in the following theorem. In the statement of the following theorem we explain the notion of evaluation at "infinity" in an elementary (and equivalent) form (see Remark 1). We also keep some repetition and present a long statement for the sake of clarity.

Theorem 2. *Let n, k be positive integers such that*

$$k \leq n, \quad 3 \leq k \text{ and } \quad n + k \leq q + 1. \tag{2}$$

Let ∞ denote the symbol at "infinity" and consider the set $\mathbb{F}_q \cup \{\infty\}$ having $q + 1$ elements. Let $\alpha_1, \alpha_2, \ldots, \alpha_n, \beta_1, \beta_2, \ldots, \beta_k \in \mathbb{F}_q \cup \{\infty\}$ be distinct elements of $\mathbb{F}_q \cup \{\infty\}$. Let $\mu(x) \in \mathbb{F}_q[x]$ be a monic irreducible polynomial of degree $k - 1$. Let $f_0(x) = \frac{1}{\mu(x)}, f_1(x) = \frac{x}{\mu(x)}, \ldots, f_{k-1}(x) = \frac{x^{k-1}}{\mu(x)} \in \mathbb{F}_q(x)$. For $\alpha \in \mathbb{F}_q$ and $0 \leq i \leq k - 1$, by $f_i(\alpha)$ we mean the evaluation of the rational function $f_i(x)$ at α. For $\alpha = \infty$ we put $f_{k-1}(\alpha) = 1$ and $f_i(\alpha) = 0$ if $0 \leq i \leq k - 2$. As in Theorem 1, let H_1 and H_2 denote the $k \times n$ and $k \times k$ matrices given by

$$H_1 = \begin{bmatrix} f_0(\alpha_1) & f_0(\alpha_2) & \cdots & f_0(\alpha_n) \\ f_1(\alpha_1) & f_1(\alpha_2) & \cdots & f_1(\alpha_n) \\ \vdots & \vdots & \ddots & \vdots \\ f_{k-1}(\alpha_1) & f_{k-1}(\alpha_2) & \cdots & f_{k-1}(\alpha_n) \end{bmatrix}$$

and

$$H_2 = \begin{bmatrix} f_0(\beta_1) & f_0(\beta_2) & \cdots & f_0(\beta_k) \\ f_1(\beta_1) & f_1(\beta_2) & \cdots & f_1(\beta_k) \\ \vdots & \vdots & \ddots & \vdots \\ f_{k-1}(\beta_1) & f_{k-1}(\beta_2) & \cdots & f_{k-1}(\beta_k) \end{bmatrix}.$$

Then H_2 is invertible. Moreover, let H be the $k \times n$ matrix given by

$$H = H_2^{-1} H_1.$$

Consider the linear code C_1 over $\mathbb{F}_q$ with generator matrix H and the linear code C_2 over $\mathbb{F}_q$ with generator matrix $[H : I_k]$. We have

1. C_1 is an $[n, k]$ MDS code over $\mathbb{F}_q$. Its dual code $C_1^\perp$ is also an MDS code over $\mathbb{F}_q$ with minimum distance $d(C_1^\perp) = k + 1$.
2. C_2 is an $[n + k, k]$ MDS code over $\mathbb{F}_q$. Its dual code $C_2^\perp$ is also an MDS code over $\mathbb{F}_q$ with the minimum distance $d(C_2^\perp) = k + 1$.

Remark 1. In Theorem 2, for $\alpha = \infty$ we put $f_i(\alpha) = 0$ if $0 \leq i \leq k - 2$, and we put $f_i(\alpha) = 1$ if $i = k - 1$ as it refers to the evaluation of the rational function $\frac{x^i}{\mu(x)}$ at "infinity". Indeed this evaluation is 0 if $0 \leq i \leq k - 2$, and for the case $i = k - 1$ this evaluation corresponds to the ratio leading coefficient of $x^{k-1}/$ leading coefficient of $\mu(x)$, which is $1/1 = 1$.

Remark 2. In Theorem 2, the extra condition $3 \leq k$ is included as we use irreducible polynomial $\mu(x)$ of degree at least 2. In fact we can easily get rid of this condition for example using two irreducible polynomials $\mu_1(x)$ and $\mu_2(x)$ of suitable degrees instead of only one.

Theorems 1 and 2 are special cases of Theorem 3 below. We give a proof of Theorem 3, which implies proofs of Theorems 1 and 2.

3   Construction of Algebraic Geometry Codes Suitable for both SCA and FIA

We will use some facts and notation from algebraic function fields. For details we refer to [19]. For example if $F/\mathbb{F}_q$ is an algebraic function field of one variable whose full constant field is $\mathbb{F}_q$, $f \in F$ and P is a place of degree one of F, then $f(P)$ denotes the evaluation of f at P. Recall that $f(P) \in \mathbb{F}_q$.

The next Lemma is used in Theorem 3 below. It shows that if $n + k$ distinct degree one places of an algebraic function field of genus g is given (under suitable conditions), then we can easily determine a subset $Q_1, \ldots, Q_k$ of this set which leads to a $k \times k$ invertible matrix. If $g = 0$, then in fact $Q_1, \ldots, Q_k$ can be chosen arbitrarily. However, if $g \geq 1$, then we should be careful but it is easy and effective to choose such a subset.

Lemma 1. *Let $F/\mathbb{F}_q$ be an algebraic function field of one variable whose full constant field is $\mathbb{F}_q$. Let g be the genus of F and $N(F)$ be the number of degree one places of F. Assume that*

$$g \leq k \leq n$$

and

$$n + k \leq N(F).$$

Let T be a set of $n + k$ distinct degree one places of F. Let G be a divisor of F such that $\deg G = k + g - 1$ and $\operatorname{supp} G \cap T = \emptyset$. Then the Riemann-Roch space $\mathcal{L}(G) = \{f \in F^\star : (f) + G \geq 0\} \cup \{0\}$ has dimension k over $\mathbb{F}_q$. Let $\{f_0, f_1, \ldots, f_{k-1}\}$ be an $\mathbb{F}_q$-basis of $\mathcal{L}(G)$.

For any subset $S \subseteq T$ of size $k + g$ there exists a subset $S_0 = \{Q_1, \ldots, Q_k\} \subseteq S$ of size k such that the matrix

$$\begin{bmatrix} f_0(Q_1) & f_0(Q_2) & \cdots & f_0(Q_k) \\ f_1(Q_1) & f_1(Q_2) & \cdots & f_1(Q_k) \\ \vdots & \vdots & \ddots & \vdots \\ f_{k-1}(Q_1) & f_{k-1}(Q_2) & \cdots & f_{k-1}(Q_k) \end{bmatrix}$$

is an invertible $k \times k$ matrix over $\mathbb{F}_q$.

Proof. As $g \leq k$, $\deg G = k + g - 1 \geq 2g - 1$ and hence the dimension $l(G)$ of $\mathcal{L}(G)$ is $\deg G + 1 - q = k$. Let $S = \{U_1, U_2, \ldots, U_{k+g}\}$ and consider the $\mathbb{F}_q$−linear map

$$\psi : \mathcal{L}(G) \to \mathbb{F}_q^{k+g}$$
$$f \mapsto (f(U_1), f(U_2), \ldots, f(U_{k+g})).$$

As $\deg(G - U_1 - U_2 - \cdots - U_{k+g}) = k + g - 1 - k - g < 0$, ψ is injective and the image is and $\mathbb{F}_q$−linear code of dimension k. The image has a generating matrix as

$$\begin{bmatrix} f_0(U_1) & f_0(U_2) & \cdots & f_0(U_{k+g}) \\ f_1(U_1) & f_1(U_2) & \cdots & f_1(U_{k+g}) \\ \vdots & \vdots & \ddots & \vdots \\ f_{k-1}(U_1) & f_{k-1}(U_2) & \cdots & f_{k-1}(U_{k+g}) \end{bmatrix}.$$

This matrix has rank k and hence there exist k linearly independent columns of this matrix. Let $Q_1, Q_2, \ldots, Q_k$ denote one of these configurations of k linearly independent columns. This completes the proof.

The following gives a construction of algebraic geometry codes suitable for both SCA and FIA.

Theorem 3. *Let $F/\mathbb{F}_q$ be an algebraic function field of one variable whose constant field is $\mathbb{F}_q$. Let g be the genus of F and $N(F)$ be the number of degree one places of F. Assume that n, k are positive integers such that*

$$g \leq k \leq n$$

and

$$n + k \leq N(F).$$

Let T be a set of $n + k$ distinct degree one places of F and G be a divisor of F such that $\deg G = k + g - 1$ and $\operatorname{supp} G \cap T = \emptyset$. Using Lemma 1 we determine k degree one places $Q_1, Q_2, \ldots, Q_k$ in T such that the $k \times k$ matrix H_2 given by

$$H_2 = \begin{bmatrix} f_0(Q_1) & f_0(Q_2) & \cdots & f_0(Q_k) \\ f_1(Q_1) & f_1(Q_2) & \cdots & f_1(Q_k) \\ \vdots & \vdots & \ddots & \vdots \\ f_{k-1}(Q_1) & f_{k-1}(Q_2) & \cdots & f_{k-1}(Q_k) \end{bmatrix}.$$

is invertible. Let $P_1, P_2, \ldots, P_n$ be the remaining degree one places in T. Let H_1 be the $k \times n$ matrix given by

$$H_1 = \begin{bmatrix} f_0(P_1) & f_0(P_2) & \cdots & f_0(P_n) \\ f_1(P_1) & f_1(P_2) & \cdots & f_1(P_n) \\ \vdots & \vdots & \ddots & \vdots \\ f_{k-1}(P_1) & f_{k-1}(P_2) & \cdots & f_{k-1}(P_n) \end{bmatrix}.$$

Moreover let H be the $k \times n$ matrix given by

$$H = H_2^{-1} H_1.$$

Consider the linear code C_1 over $\mathbb{F}_q$ with generator matrix H and the linear code C_2 over $\mathbb{F}_q$ with generator matrix $[H : I_k]$. We have

1. *C_1 is an $[n, k]$ linear code over $\mathbb{F}_q$. Its dual code $C_1^\perp$ has minimum distance $d(C_1^\perp)$ satisfying $d(C_1^\perp) \geq k - g + 1$.*
2. *C_2 is an $[n+k, k]$ linear code over $\mathbb{F}_q$. Its dual code $C_2^\perp$ has minimum distance $d(C_2^\perp)$ satisfying $d(C_1^\perp) \geq k - g + 1$.*

Proof. There exists a (Weil) differential η of F such that $v_{P_i}(\eta) = -1$, $v_{Q_j}(\eta) = -1$, $\mathrm{res}_{P_i}(\eta) = 1$, $\mathrm{res}_{Q_i}(\eta) = 1$ for $1 \leq i \leq n$ and $1 \leq j \leq k$ ([19, Lemma 2.2.9]). Let $D = P_1 + P_2 + \cdots + P_n$. Recall that $C_{\mathcal{L}}(D, D - G + (\eta))$ denotes the geometric Goppa code obtained by taking evaluations $(f(P_1), \ldots, f(P_n))$ as f runs through $\mathcal{L}(D - G + (\eta))$ (see [19, Definition 2.2.1]).

Using [19, Theorem 2.2.2] we conclude that

$$d(C_1^\perp) \geq n - \deg(D - G + (\eta)) = k - g + 1,$$

where we use the fact that $\deg(\eta) = 2g - 2$. Let $E = Q_1 + Q_2 + \cdots + Q_k$. Similarly we have

$$C_2^\perp = C_{\mathcal{L}}(D + E, D + E - G + (\eta)),$$

and hence $d(C_2^\perp) \geq k - g + 1$ as well.

Remark 3. Note that both Theorems 1 and 2 are special cases of Theorem 3 obtained by taking $g = 0$. The corresponding codes in Theorems 1 and 2 are MDS because of the Singleton bound.

4   An Expansion (Concatenation) Construction Suitable for both SCA and FIA

Let $m \geq 2$ be an integer and $(\epsilon_1, \ldots, \epsilon_m)$ be an $\mathbb{F}_q$ basis of $\mathbb{F}_{q^m}$. Let $\pi : \mathbb{F}_{q^m} \to \mathbb{F}_q^m$ be the $\mathbb{F}_q$-linear map sending $\alpha \in \mathbb{F}_q^m$ to $(a_1, \ldots, a_m)$ where $\alpha = a_1 \epsilon_1 + \cdots + a_m \epsilon_m$. For an integer $n \geq 1$ we also denote the map $\pi : \mathbb{F}_{q^m}^n \to \mathbb{F}_{q^{mn}}$ defined by

$$\pi(\alpha_1, \ldots, \alpha_n) = (\pi(\alpha_1), \ldots, \pi(\alpha_n)).$$

Let $(\epsilon_1^*, \ldots, \epsilon_m^*)$ be the trace dual basis of $\mathbb{F}_{q^m}$ over $\mathbb{F}_q$. Similarly let $\pi' : \mathbb{F}_{q^m} \to \mathbb{F}_q^m$ be the $\mathbb{F}_q$-linear map sending $\alpha \in \mathbb{F}_q^m$ to $(a_1, \ldots, a_m)$ where $\alpha = a_1 \epsilon_1^* + \cdots + a_m \epsilon_m^*$. For an integer $n \geq 1$ we again denote the map $\pi' : \mathbb{F}_{q^m}^n \to \mathbb{F}_{q^{mn}}$ defined by

$$\pi'(\alpha_1, \ldots, \alpha_n) = (\pi'(\alpha_1), \ldots, \pi'(\alpha_n)).$$

If C is an $\mathbb{F}_{q^m}$-linear code in $\mathbb{F}_{q^m}^n$ then $\pi(C)$ and $\pi'(C)$ are $\mathbb{F}_q$-linear codes in $\mathbb{F}_q^{mn}$. The following is an easy but useful result that we use. We prefer to give a short proof for completeness.

Lemma 2. *Under notation as above let C be an $\mathbb{F}_{q^m}$-linear code in $\mathbb{F}_{q^m}^n$ and $C^\perp$ be its dual in $\mathbb{F}_{q^m}^n$. Then for the dual of $\pi(C)$ in $\mathbb{F}_q^{mn}$ we have*

$$\pi(C)^\perp = \pi'(C^\perp).$$

Proof. It is clear that

$$\dim_{\mathbb{F}_q} \pi(C)^\perp = mn - m\dim_{\mathbb{F}_{q^m}} C = m(n - \dim_{\mathbb{F}_{q^m}} C) = m\dim_{\mathbb{F}_{q^m}} C^\perp = \dim_{\mathbb{F}_q} \pi'(C^\perp).$$

Hence it is enough to show that $\pi'(C^\perp) \subseteq \pi(C)^\perp$. Let $(\beta_1, \ldots, \beta_n) \in C^\perp$. We have

$$0 = (\beta_1, \ldots, \beta_n) \cdot (\alpha_1, \ldots, \alpha_n) = \sum_{l=1}^{n} \alpha_l \beta_l = \sum_{l=1}^{n} (\sum_{i=1}^{m} a_{l,i}\epsilon_i)(\sum_{j=1}^{m} b_{l,j}\epsilon_j^*)$$

for all $(\alpha_1, \ldots, \alpha_n) \in C$. Taking the trace of both sides we obtain

$$0 = \sum_{l=1}^{n} \sum_{i=1}^{m} \sum_{j=1}^{m} a_{l,i} b_{l,j} \mathrm{Tr}(\epsilon_i \epsilon_j^*)$$

$$= \pi(\alpha_1, \ldots, \alpha_n) \cdot \pi'(\beta_1, \ldots, \beta_n).$$

Hence $\pi'(\beta_1, \ldots, \beta_n) \in \pi(C)^\perp$, which completes the proof.

The following theorem gives an expansion (concatenation) construction allowing to go to a subfield. Namely if C_1 and C_2 are $[n, k]$ and $[n+k, k]$ linear codes over $\mathbb{F}_{q^m}$ giving a solution to the Problem posed in Sect. 2 over $\mathbb{F}_{q^m}$, then using an expansion (concatenation) we obtain $[nm, km]$ and $[nm + km, km]$ linear codes $\pi(C_1)$ and $\pi(C_2)$ over $\mathbb{F}_q$ giving a solution to the same problem over $\mathbb{F}_q$. Moreover $d(\pi(C_1)^\perp) \geq d(C_1^\perp)$ and $d(\pi(C_2)^\perp) \geq d(C_2^\perp)$. This is similar to the analogous construction in [4] and [12] (see also [7]). We further note that π exists for all q and m, which is not the case for the corresponding construction in [4] and [12].

Theorem 4. *Let $m \geq 2$ and $1 \leq k \leq n$ be positive integers. Let $(\epsilon_1, \ldots, \epsilon_m)$ be an $\mathbb{F}_q$ basis of $\mathbb{F}_{q^m}$ and $\pi : \mathbb{F}_{q^m}^n \to \mathbb{F}_{q^{mn}}$ be the corresponding $\mathbb{F}_q$-linear map defined above.*

Let G be a $k \times n$ full rank matrix over $\mathbb{F}_{q^m}$. Let C_1 and C_2 be the $[n, k]$ and $[n + k, k]$ linear codes over $\mathbb{F}_{q^m}$ given by

– *G is a generator matrix of C_1 and*
– *$[G : I_k]$ is a generator matrix of C_2.*

Let $G_1, \ldots, G_k \in \mathbb{F}_{q^m}^{1 \times n}$ be the rows of G so that

$$G = \begin{bmatrix} G_1 \\ G_2 \\ \vdots \\ G_k \end{bmatrix}.$$

Let $\overline{G}$ be the $km \times mn$ matrix over $\mathbb{F}_q$ defined as

$$\overline{G} = \begin{bmatrix} \pi(\epsilon_1 G_1) \\ \vdots \\ \pi(\epsilon_m G_1) \\ \vdots \\ \pi(\epsilon_1 G_k) \\ \vdots \\ \pi(\epsilon_m G_k) \end{bmatrix}.$$

Then $\pi(C_1)$ is an $[nm, km]$ linear code over $\mathbb{F}_q$ and $\pi(C_2)$ is an $[nm + km, km]$ linear code over $\mathbb{F}_q$ such that

- *$\overline{G}$ is a generator matrix of $\pi(C_1)$ and*
- *$[\overline{G} : I_{km}]$ is a generator matrix of $\pi(C_2)$.*

Moreover $d(\pi(C_1)^{\perp}) \geq d(C_1^{\perp})$ and $d(\pi(C_2)^{\perp}) \geq d(C_2^{\perp})$.

Proof. It is clear that $\overline{G}$ is a generator matrix of $\pi(C_1)$. Let $\epsilon_1, \ldots, \epsilon_m$ be the rows of I_k. It is not difficult to observe that

$$\begin{bmatrix} \pi(\epsilon_1) & 0 & \cdots & 0 \\ \pi(\epsilon_2) & 0 & \cdots & 0 \\ \vdots & \vdots & \cdots & \vdots \\ \pi(\epsilon_m) & 0 & \cdots & 0 \\ 0 & \pi(\epsilon_1) & \cdots & 0 \\ 0 & \pi(\epsilon_2) & \cdots & 0 \\ \vdots & \vdots & \cdots & \vdots \\ 0 & \pi(\epsilon_m) & \cdots & 0 \\ \vdots & \vdots & \cdots & \vdots \\ 0 & 0 & \cdots & \pi(\epsilon_1) \\ 0 & 0 & \cdots & \pi(\epsilon_2) \\ \vdots & \vdots & \cdots & \vdots \\ 0 & 0 & \cdots & \pi(\epsilon_m) \end{bmatrix} = I_{km}.$$

This implies that $[\overline{G} : I_{km}]$ is a generator matrix of $\pi(C_2)$. Using Lemma 2 we have

$$\pi(C_1)^{\perp} = \pi'(C_1^{\perp}) \text{ and } \pi(C_2)^{\perp} = \pi'(C_2^{\perp}).$$

As π' gives a concatenated code such that the inner code is an $[m, m, 1]$ linear code over $\mathbb{F}_q$ we obtain that

$$d(\pi(C_1)^{\perp}) \geq d(C_1^{\perp}) \text{ and } d(\pi(C_2)^{\perp}) \geq d(C_2^{\perp}).$$

5   Applications

In this section we give some applications of our constructions, providing a solution to the main problem of this article stated in Sect. 2.

Example 1. Let $F = \mathbb{F}_q(x)$. Then the genus of F is 0 and F has $q+1$ degree one places. Let k and n be integers such that $1 \leq k \leq n$ and $k + n \leq q + 1$. Using Theorem 3 (see also Theorems 1 and 2) we construct

1. $[n, k]$ linear code C_1 over $\mathbb{F}_q$ and
2. $[n + k, k]$ linear code C_2 over $\mathbb{F}_q$

such that

$$d(C_1^{\perp}) = k + 1 \text{ and } d(C_2^{\perp}) = k + 1.$$

Note that C_1, $C_1^{\perp}$, C_2 and $C_2^{\perp}$ are MDS codes.

Example 2. Let $m \geq 2$ be an integer and $F = \mathbb{F}_{q^m}(x)$. Let k and n be integers such that

$$1 \leq k \leq n \text{ and } k + n \leq q^m + 1.$$

Using Theorems 3 and 4 we construct

1. $[nm, km]$ linear code $\pi(C_1)$ over $\mathbb{F}_q$ and
2. $[nm + km, km]$ linear code $\pi(C_2)$ over $\mathbb{F}_q$

such that

$$d(\pi(C_1)^{\perp}) \geq k + 1 \text{ and } d(\pi(C_2)^{\perp}) \geq k + 1.$$

Example 3. Let $F = \mathbb{F}_{q^2}(x, y)$ where $y^q + y = x^{q+1}$. Note that F is the Hermitian function field over $\mathbb{F}_{q^2}$. The genus of F is $\frac{q(q-1)}{2}$ and the number of degree one places of F is $q^3 + 1$. Let k and n be integers such that

$$\frac{q(q-1)}{2} \leq k \leq n \text{ and } k + n \leq q^3 + 1.$$

Using Theorem 3 we construct

1. $[n, k]$ linear code over $\mathbb{F}_{q^2}$ and
2. $[n + k, k]$ linear code over $\mathbb{F}_{q^2}$

such that

$$d(\pi(C_1)^{\perp}) \geq k + 1 - \frac{q(q-1)}{2} \text{ and } d(\pi(C_2)^{\perp}) \geq k + 1 - \frac{q(q-1)}{2}.$$

Example 4. Let $m \geq 2$ be an integer. Let $F = \mathbb{F}_{q^{2m}}(x, y)$ where $y^{q^m} + y = x^{q^m+1}$. Then F is the Hermitian function field over $\mathbb{F}_{q^{2m}}$, the genus of F is $\frac{q^m(q^m-1)}{2}$ and the number of degree one places of F is $q^{3m} + 1$. Let k and n be integers such that

$$\frac{q^m(q^m - 1)}{2} \leq k \leq n \text{ and } k + n \leq q^{3m} + 1.$$

Using Theorems 3 and 4 we construct

1. $[nm, km]$ linear code $\pi(C_1)$ over $\mathbb{F}_{q^2}$ and
2. $[nm + km, km]$ linear code $\pi(C_2)$ over $\mathbb{F}_{q^2}$

such that

$$d(\pi(C_1)^\perp) \geq k + 1 - \frac{q^m(q^m - 1)}{2} \text{ and } d(\pi(C_2)^\perp) \geq k + 1 - \frac{q^m(q^m - 1)}{2}.$$

Next we give the numerical examples.

Example 5. Let $q = 2$ and $m = 5$ in Example 2. Let $k = 16$, $n = 17$. Then we obtain

1. $[85, 80]$ linear code $\pi(C_1)$ over $\mathbb{F}_2$ and
2. $[165, 80]$ linear code $\pi(C_2)$ over $\mathbb{F}_2$

such that
$$d(\pi(C_1)^\perp) \geq 17 \text{ and } d(\pi(C_2)^\perp) \geq 17.$$

Example 6. Let $q = 2$ and $m = 3$ in Example 4. Let $k = 256$, $n = 257$. Then we obtain

1. $[771, 768]$ linear code $\pi_1(C_1)$ over $\mathbb{F}_4$ and
2. $[1539, 768]$ linear code $\pi_1(C_2)$ over $\mathbb{F}_4$

such that $d(\pi(C_1)^\perp) \geq 229$ and $d(\pi(C_2)^\perp) \geq 229$. Using Theorem 4 via $\pi : \mathbb{F}_4 \to \mathbb{F}_2^2$, we further obtain

(3) $[1542, 1536]$ linear code $\pi_2(\pi_1(C_1))$ over $\mathbb{F}_2$ and
(4) $[3078, 1536]$ linear code $\pi_2(\pi_1(C_2))$ over $\mathbb{F}_2$

such that $d(\pi_2(\pi_1(C_1))^\perp) \geq 229$ and $d(\pi_2(\pi_1(C_2))^\perp) \geq 229$.

Acknowledgement. Güneri and Özbudak are supported by the TÜBİTAK project 215E200, which is associated with the SECODE project in the scope of the CHIST-ERA Program. Carlet and Mesnager are also supported by the SECODE Project.

References

1. Bhasin, S., Danger, J.-L., Guilley, S., Najm, Z., Ngo, X.T.: Linear complementary dual code improvement to strengthen encoded circuit against hardware Trojan horses. In: IEEE International Symposium on Hardware Oriented Security and Trust (HOST), 5–7 May 2015
2. Bringer, J., Carlet, C., Chabanne, H., Guilley, S., Maghrebi, H.: Orthogonal direct sum masking. In: Naccache, D., Sauveron, D. (eds.) WISTP 2014. LNCS, vol. 8501, pp. 40–56. Springer, Heidelberg (2014). https://doi.org/10.1007/978-3-662-43826-8_4
3. Carlet, C., Daif, A., Guilley, S., Tavernier, C.: Polynomial direct sum masking to protect against both SCA and FIA. J. Cryptogr. Eng. (2018). https://doi.org/10.1007/s13389-018-0194-9

4. Carlet, C., Guilley, S.: Complementary dual codes for counter-measures to side-channel attacks. Adv. Math. Commun. **10**(1), 131–150 (2016)
5. Carlet, C., Guilley, S.: Satatistical properties of side-channel and fault injection attacks using coding theory. Cryptogr. Commun. **10**, 909–933 (2018)
6. Carlet, C., Güneri, C., Özbudak, F., Özkaya, B., Solé, P.: On linear complementary pairs of codes. IEEE Trans. Inf. Theory, to appear
7. Carlet, C., Güneri, C., Özbudak, F., Solé, P.: A new concatenated type construction for LCD codes and isometry codes. Discrete Math. **341**, 830–835 (2018)
8. Carlet, C., Mesnager, S., Tang, C., Qi, Y.: Euclidean and Hermitian LCD MDS codes. Des. Codes Cryptogr. **86**, 1–4 (2018). https://doi.org/10.1007/s10623-018-0463-8
9. Carlet, C., Mesnager, S., Tang, C., Qi, Y.: New characterization and parametrization of LCD codes. IEEE Trans. Inf. Theory, vol. To appear. https://arxiv.org/abs/1709.03217
10. Carlet, C., Mesnager, S., Tang, C., Qi, Y., Pellikaan, R.: Linear codes over $\mathbb{F}_q$ are equivalent to LCD codes for $q>3$. IEEE Trans. Inf. Theory **64**(4), 3010–3017 (2018)
11. Ding, C., Li, C., Li, S.: LCD Cyclic codes over finite fields. arXiv:1608. 0217v1 [cs.IT]
12. Güneri, C., Özkaya, B., Solé, P.: Quasi-cyclic complementary dual codes. Finite Fields Appl. **42**, 67–80 (2016)
13. Güneri, C., Özbudak, F., Özkaya, B., Saçıkara, E., Sepasdar, Z., Solé, P.: Structure and performance of generalized quasi-cyclic codes. Finite Fields Appl. **47**, 183–202 (2017)
14. Li, S., Ding, C., Liu, H.: A family of reversible BCH codes. arXiv:1608.02169v1 [cs.IT]
15. Li, S., Ding, C., Liu, H.: Parameters of two classes of LCD BCH codes. arXiv:1608.02670 [cs.IT]
16. Mesnager, S., Tang, C., Qi, Y.: Complementary dual algebraic geometry codes. IEEE Trans. Inf. Theory **64**(4), 2390–2397 (2018)
17. Jin, L.: Construction of MDS codes with complementary duals. IEEE Trans. Inf. Theory **63**(5), 2843–2847 (2016)
18. SECODE Project Report: Preliminary assesment of the candidate codes with respect to fault injection attacks, December 2017
19. Stichtenoth, H.: Algebraic Function Fields and Codes. Springer, Heidelberg (1993)
20. Yang, X., Massey, J.L.: The condition for a cyclic code to have a complementary dual. J. Discrete Math. **126**, 391–393 (1994)

Hardware Implementations

On Hardware Implementation
of Tang-Maitra Boolean Functions

Mustafa Khairallah[1]([⊠]), Anupam Chattopadhyay[1], Bimal Mandal[2],
and Subhamoy Maitra[2]

[1] Nanyang Technological University, Singapore, Singapore
mustafam001@e.ntu.edu.sg, anupam@ntu.edu.sg
[2] Indian Statistical Institute, Kolkata, India
bimalmandal90@gmail.com, subho@isical.ac.in

Abstract. In this paper, we investigate the hardware circuit complexity of the class of Boolean functions recently introduced by Tang and Maitra (IEEE-TIT 64(1): 393–402, 2018). While this class of functions has very good cryptographic properties, the exact hardware requirement is an immediate concern as noted in the paper itself. In this direction, we consider different circuit architectures based on finite field arithmetic and Boolean optimization. An estimation of the circuit complexity is provided for such functions given any input size n. We study different candidate architectures for implementing these functions, all based on the finite field arithmetic. We also show different implementations for both ASIC and FPGA, providing further analysis on the practical aspects of the functions in question and the relation between these implementations and the theoretical bound. The practical results show that the Tang-Maitra functions are quite competitive in terms of area, while still maintaining an acceptable level of throughput performance for both ASIC and FPGA implementations.

Keywords: Boolean functions · Bent functions · Cryptology
Finite fields · Hardware implementation · Stream cipher

1 Introduction

Boolean functions are used in many domains such as sequences, cryptography, coding theory and combinatorics. In many cryptosystems, for example, linear/non-linear feedback shift register (LFSR and NFSR) based stream ciphers, a Boolean function is used to combine the outputs of several LFSRs/NFSRs. A special class of Boolean functions, having the maximum distance from the set of all affine functions, are known as bent functions. Introduced by Rothaus in 1976 [Rot76], these functions maximally resist any kind of affine approximations.

However, bent functions are not directly used as cryptographic primitives, since they are not balanced. Moreover, bent functions $f \in \mathcal{B}_n$ (we denote the set of n-variable Boolean functions as $\mathcal{B}_n$) exist only for even number of variables and its degree is at most $\frac{n}{2}$. Dillon [Dil74] constructed a class of bent

© Springer Nature Switzerland AG 2018
L. Budaghyan and F. Rodríguez-Henríquez (Eds.): WAIFI 2018, LNCS 11321, pp. 111–127, 2018.
https://doi.org/10.1007/978-3-030-05153-2_6

functions, which is called the partial spread ($\mathcal{PS}$) class of bent functions, and the bent functions in $\mathcal{PS}_{ap}$ is a subclass of $\mathcal{PS}$. Another generic class of bent functions, called Maiorana–McFarland class (denoted by $\mathcal{M}$), was introduced in [McF73] and further investigated in [Dil74]. Further, Dobbertin [Dob94] and Carlet [Car93] constructed different classes of bent functions. For more details of bent Boolean functions, we refer to [MM16, CS09]. Recently, Tang and Maitra [TM18, Construction 1] constructed a class of cryptographically significant balanced Boolean functions by modifying a special type of bent functions in $\mathcal{PS}_{ap}$. Such functions have very good nonlinearity and autocorrelation at the same time. Further research in this direction has been reported in [KMT18].

Since the functions of [TM18] are derived from Dillon type bent functions (related to Maiorana-McFarland type functions), the actual hardware should follow from the finite field implementation ideas. This is the task that we take up here. Noting that such functions might be useful in lightweight stream ciphers, we try to see how efficiently one can actually implement such functions. In fact, we note that the 12-variable function requires area less than 500 GE (gate equivalent) which may be embedded in a lightweight stream cipher circuit of 1000 GE. Functions on 22-variables can be implemented with little more than 3000 GE. This underlines that such implementations might be of interest as primitives in stream cipher design.

Our Contributions. In this paper, we study the circuit complexity of the Tang-Maitra class of Boolean functions. Since the construction is based on finite field arithmetic, we consider different representations for finite fields $\mathbb{F}_{2^n}$. The first two representations are the polynomial and normal bases representations. The third representation is the discrete-log representation of the finite field elements (though there are some additional complexities that we will discuss later). The results are summed up in the following three lemmas:

1. Lemma 1: The polynomial basis implementation of a Tang-Maitra function on n variables, where n is even, has the circuit complexity bounded by $\mathcal{O}(2^k + k^2)$ and depth of $\mathcal{O}(k)$, where $n = 2k$.
2. Lemma 2: The normal basis implementation of a Tang-Maitra function on n variables, where n is even, has the circuit complexity bounded by $\mathcal{O}(2^k + k^3)$, where $n = 2k$.
3. Lemma 4: The discrete-log implementation of a Tang-Maitra function on n variables, where n is even, has the circuit complexity bounded by $\mathcal{O}(2^k + k^2 + 2^k/k^2)$, where $n = 2k$.

We also propose a general circuit construction for Rotation Symmetric Boolean Functions (RSBF), bounded by $\mathcal{O}(k^2 + 2^k/k^2)$ (Lemma 3). Further, we briefly discuss the relationship between these different representations (Sect. 3.3) to clarify the well-known understanding of polynomial and normal bases representations related to certain cryptographic properties. Finally, we practically implement several instances of the Tang-Maitra functions for different values of n, where n is the number of variables. We show the implementation results for both ASIC and FPGA, providing a comparison with the hardware implementation GMM class of Boolean functions (Sect. 4). The results are based on the polynomial representation, which has the lowest asymptotic circuit complexity.

2 Preliminaries

2.1 Finite Field $\mathbb{F}_{2^n}$ Arithmetic

The finite field $\mathbb{F}_{2^n}$ is an extension field of $\mathbb{F}_2$ of degree n, where $\mathbb{F}_2$ is a prime field of two elements $\{0, 1\}$. In other words, $\mathbb{F}_{2^n}$ is the field of polynomials of degree at most $n-1$ over $\mathbb{F}_2$. It consists of 2^n elements and two basic operations are defined for it:

1. Addition ($\oplus$): polynomial addition modulo 2.
2. Multiplication ($\odot$): polynomial multiplication modulo $F[x]$, where $F[x]$ is an irreducible polynomial of degree n over $\mathbb{F}_2$.

There are three more operations of interest: Squaring, Inversion and the Trace Function, but they can be defined using the former two basic operations. Since there can be more than one irreducible polynomial of degree n, $\mathbb{F}_{2^n}$ according to this definition is not unique. However, all the fields generated by different irreducible polynomial choice are isomorphic, i.e., a certain field can be changed to another one by a permutation of the elements.

The trace function is a frequently referred to, in finite field theory. It is defined as $\mathrm{Tr}_1^n : \mathbb{F}_{2^n} \to \mathbb{F}_2$,

$$\mathrm{Tr}_1^n(\alpha) = \alpha \oplus \alpha^2 \oplus \alpha^{2^2} \oplus \ldots \oplus \alpha^{2^{n-1}}, \text{ for all } \alpha \in \mathbb{F}_{2^n} \tag{1}$$

We list here certain properties of the trace function Tr_1^n that are important for our work. An interested reader may refer to [LN94] for a complete discussion related to finite fields $\mathbb{F}_{2^n}$. These properties are:

1. $\mathrm{Tr}_1^n(\alpha \oplus \beta) = \mathrm{Tr}_1^n(\alpha) \oplus \mathrm{Tr}_1^n(\beta)$, for all $\alpha, \beta \in \mathbb{F}_{2^n}$.
2. $\mathrm{Tr}_1^n(\alpha^2) = \mathrm{Tr}_1^n(\alpha)$, for all $\alpha \in \mathbb{F}_{2^n}$.

2.2 Boolean Functions

Let $\mathbb{F}_2^n$ be the vector space of dimension n over $\mathbb{F}_2$ and any element $x \in \mathbb{F}_2^n$ can be written as $x = (x_{n-1}, \ldots, x_1, x_0)$, where $x_i \in \mathbb{F}_2$, $0 \leq i \leq n-1$. Any function f from $\mathbb{F}_2^n$ (or $\mathbb{F}_{2^n}$) to $\mathbb{F}_2$ is called Boolean function in n variables. The set of n-variable Boolean functions is denoted by $\mathcal{B}_n$. Any Boolean function $f \in \mathcal{B}_n$ can be represented in a unique way as

$$f(x) = \bigoplus_{\alpha \in \mathbb{F}_2^n} \mu_\alpha x_0^{\alpha_0} x_1^{\alpha_1} \ldots x_{n-1}^{\alpha_{n-1}},$$

for all $x \in \mathbb{F}_2^n$, where $\mu_\alpha \in \mathbb{F}_2$. This polynomial representation is called the algebraic normal form (ANF) of $f \in \mathcal{B}_n$. The algebraic degree of a Boolean function $f \in \mathcal{B}_n$ is defined as $deg(f) = \max_{\alpha \in \mathbb{F}_2^n}\{wt(\alpha) : \mu_\alpha \neq 0\}$, where $wt(\alpha)$ is the Hamming weight of $\alpha \in \mathbb{F}_2^n$, defined as $wt(\alpha) = \sum_{i=0}^{n-1} \alpha_i$ (the sum is over the ring of integers). Further more details, we refer to [MM16, CS09].

2.3 $\mathbb{F}_{2^n}$ Practical Representations

In order to perform operations on finite field elements in practice, we need to represent the elements and operations in terms of binary representations and circuits/algorithms, respectively. In this section, we describe three possible representations, with the advantages and disadvantages of each. An inquisitive reader can find more details in [DIS09].

Polynomial Basis. Let $\alpha \in \mathbb{F}_{2^n}$ be a root of the irreducible polynomial $F[x]$ used to define $\mathbb{F}_{2^n}$. Hence, $\mathbb{F}_{2^n}$ can be defined as $\{a(\alpha)|a(\alpha) = \bigoplus_{i=0}^{n-1} a_i\alpha^i\}$, where $a_i \in \mathbb{F}_2$. Therefore, $x(\alpha)$ can be represented by the binary string $(x_{n-1}, \ldots, x_1, x_0)$. Operations are performed as follows:

1. *Addition:* Addition is performed using coefficient-wise XOR.
2. *Multiplication:* Multiplication is more complicated. Since multiplication is defined as polynomial multiplication, the circuit complexity is $\mathcal{O}(n^{1+\epsilon})$, where $\epsilon > 0.5$ for the available circuits, as opposed to $\mathcal{O}(n)$ in case of addition. However, we explain the matrix-vector multiplication method, which has a complexity of $\mathcal{O}(n^2)$, as it is useful in discussing the complexity of squaring and the trace function. In the first step, we compute the polynomial $d(x) = a(x) \cdot b(x)$ of degree at most $2n - 2$. This is performed by the equation

$$
\begin{bmatrix} d_0 \\ d_1 \\ \vdots \\ d_{n-1} \\ \vdots \\ d_{2n-2} \end{bmatrix} = \begin{bmatrix} a_0 & 0 & 0 & \ldots & 0 & 0 \\ a_1 & a_0 & 0 & \ldots & 0 & 0 \\ \vdots & \vdots & \vdots & \ddots & \vdots \\ a_{n-1} & a_{n-2} & a_{n-3} & \ldots & a_1 & a_0 \\ 0 & a_{n-1} & a_{n-2} & \ldots & a_2 & a_1 \\ \vdots & \vdots & \vdots & \ddots & \vdots \\ 0 & 0 & 0 & \ldots & 0 & a_{n-1} \end{bmatrix} \begin{bmatrix} b_0 \\ b_1 \\ \vdots \\ b_{n-1} \end{bmatrix} \tag{2}
$$

This step requires n^2 AND gates and $(n - 1)^2$ XOR gates. The second step is to compute $c(x) \equiv d(x) \ mod \ F[x]$, which is performed by the equation

$$
\begin{bmatrix} c_0 \\ c_1 \\ \vdots \\ c_{n-1} \end{bmatrix} = \begin{bmatrix} I \ R \end{bmatrix} \begin{bmatrix} d_0 \\ d_1 \\ \vdots \\ d_{n-1} \\ \vdots \\ d_{2n-2} \end{bmatrix} \tag{3}
$$

where I is an $n \times n$ identity matrix and R is a matrix whose elements are functions of $F[x]$. This step requires $wt(R)$ XOR gates and it can be lowered by choosing a suitable $F[x]$, e.g., a trinomial.

3. *Squaring:* It can be implemented using only reduction (and wiring). This is due to the fact that $a(x) \cdot a(x) = a_{n-1}x^{2n-2} \oplus a_{n-2}x^{2n-4} \oplus \ldots \oplus a_1x^2 \oplus a_0$. Hence, only the last step of the matrix-vector multiplication method is required, leading to a circuit with $wt(\boldsymbol{R})$ XOR gates.
4. *Inversion:* Inversion is the most complex operation in polynomial basis. It requires $\mathcal{O}(n^2)$ gates. For example, a straightforward implementation of the Extended Euclidean Algorithm for Inversion requires around $(24n^2 + 24n)$ MUXes, $(n^2 + n)$ AND gates and $(5n^2 + 5n)$ XOR gates. Complex Boolean optimization heuristics are usually used to design more efficient circuits.
5. *Trace Function* Tr_1^n: Requires $n - 1$ squarings and $n - 1$ XOR gates, as only the constant coefficients of α^{2^i} need to be added. The overall complexity is $(wt(\boldsymbol{R}) + 1)(n - 1)$ XOR gates.

Normal Basis. The polynomial basis representation can be viewed as a vector space representation with the basis $\{1, \alpha, \alpha^2, \alpha^3, \ldots, \alpha^{n-1}\}$. Since it is known that the finite field representation is not unique, it is useful to look for other suitable bases that can be used. One special basis is the normal basis, which is defined as $\{\beta, \beta^2, \beta^{2^2}, \ldots, \beta^{2^{n-1}}\}$, where $\beta^{2^n} = \beta$ and $\beta^i \neq \beta$ for all $1 < i < 2^n$, with β is a primitive element of $\mathbb{F}_{2^n}$. $\mathbb{F}_{2^n}$ is defined as $\{a(\beta) | a(\beta) = \bigoplus_{i=0}^{n-1} a_i \beta^{2^i}\}$, where $a_i \in \mathbb{F}_2$. Since the vectors of the normal basis are linearly independent, all the elements of $\mathbb{F}_{2^n}$ can be generated as linear combinations of the basis vectors.

Again, $x(\beta)$ can be represented by the binary string $(x_{n-1}, \ldots, x_1, x_0)$ and addition is the same as in the case of polynomial basis. While multiplication is even more complex than in the case of polynomial basis ($\mathcal{O}(n^{1+\epsilon})$, where $\epsilon > 0.6$), two of the operations we are interested in are very simple using the normal basis representation; Squaring and Tr_1^n. Equation (4) shows that squaring in $\mathbb{F}_{2^n}$ is a linear operation, while Eq. (5) is an application of Fermat's Little Theorem to $\mathbb{F}_{2^n}$. Using these two properties, it can be shown that in the normal basis representation x^2 is just a cyclic shift of the bit representation of x and $\mathrm{Tr}_1^n(x) = \bigoplus_{i=0}^{n-1} x_i$.

$$(\beta \oplus \gamma)^2 = \beta^2 \oplus \gamma^2, \text{ for all } \beta, \gamma \in \mathbb{F}_{2^n} \tag{4}$$

$$\beta^{2^n} = \beta, \text{ for all } \beta \in \mathbb{F}_{2^n} \tag{5}$$

Inversion, however, is not efficient in the normal basis. Using Fermat's little theorem, it requires circuit complexity of $\mathcal{O}(n^{2+\epsilon})$, where $\epsilon > 0.6$, since exponentiation requires $n - 2$ multiplication operations. On the other hand, for special choices of n, more efficient circuits can be implemented. For example, for $n = 2^r + 1$ only $\log(n-1)$ multiplications are required. Similar results are available for the cases when $n = m * k$, using Tower Fields [IT88].

Discrete-Log Representation. Another interesting representation of finite field elements is the discrete logarithmic representation. Let α be a primitive element of $\mathbb{F}_{2^n}$. Then $\alpha^{2^n} = \alpha$, $\alpha^i \neq \alpha$, for all $0 \leq i \leq 2^n - 2$ and any nonzero element $x \in \mathbb{F}_{2^n}$ is represented as $x = \alpha^i$ for a $0 \leq i \leq 2^n - 2$. A special representation is used for $x = 0$. Hence, only the exponent i needs to be stored. Since α has order of $2^n - 1$, each of the $2^n - 1$ non-zero field elements have unique representation $\in \mathbb{Z}_{2^n - 1}$. Besides, $i = 2^n - 1$ is used to represent $x = 0$. This representation is very useful for applications that include a lot of multiplication, squaring and inversion operations.

However, the transformation to/from discrete-log is a non-linear operation, and the relation between this representation and the polynomial basis representation is non-linear. Hence, it cannot be used directly for the Tang-Maitra functions without implementing this non-linear transformation increasing the circuit complexity (details are given in Sect. 3.3). This representation is discussed in details in Appendices A and B as the cost of such non-linear transformations cannot be estimated immediately.

2.4   The Tang-Maitra Class of Functions

Tang and Maitra [TM18] constructed a class of Boolean functions with good cryptographic properties by modifying $\mathcal{PS}_{ap}$, a subclass of partial spread, as follow:

Construction 1 [TM18, Construction 1]. *Let $n = 2k$ and $\lambda, \mu \in \mathbb{F}_{2^n}^*$, where $k \geq 9$ is an odd integer. An n-variable Boolean function over $\mathbb{F}_{2^n}$ is defined as follows*

$$f(x, y) = \begin{cases} h_0(y) & \text{if } x = 0 \\ h_1(y) & \text{if } x = \mu \\ \mathrm{Tr}_1^k(\frac{\lambda x}{y}) & \text{otherwise} \end{cases} \qquad (6)$$

Here we assume that $\mathrm{Tr}_1^k(\frac{\lambda x}{0}) = 0$, for all $x \in \mathbb{F}_{2^k}$. The functions h_0 and h_1 must satisfy some cryptographic properties, which can be found in [TM18, KMT18]. In Sect. 3, we provide circuits for Construction 1 using different finite field representations and derive an estimation of its circuit complexity.

3   Circuit Architectures of the Tang-Maitra Class of Functions

In order to study the hardware complexity of the Tang-Maitra class of functions, we need to divide it into its smaller components, which are:

1. Three equality comparators ($x = 0$, $x = \mu$, $y = 0$).
2. Finite field inverter y^{-1}.
3. Constant multiplier λx.
4. Finite field multiplier $(\lambda x) \cdot y^{-1}$.
5. The trace function Tr_1^k.
6. The circuits for h_0 and h_1.
7. A 4×1 multiplexer.
8. A 3×2 encoder to convert the output of the three comparator into 2 selection bits.

From the previous decomposition, we can already observe some of the properties of the hardware combinational circuit and also some simple optimizations. First, the cost of the multiplexer and encoder is small, constant and does not depend on k. Precisely, for any k, 6 AND gates and 3 OR gates are needed for the 4×1 multiplexer and the required encoder can be implemented using 3 AND gates and 2 OR gates. Second, an obvious optimization is to choose $\lambda \equiv e$ where e is the multiplicative identity element of $\mathbb{F}_{2^k}$. Hence, regardless of the finite field representation use, the cost for the constant multiplication λx is zero. Third, while the comparator cost is linear in k, precisely k XNOR gates and $k - 1$ AND gates, we can choose μ in a way that reduces the overall cost of the three comparators. We consider, μ such that consider $wt(\mu \oplus 0) = 1$. The two k-bit comparators $x = 0$ and $x = \mu$ can be implemented using only one $(k - 1)$-bit comparators, 2 single bit comparators and 2 extra AND gates. Overall, the three comparators will require $2k + 1$ XNOR gates and $2k - 1$ AND gates.

In the rest of this section, we discuss the overall circuit using different finite field representations, providing estimations for the cost using each of them. Since h_0 and h_1 are constructed as Boolean functions and not as finite field functions, we consider their cost to be roughly the same for all representations. In the next section, we discuss their cost more precisely. For all figures, thick arrows represent k-bit buses, blue blocks have $poly(k)$ cost, green blocks have constant cost and red boxes represent blocks of unknown cost.

3.1   Circuit Based on Polynomial Basis

Lemma 1. *The polynomial basis implementation of the Tang-Maitra function of n variables, where n is even, has a circuit complexity bounded by $\mathcal{O}(2^k + k^2)$ and depth of $\mathcal{O}(k)$, where $n = 2k$.*

Figure 1 shows the polynomial basis circuit. The first branch for the multiplexer is the default option, which computes $\mathrm{Tr}_1^k(\frac{\lambda x}{y})$. Addition over $\mathbb{F}_{2^k}$ using polynomial basis is carry-free. Besides, 0 is represented as the binary string 0^k and 1 is represented the binary string $0^{k-1}1$. Hence, the addition part of Tr_1^k can be performed by adding only the least significant bits, requires $k - 1$ XOR gates, instead of $k(k-1)$. Choosing $\lambda \equiv e$, the cost of this branch is the cost of 1 inversion, 1 multiplication, $k - 1$ squaring and $k - 1$ XOR gates. Multiplication costs $\mathcal{O}(k^{1+\epsilon})$, where $0.5 \leq \epsilon \leq 1$. The cost for every squaring is $\mathcal{O}(k)$, and inversion has cost of $\mathcal{O}(k^2)$. The overall complexity of this branch is $\mathcal{O}(k^2)$ and depth of $\mathcal{O}(k)$. However, as explained in Sect. 2.3, this complexity can be bad in practice due to large coefficients. Besides, the circuit complexity of the h_0 and h_1 is bounded by $\mathcal{O}(2^k)$.

3.2   Circuit Based on Normal Basis

Lemma 2. *The normal basis implementation of the Tang-Maitra function of n variables, where n is even, has a circuit complexity bounded by $\mathcal{O}(2^k + k^3)$, where $n = 2k$.*

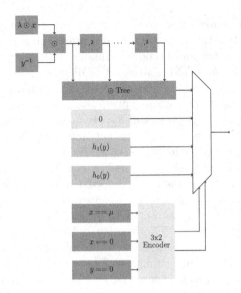

Fig. 1. Polynomial basis circuit for a Tang-Maitra function

Figure 2 shows the normal basis circuit, it is similar to the polynomial basis circuit, from a high-level point of view. However, since the trace function is implemented as the XOR of all input bits, squaring is not required. The complexity of the first branch is $\mathcal{O}(k^3)$, but it should be either smaller than or comparable to the polynomial basis circuit in practice. Besides, for certain choices of k (e.g. $k = 2^r + 1$ or $k = m(2^r + 1)$), the complexity can be in the order of $\mathcal{O}(k^2 \cdot \log(k))$. λ is chosen as e and $\mu = \beta$.

3.3   Comments on Different Representations

The Tang-Maitra functions are defined as finite field functions. However, the cryptographic properties of interest were evaluated for the Boolean circuit generated by implementing the polynomial representation of the underlying field. An important question is: are the cryptographic properties of the Tang-Maitra functions preserved under the change of basis/representation? The answer to that question is that if the change of representation operation is a linear (or affine) operation, they are preserved. Hence, both the polynomial and normal bases representations have the same cryptographic properties. However, since the transformation to/from discrete-log is a non-linear operation, it cannot be used directly without adjusting the input.

Let α be a primitive element of $\mathbb{F}_{2^k}$, and B_p^k and B_n^k be the polynomial and normal bases of $\mathbb{F}_{2^k}$ over $\mathbb{F}_2$, respectively. Then

$$B_p^k = \{1, \alpha, \alpha^2, \ldots, \alpha^{k-1}\}, \quad B_n^k = \{\alpha, \alpha^2, \ldots, \alpha^{2^{k-1}}\}.$$

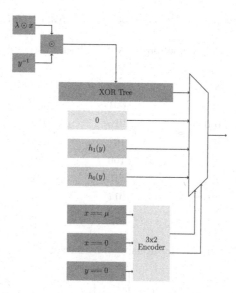

Fig. 2. Normal basis circuit for a Tang-Maitra function

Thus, any element $x \in \mathbb{F}_{2^k}$ can be written as $x = \bigoplus_{j=0}^{k-1} a_j \alpha^j$ and also $x = \bigoplus_{j=0}^{k-1} b_j \alpha^{2^j}$, where $a_j, b_j \in \mathbb{F}_2$, $0 \leq j \leq k - 1$. The binary strings $(a_{k-1}, \ldots, a_1, a_0)$ and $(b_{k-1}, \ldots, b_1, b_0)$ are called the binary representation of x with respect to polynomial and normal bases, respectively. It is clear that two binary representations of $\mathbb{F}_{2^k}$ using two bases, normal and polynomial bases, are related by a linear nonsingular mapping.

For discrete-log representation of $\mathbb{F}_{2^k}$, let $0 = (1, 1, \ldots, 1)$, all one vector of $\mathbb{F}_2^k$, and $\alpha^i = (c_{k-1}, \ldots, c_1, c_0)$, where $i = \sum_{j=0}^{k-1} c_j 2^j$ and $c_j \in \mathbb{F}_2$, $0 \leq j \leq k-1$, for all $0 \leq i \leq 2^k - 2$. In this representation, it is not possible to write all elements of $\mathbb{F}_{2^k}$ in the linear combination of k linearly independent elements and there is no linear (or affine) mapping between discrete-log and normal (or polynomial) basis representations.

We know that cryptographic properties of a Boolean function such as algebraic degree, balancedness, Walsh–Hadamard spectra, autocorrelation spectra, nonlinearity are invariant under the nonsingular affine transformations. But, if the transformations are not affine, then these properties may or may not be same.

For example let $k = 3$ and α be a primitive element of $\mathbb{F}_{2^3}$ such that $\alpha^3 \oplus \alpha^2 \oplus 1 = 0$. Then $B_p^3 = \{1, \alpha, \alpha^2\}$ and $B_n^3 = \{\alpha, \alpha^2, \alpha^4\}$ are the polynomial and normal bases of $\mathbb{F}_2^3$ over $\mathbb{F}_2$, respectively. For discrete-log representation of $\mathbb{F}_{2^3}$, let $0 = (1, 1, 1)$ and $\alpha^i = (c_2, c_1, c_0)$, where $i = \sum_{j=0}^{2} c_j 2^j$ and $c_j \in \mathbb{F}_2$, $0 \leq j \leq 2$, for all $0 \leq i \leq 6$. The binary representations of $\mathbb{F}_{2^3}$ with respect to B_p^3, B_n^3 and discrete-log are given in Table 1.

Table 1. Binary representations of $\mathbb{F}_{2^3}$ with respect different bases

$\mathbb{F}_{2^3}$	$\mathrm{Tr}_1^3(x)$	$\mathrm{Tr}_1^3(x^3)$	Polynomial basis	Normal basis	Discrete-log representation
0	0	0	000	000	111
1	1	1	001	111	000
α	1	0	010	001	001
α^2	1	0	100	010	010
α^3	0	1	101	101	011
α^4	1	0	111	100	100
α^5	0	1	011	110	101
α^6	0	1	110	011	110

One can check that using the nonsingular matrix

$$A = \begin{bmatrix} 1\,1\,1 \\ 1\,0\,0 \\ 0\,1\,0 \end{bmatrix}, \tag{7}$$

the binary representation of $\mathbb{F}_{2^3}$ with respect to normal and polynomial bases are related. There is no such linear (or affine) transformation that maps between discrete-log and normal (or polynomial) basis. This we explain by an example here.

Suppose $f, g \in \mathcal{B}_3$ are defined as $f(x) = \mathrm{Tr}_1^3(x)$ and $g(x) = \mathrm{Tr}_1^3(x^3)$, for all $x \in \mathbb{F}_{2^3}$ (defined as in Table 1). The algebraic degrees of f and g are 1 and 2 respectively. Then the algebraic normal form (ANF) of $f(x_1, x_2, x_3)$ over $\mathbb{F}_2^3$ with respect to normal and polynomial bases is $x_1 \oplus x_2 \oplus x_3$, but with respect to discrete-log representation, it becomes $x_1 x_2 \oplus x_2 x_3 \oplus x_1 x_3 \oplus 1$. Moreover, the ANF of $g(x_1, x_2, x_3)$ over $\mathbb{F}_2^3$ with respect to normal an polynomial bases are $x_1 x_2 \oplus x_2 x_3 \oplus x_1 x_3$ and $x_1 \oplus x_2 x_3$ respectively, but with respect to discrete-log representation, it is $x_1 \oplus x_2 \oplus x_3 \oplus 1$. Thus, while there are certain advantages in the discrete-log representation, unless the proper nonlinear transformation cannot be decided, the implementation is not complete. Still we explain the implementation in discrete-log domain in Appendices A and B.

3.4   Comparison

We assume the complexity of h_0 and h_1 is the same for the three representations. Comparing the asymptotic complexity of $\mathrm{Tr}_1^k(\frac{\lambda x}{y})$, polynomial basis has the smallest area complexity, while the normal basis has the smallest logical Depth. Moreover, in practice, sometimes the normal basis circuit can have a better area compared to the polynomial basis circuit, especially for good choices of k. The discrete-log circuit (Appendix B) has good parameters for small n, in practice, but due to its non-linearity with respect to the polynomial basis, it needs input adjustment, which can be costly.

4   Practical Implementations

In order to verify the theoretical bounds discussed in the paper, we have implemented the polynomial basis construction for the cases of $n \in \{12, 14, 16, 18, 20, 22\}$ for both FPGA and ASIC.

Table 2. Implementation results on the Virtex-7 FPGA

n	LUTs	Critical path (ns)	Logic levels
12	16	2.25	5
12 [PCZ17]	115	3.47	9
14	42	2.98	6
16	89	3.53	8
16 [PCZ17]	828	4.73	11
18	196	5.13	11
20	640	18.86	33
22	813	23.6	40

For FPGA, the design have been synthesized on Virtex-7 using Xilinx ISE 14.7 design flow, given in Table 2. The results show quadratic and linear increase in the number of LUTs and the critical path, respectively. This conforms with the theoretical estimations in the paper. The technology mapping techniques from [KCP17] have been used to reduce the area, specially for h_0 and h_1 functions. In comparison with the implementations of the GMM functions in [PCZ17], the area and performance on FPGA are both better. Moreover, while the complexity h_0 and h_1 is theoretically exponential, the results show that the dominant factor of the circuit cost in practice is the cost of the trace function. The results also show that the circuit is relatively more costly for $n > 18$.

For ASIC, we have used Synopsys Design Compiler for both the Open Source Nangate 45 nm and TSMC 65 nm Standard Cell Libraries, given in Tables 3 and 4. The implementations from [PCZ17] have also been synthesized for the same technology. The ASIC results also follow the theoretical estimations. However, compared to [PCZ17], the cost of the Tang-Maitra functions is higher in terms of performance, due to the depth of linear order. With respect to area, the Tang-Maitra functions show a quadratic growth rate, compared to the sub-exponential rate in [PCZ17]. In addition, while the Tang-Maitra ASIC implementations are slower than the GMM ASIC implementations, the clock frequency for 22 variables is still more than 100 MHz, which is faster than the speed required by many practical applications. The hardware results in this paper combined with the cryptographic properties of the Tang-Maitra functions show that they can be a building block for promising cryptographic primitives. Besides, we have not studied the effect of advanced circuit optimization techniques on the ASIC implementations, which can further improve both the area and performance.

Table 3. Implementation results on the Nangate 45 nm ASIC technology library

n	Area (GE)	Critical path (ns)	Clock frequency (MHz)
12	580	2.9	344
12 [PCZ17]	381	0.95	1052
14	898	3.49	286
16	1328	4.56	219
16 [PCZ17]	1778	1.47	680
18	1657	5.66	176
20	2602	6.73	148
22	3501	7.95	126

Table 4. Implementation results on the TSMC 65 nm ASIC technology library

n	Area (GE)	Critical path (ns)	Clock frequency (MHz)
12	477	3.46	289
12 [PCZ17]	378	1.51	1052
14	802	4.79	208
16	1250	5.62	177
16 [PCZ17]	1510	1.64	680
18	1503	6.82	147
20	2366	8.37	119
22	3273	9.84	102

5   Conclusion

In this paper, we consider how a recently proposed construction of cryptographically significant Boolean functions in [TM18] can be efficiently implemented. Such functions are derived from Dillon type bent functions and can be interpreted as Maiorana-McFarland bent functions as well. Given that the functions have very good nonlinearity and autocorrelation properties, they may be useful as primitives in hardware stream cipher design. In particular, such functions can be exploited for lightweight stream ciphers too. The implementation methods follow different ideas of finite field representation and the actual implementation results are explained. One may note that in truth table domain (considering the Boolean functions as a mapping from $\{0,1\}^n \rightarrow \{0,1\}$), the Maiorana-McFarland type bent functions can be seen as the concatenation of small affine functions. A theoretical view from this direction is being explored in an independent and parallel work [TKMM18] recently.

In a stream cipher, other than the Boolean function the main circuit component is LFSR or NFSR and the state size is decided by that. Thus, with a Boolean function with 500 GEs may be accommodated in a complete stream cipher circuit of 1000 GEs. Further, the Boolean function that we use have very good autocorrelation spectra, which may resist against proper signature generation in Differential Fault Attack. The complete design of stream cipher using such functions and the resistance against different attacks will be presented in the journal version of this paper.

A   Discrete-Log Representation of $\mathbb{F}_{2^n}$ Arithmetic

The Discrete-Log representation is described in Sect. 2.3. Multiplication can be defined as

$$x_1 \odot x_2 = \begin{cases} 2^n - 1 & \text{if } x_1 = 2^n - 1 \text{ or } x_2 = 2^n - 1 \\ x_1 + x_2 \pmod{2^n - 1} & \text{otherwise} \end{cases} \tag{8}$$

While inversion can be defined as

$$x^{-1} = \begin{cases} 2^n - 1 & \text{if } x = 2^n - 1 \\ -x \pmod{2^n - 1} & \text{otherwise} \end{cases} \tag{9}$$

Both operations require circuit complexity of $\mathcal{O}(n)$, which is smaller than the corresponding circuits for both normal and polynomial bases. While the same can be said about squaring, we show now that it can be implemented as a cyclic shift operation (similar to the case of normal basis). Squaring can be written in terms of multiplication as follows, where $\times$ is used for integer multiplications as opposed to finite field multiplication $\odot$,

$$x^2 = \begin{cases} 2^n - 1 & \text{if } x = 2^n - 1 \\ 2 \times x \pmod{2^n - 1} & \text{otherwise} \end{cases} \tag{10}$$

and

$$2 \times x \pmod{2^n - 1} = \begin{cases} x \ll 1 & \text{if } 2 \times x < 2^n - 1 \\ (x \ll 1) - (2^n - 1) & \text{otherwise} \end{cases} \tag{11}$$

Using the two's complement representation of integer arithmetic, Eq. (11) can be written as

$$2 \times x \pmod{2^n - 1} = \begin{cases} x \ll 1 & \text{if } 2 \times x < 2^n - 1 \\ (x \ll 1) + 2^n + 1 \pmod{2^n} & \text{otherwise} \end{cases} \tag{12}$$

Equation (12) means that the squaring operation in the discrete-log representation is a left shift operation with the most significant bit of x becoming the least significant bit, i.e., a cyclic shift of x.

In addition, however, in the discrete-log representation is complicated. It can be implemented by using look-up tables or by conversion to another representation. Hence, studying the complexity of trace function is this representation without using addition is an interesting problem. Using property 2 of trace function in Sect. 2.1 and Eq. (12), we can conclude, as in the case of normal basis, that trace function is a Rotation Symmetric Boolean Function (RSBF). Now we define the rotation symmetric Boolean functions. Let $x_i \in \mathbb{F}_2$ for $0 \leq i \leq n - 1$. We define

$$\rho_n^r(x_i) = x_{(i+r) \bmod n} = \begin{cases} x_{i+r}, & \text{if } i+r \leq n-1; \\ x_{i+r-n}, & \text{if } i+r \geq n. \end{cases}$$

Let $P_n = \{\rho_n^0, \rho_n^1, \ldots, \rho_n^{n-1}\}$ be the permutation group which contains the rotations of n symbols, defined as

$$\rho_n^i(x) = \rho_n^i(x_{n-1}, x_{n-2}, \ldots, x_0) = (x_{(n-1+i) \bmod n}, x_{(n-2+i) \bmod n}, \ldots, x_{(i) \bmod n}).$$

Definition 1. *A Boolean function f in n variables is said to be rotation symmetric if and only if for any $x \in \mathbb{F}_2^n$, $f(\rho_n^i(x)) = f(x)$, for all $0 \leq i \leq n - 1$.*

The problem of defining an RSBF is related to the problem of necklace equivalence in combinatorics. This helps to derive an upper bound on the circuit complexity of a trace function in the discrete-log representation.

Definition 2. *A binary necklace of length n is an equivalence class of n-character strings over the alphabet $\{0, 1\}$, where two arrangements are equivalent if one can be obtained from the other by applying cyclic rotations.*

Definition 3. *The lexicographical representation of a binary necklace N is the member of $[N]$ with the maximum number of leading 0's.*

B   Circuit for the Tang-Maitra Functions Based on Discrete-Log Representation

The circuit in Fig. 3 can be used to compute the Tang-Maitra function when the inputs are in the discrete-log representation. The operation $\frac{x}{y}$ is computed as $x - y \pmod{2^k - 1}$, with complexity $\mathcal{O}(k)$. After that, Tr_1^k is computed as an RSBF. In this Section, we give a circuit for any RSBF, with sub-exponential complexity $\mathcal{O}(k^2 + 2^k/k^2)$.

Rotation Symmetric Boolean Function Circuits. Let f be a rotation symmetric Boolean function in k variables, i.e., $f(\rho_k^i(x)) = f(x)$, for all $0 \leq i \leq k - 1$. Hence, $[x]$ is an equivalence class (orbit) that includes all the rotations of x, i.e., $[x] = \{\rho_k^i(x) | 0 \leq i \leq k - 1\}$. We choose the representative of that class to be $\rho_k^r(x)$, such that $\rho_k^r(x) \geq \rho_k^i(x)$, for all $0 \leq i \leq k - 1$. In other words, it is the rotation of x that has the maximum integer value. For more details of rotation symmetric Boolean function we refer to [FF98, Fon99]. This is the lexicographical representation of $[x]$ based on the alphabet $\{0, 1\}$.

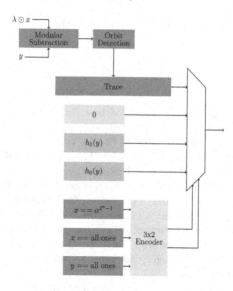

Fig. 3. Discrete-log circuit for a Tang-Maitra function

Lemma 3. *A rotation symmetric Boolean function (RSBF) of k variables has a circuit complexity bounded by $\mathcal{O}(k^2 + 2^k/k^2)$.*

Lemma 4. *The discrete-log implementation of the Tang-Maitra function of n variables, where n is even, has a circuit complexity bounded by $\mathcal{O}(2^k + k^2 + 2^k/k^2)$, where $n = 2k$.*

Proof. In order to convert any x to its lexicographical orbit representation, the orbit detection circuit generates all the k rotations of x, then chooses the value of x that has the maximum integer value using a selection tree that consists of $k-1$ two-input MAX circuits. Every two-input MAX circuit consists of $k+1$ integer subtractor ($6k+6$ gates) and k 2×1 MUXes, $3K$ gates. Hence, the orbit detection circuit has a complexity of around $9k^2 - 3k - 6$ gates. After the lexicographical orbit representation has been detected, a circuit decides whether the given orbit functional value is 0 or 1. This circuit expects only 1 of the lexicographical representations, which, according to Burnside's Lemma and [SM08, Theorem 3], are $N_O = \frac{1}{k} \sum_{d|k} \phi(d) 2^{\frac{k}{d}}$, where ϕ is Euler's phi-function. Hence, $n_x = 2^k - N_O$ values in the Truth table of such circuit can be set as DON'T CARES 'X'. In [Spi80], the author gave an analysis of the circuit complexity of combinational circuits with a large number of DON'T CARES. The number of AND/OR/NOT gates was given by

$$L_\infty = (1 - d)H(p)L_\infty(G),$$

where $d = \frac{n_x}{2^k}, p = \frac{n_1}{(1-d)2^k}, H(p) = -p\log(p) - (1-p)\log(1-p)$ and $L_\infty(G) = \frac{2^k}{k}$. By substitution for the case of the trace circuit, the number of gates is $\frac{N_O}{n} H(p)$, where $H(p) \leq 1$. Hence, the circuit complexity is $\mathcal{O}(\frac{N_O}{n})$, and from Burnside's

Lemma, it can be expressed as $\mathcal{O}(\frac{2^k}{k})$. Hence, the overall complexity of this construction is $\mathcal{O}(k^2 + 2^k/k^2)$. ☐

References

[Car93] Carlet, C.: Two new classes of bent functions. In: Helleseth, T. (ed.) EUROCRYPT 1993. LNCS, vol. 765, pp. 77–101. Springer, Heidelberg (1994). https://doi.org/10.1007/3-540-48285-7_8

[CS09] Thomas, W., Cusick, W., Stănică, P.: Cryptographic Boolean Functions and Applications. Academic Press, Cambridge (2009)

[Dil74] Dillon, J.F.: Elementary Hadamard difference sets. Ph.D. thesis (1974)

[DIS09] Deschamps, J.-P., Imana, J.L., Sutter, G.D.: Hardware Implementation of Finite-Field Arithmetic. McGraw-Hill, New York (2009)

[Dob94] Dobbertin, H.: Construction of bent functions and balanced Boolean functions with high nonlinearity. In: Preneel, B. (ed.) FSE 1994. LNCS, vol. 1008, pp. 61–74. Springer, Heidelberg (1995). https://doi.org/10.1007/3-540-60590-8_5

[FF98] Filiol, E., Fontaine, C.: Highly nonlinear balanced Boolean functions with a good correlation-immunity. In: Nyberg, K. (ed.) EUROCRYPT 1998. LNCS, vol. 1403, pp. 475–488. Springer, Heidelberg (1998). https://doi.org/10.1007/BFb0054147

[Fon99] Fontaine, C.: On some cosets of the first-order Reed-Muller code with high minimum weight. IEEE Trans. Inf. Theory 45(4), 1237–1243 (1999)

[IT88] Itoh, T., Tsujii, S.: A fast algorithm for computing multiplicative inverses in GF (2m) using normal bases. Inf. comput. 78(3), 171–177 (1988)

[KCP17] Khairallah, M., Chattopadhyay, A., Peyrin, T.: Looting the LUTs: FPGA optimization of AES and AES-like ciphers for authenticated encryption. In: Patra, A., Smart, N.P. (eds.) INDOCRYPT 2017. LNCS, vol. 10698, pp. 282–301. Springer, Cham (2017). https://doi.org/10.1007/978-3-319-71667-1_15

[KMT18] Kavut, S., Maitra, S., Tang, D.: Searching balanced Boolean functions on even number of variables with excellent autocorrelation profile. In: Tenth International Workshop on Coding and Cryptography, Saint-Petersburg, Russia, 18–22 September 2017

[LN94] Lidl, R., Niederreiter, H.: Introduction to Finite Fields and Their Applications. Cambridge University Press, Cambridge (1994)

[McF73] McFarland, R.L.: A family of difference sets in non-cyclic groups. J. Comb. Theory Ser. A 15(1), 1–10 (1973)

[MM16] Mesnager, S.: Bent Functions. Springer, Cham (2016). https://doi.org/10.1007/978-3-319-32595-8

[PCZ17] Pasalic, E., Chattopadhyay, A., Zhang, W.: Efficient implementation of generalized Maiorana-McFarland class of cryptographic functions. J. Cryptogr. Eng. 7(4), 287–295 (2017)

[Rot76] Rothaus, O.S.: On "bent" functions. J. Comb. Theory Ser. A 20(3), 300–305 (1976)

[SM08] Stănică, P., Maitra, S.: Rotation symmetric Boolean functions-count and cryptographic properties. Discrete Appl. Math. 156(10), 1567–1580 (2008)

[Spi80] Spillman, R.J.: The effect of DON'T CARES on the complexity of combinational circuits. Proc. IEEE 68(8), 1021–1022 (1980)

[TM18] Tang, D., Maitra, S.: Construction of n-variable ($n \equiv 2 \bmod 4$) balanced Boolean functions with maximum absolute value in autocorrelation spectra $< 2^{n/2}$. IEEE Trans. Inf. Theory **64**(1), 393–402 (2018)

[TKMM18] Tang, D., Kavut, S., Mandal, B., Maitra, S.: Modifying Maiorana-McFarland type bent functions for good cryptographic properties, April 2018 (preprint)

Rapid Hardware Design
for Cryptographic Modules with Filtering
Structures over Small Finite Fields

Nusa Zidaric$^{(\boxtimes)}$ (iD), Mark Aagaard (iD), and Guang Gong (iD)

University of Waterloo, Waterloo, ON N2L 3G1, Canada
{nzidaric,maagaard,ggong}@uwaterloo.ca

Abstract. This paper presents a design automation toolkit for hardware implementations of linear and non-linear feedback shift registers (FSRs). The toolkit is implemented in the GAP computer algebra system and generates both executable GAP code and VHDL for synthesizable hardware. To design an FSR, the user needs only to provide a template and instantiate a few parameters. The primary objects are LFSRs; NLFSRs; and arbitrary combinational functions, which are modelled as FILFUNs, for "filtering functions". Conventional feedback functions are modelled as univariate or multivariate polynomials. More complex functions can be modelled as FILFUNs. The paper demonstrates the capabilities of the toolkit using the WG-7 and WG-8 keystream generators and the Grain v1 stream cipher. Less than 30 lines of GAP code are required to generate a complete datapath in VHDL.

Keywords: Feedback shift registers · Filtering generators
Rapid hardware design · Stream ciphers · GAP · VHDL

1 Introduction

Feedback shift registers (FSRs) play an important role in stream cipher design. A milestone in stream cipher design is the eSTREAM project [1], launched in 2004. All 3 hardware portfolio ciphers, Grain, MICKEY and Trivium, as well as the software portfolio cipher Sosemanuk, use FSRs. The stream cipher ACORN [2], a round 3 CAESAR candidate [3], is based on 6 LFSRs. Last but not least, the two stream ciphers used for encryption and integrity of communications in mobile networks, Snow3G and ZUC [4,5], both use LFSRs. Another application area for LFSRs are the cyclic redundancy codes (CRC) used in many communication and data storage devices for error-detection. Less noticeable is the use of LFSRs in algorithms for finite field arithmetic: e.g. a serial circuit implementing multiplication by x followed by reduction modulo the field defining polynomial $f(x)$ can be implemented as an LFSR with the feedback $f(x)$.

The authors would like to thank Dr. Alexander Konovalov from University of St. Andrews for his advice during the FSR package implementation.

L. Budaghyan and F. Rodríguez-Henríquez (Eds.): WAIFI 2018, LNCS 11321, pp. 128–145, 2018.
https://doi.org/10.1007/978-3-030-05153-2_7

This work presents a toolkit for rapid hardware design of modules composed of feedback shift registers and filtering functions. The toolkit consists of two packages written in GAP [6]. The first package is called FSR, and allows the creation, initialization and running of both LFSRs and NLFSRs. In addition, it implements filtering functions FILFUNs (multivariate polynomials). The generality of the FSR package enables the FSRs to be used directly as building blocks for many cryptographic modules. The second package is called FSRtoVHDL, and as its name suggests, it creates hardware modules using the VHDL language. By the design of the two packages, each FSR object corresponds to a hardware entity, and the FSR objects themselves contain all the information needed for both their execution in GAP and their generation in hardware. Hardware is generated from a template, and for a simple cipher, such as WG-7, the entire VHDL datapath was created from just 30 lines of GAP code. The FSR package can be used to implement arbitrary primitives in GAP, which can operate e.g. as random number generators.

Two critical points in the design of this toolkit were modular thinking, inherent to hardware designs, and recognition and exploitation of structural similarities between LFSRs, NLFSRs and filters, from both mathematical and hardware perspectives. A modular approach to design and implementation can open new perspectives and improve the initial design. Cryptographic primitives are always carefully selected to meet certain security requirements and an appropriate trade-off between security and hardware efficiency is desirable. It is thus imperative to be able to estimate the hardware cost of the design early on, and having the ability to quickly generate and synthesize hardware modules is very beneficial. Theoretical estimates of hardware cost, such as Hamming weights, often do not reflect the actual area and delay in hardware. Modern synthesis tools are powerful and quite successful at optimizing combinational logic, especially for small designs. Furthermore, they are aware of hardware resources available for the chosen target technology and can be instructed to optimize for a specific performance goal, e.g. high speed or small area. These optimizations can cause discrepancies between theoretical estimates and actual hardware, and the results can be quite surprising.

The toolkit is able to define (N)LFSRs and filters described by multivariate polynomials. It can execute these (N)LFSRs in GAP, generate traces, and generate VHDL code. Furthermore, it can define, execute, and generate VHDL datapaths for many hierarchical modules constructed from a set of (N)LFSRs and FILFUN filters.

2   Background and Related Work

2.1   Basic Terminology

To keep this section short, many details are omitted and the reader can refer to a number of sources such as [7–9].

Multivariate Polynomials. Let $\mathcal{F} \equiv \mathbb{F}_q$ where q is a prime or a prime power. A multivariate function in t variables $x_0, x_1, \ldots, x_{t-1}$ is defined as follows:

$$f : \mathcal{F}^t \to \mathcal{F}$$
$$f(x_0, x_1, \ldots, x_{t-1}) = \sum_{\forall (i_0, i_1, \ldots, i_{t-1}) \in Z_q^t} c_{i_0, i_1, \ldots, i_{t-1}} x_0^{i_0} x_1^{i_1} \ldots x_{t-1}^{i_{t-1}} \qquad (1)$$

with coefficients $c_{i_0, i_1, \ldots, i_{t-1}} \in \mathcal{F} \equiv \mathbb{F}_q$ and where $i_j \in Z_q$ for $0 \le j < t$. The sum in Eq. (1) runs over all possible monomials $x_0^{i_0} x_1^{i_1} \ldots x_{t-1}^{i_{t-1}}$, where $\forall x \in \mathcal{F}$: $x^q = x$. The expression on the r.h.s. of Eq. (1) describes an univariate polynomial when $t = 1$, and a multivariate polynomial when $t > 1$.

The degree of a monomial is defined as the sum of all its exponents (Eq. (2)) and the degree of the polynomial as the maximum degree of all its monomials (Eq. (3)). For readability, the notation $m_{i_0, i_1, \ldots, i_{t-1}}$ is introduced for monomials:

$$m_{i_0, i_1, \ldots, i_{t-1}} = m(x_0, x_1, \ldots, x_{t-1}) = x_0^{i_0} x_1^{i_1} \ldots x_{t-1}^{i_{t-1}}$$
$$\deg(m(x_0, x_1, \ldots, x_{t-1})) = \sum_{j=0}^{t-1} i_j \qquad (2)$$
$$\deg(f(x_0, x_1, \ldots, x_{t-1})) = \max_{\forall (i_0, i_1, \ldots, i_{t-1}) \in Z_q^t} \left\{ \deg(m_{i_0, i_1, \ldots, i_{t-1}}) \right\} \qquad (3)$$

Based on the degree of the polynomial function, given by Eq. (3), a multivariate polynomial is classified as constant for $\deg(f) = 0$, linear for $\deg(f) = 1$, and nonlinear function for $\deg(f) > 1$. Futhermore, when $q = 2$, $f(x_0, x_1, \ldots, x_{t-1})$ is a Boolean function in t variables.

Feedback Shift Registers (FSR). An n-stage shift register over a finite field $\mathcal{F}$ is an array of n registers (stages), denoted S_i, $i = n-1, \ldots, 0$. Each stage holds a value from the underlying finite field $\mathcal{F}$. The parameter n is also referred to as the *length of the FSR*. This memory array is shifted with each step $S_i \to S_{i-1}$ for $i = n-1, \ldots, 1$, and the vacant register S_{n-1} is updated with a new value obtained from the feedback function $f(x_0, \cdots, x_{n-1})$, a multivariate polynomial function in n variables, hence the name *feedback shift register* (FSR). One of the stages is used to generate the output and each time the FSR is shifted, that stage produces a new element $s_i \in \mathcal{F}$. In this way, the FSR produces a sequence of elements:

$$\underline{s} = \{s_k\} = s_0, s_1, s_2, \ldots \qquad (4)$$

where s_k's satisfy the following recursive relation

$$s_{k+n} = f(s_k, s_{k+1}, \cdots, s_{k+n-1}), k = 0, 1, \cdots$$

A simple schematic of an n-stage FSR is shown in Fig. 1(a), with the output sequence produced by stage S_0.

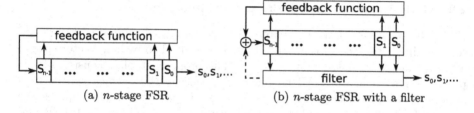

(a) n-stage FSR   (b) n-stage FSR with a filter

Fig. 1. Top level schematic of an n-stage FSR

The feedback f of the FSR is a function in $t = n$ variables $x_0, x_1, \ldots, x_{n-1}$ as defined in Eq. (1), whereby n is a positive integer and the variable x_i corresponds to the stage S_i, $i \in Z_n$, the residue ring modulo n. A linear feedback yields a linear and a nonlinear feedback a nonlinear feedback shift register, which is an LFSR and an NLFSR, respectively.

In case of the LFSR, the feedback is given by $f(x_0, \cdots, x_{n-1}) = \sum_{i=0}^{n-1} c_i x_i$ which can be represented with an univariate polynomial

$$h(y) = y^n + \sum_{j=0}^{n-1} c_j y^j \tag{5}$$

where y^j corresponds to the stage S_j for $j \in Z_n$, and y^n to the new value computed by the feedback. Coefficients c_j, $j \in Z_n$, of the polynomial in Eq. (5) belong to the underlying field $\mathcal{F}$.

At any given moment, the stages of the FSR hold n values from the underlying finite field, and can be written as a vector of length n: $(s_0, s_1, \ldots, s_{n-1}) \in \mathcal{F}^n$. This vector is called the *state* of the FSR and the state right after loading is called the *initial state*. The output sequence $\underline{s}$ is completely determined by the feedback f and the initial state.

Filtering Generators. A typical structure of a filtering generator is shown in Fig. 1(b): it consists of a filter, i.e. a nonlinear multivariate polynomial function, applied to an LFSR with n stages. Let $(s_k, s_{k+1}, \ldots, s_{k+n-1}) \in \mathcal{F}^n$ be the kth state of the LFSR, $g(x_0, \cdots, x_{t-1})$, a multivariate polynomial in t variables, where $t \leq n$, and $(d_0, \cdots, d_{t-1})$, a selection of t tap positions in the state, i.e., $0 \leq d_0 < d_1 < \cdots < d_{t-1} < n$. The output sequence $\underline{a} = \{a_k\}$ is given by

$$a_k = g(s_{k+d_0}, \cdots, s_{k+d_{t-1}}), \quad k = 0, 1, \cdots.$$

This is referred to as a filtering generator where g is a called a filtering function, or simply filter, and $\underline{a} = \{a_k\}$, a filtering sequence.

Example 1. For $q = 2$, $n = 5$, $t = 3$, let $\{s_k\}$ be a binary sequence generated by the LFSR with the feedback $y^5 + y^3 + 1$, a selection of tap positions $(d_0, d_1, d_2) = (0, 2, 3)$, and filtering function $g(x_0, x_1, x_2) = x_0 + x_1 x_2$. Then the output sequence, i.e., a filtering sequence, is given by

$$a_k = s_k + s_{k+2} s_{k+3}, \quad k = 0, 1, \cdots.$$

2.2   GAP, Hardware Synthesis and Related Work

GAP (Groups, Algorithms, Programming) is a specialized computer algebra system, originally intended for group theory, but evolved to include vector spaces, algebras, matrices, polynomials, etc. [6]. The proposed toolkit consists of two packages written in the GAP language: the package FSR, which can be used as a stand-alone package, and the FSRtoVHDL package, which requires the package FSR. GAP is included in SageMath [10], which allows both FSR and FSRtoVHDL to be loaded and used in SageMath as well.

A SageMath package *Cryptography* [11], implements *LFSRCryptosystem* over the finite field $\mathbb{F}_2$, but it does not support extension fields. A simple Mathematica package, called Symbolic Linear Feedback Shift Registers [12] can generate bit sequences from LFSRs. The FSR package presented in this work is capable of working with LFSRs, NLFSRs and filtering functions, defined over both prime and extension fields. To the best of authors knowledge, this toolkit is unique in its ability to generate VHDL from a mathematical description containing finite field arithmetic. For integer, fixed point, and floating point arithmetic, MathWorks [13] offers an extensive range of embedded software and hardware support, from ARM microprocessors, Altera and Xilinx FPGAs to PARROT Minidrones. Similarly, tools like MATLAB Coder, Simulink Coder, Embedded Coder generate C and C++ code and HDL Coder generates synthesizable Verilog and VHDL code.

The FSRtoVHDL package is situated on top of Register-Transfer-Level, but lower than High-Level Synthesis, which transform a behavioural description into RTL designs [14]. The presented toolkit generates synthesizable VHDL modules suitable for implementation on FPGAs or ASICs. The ASIC implementation results for the examples shown in this work were obtained for a 65 nm CMOS ASIC technology using Synopsis Design Compiler for synthesis. The FPGA results are obtained for Xilinx Spartan 3 devices using Xilinx-ISE. The FPGA and ASIC implementation results were obtained post place-and-route.

3   Toolkit for Generating Hardware Modules

This section begins with an overview of the toolkit and hardware-design workflow (Sect. 3.1), then describes the FSR package (Sect. 3.2), the top-level modules (Sect. 3.3), and finally the FSRtoVHDL package (Sect. 3.4).

3.1   Overview and Workflow

The toolkit consists of two GAP packages: FSR and FSRtoVHDL. The FSR package is used to create various FSRs and filters, which are then translated to VHDL modules using the FSRtoVHDL package, i.e. the FSRtoVHDL package takes care of the design entry. The produced VHDL code is used as input to synthesis tools to evaluate the design for different metrics. This design flow is captured in Fig. 2(a). The dashed horizontal line marks the end of design entry.

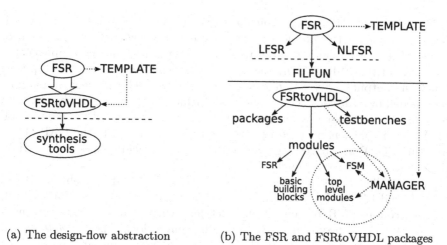

(a) The design-flow abstraction   (b) The FSR and FSRtoVHDL packages

Fig. 2. Toolkit overview

The toolkit significantly reduces the amount of human effort for both software implementation of a cipher using GAP and the design entry in VHDL. The design of the toolkit was guided by the following principles:

1. The toolkit lies in the intersection of finite field arithmetic and hardware design (with VHDL as the preferred choice for design entry).
2. Hardware-style thinking involves a modular approach: a cipher can be implemented as a collection of basic modules, which were identified as LFSRs, NLFSRs and FILFUN filters.
   (a) For most (N)LFSRs used in practice, the feedback function can be modelled by a polynomial (Example 3, Case studies 1 and 2).
   (b) Arbitrarily complicated NLFSRs can be implemented by connecting an (N)LFSR and FILFUN(s) (Example 4).
   (c) Complex FILFUNs can be implemented by connecting multiple simple FILFUNs (Case studies 1 and 2).
3. Recognizing and exploiting structural similarities between LFSRs, NLFSRs and FILFUNs, from both mathematical and hardware perspective, reduces the number of implemented objects, functions and methods.
4. A cipher as a collection of basic modules must be represented in a well structured manner with sufficient information for generation of the top-level VHDL module (Case studies 1 and 2).
5. Highly structural FSR package design is mandatory for VHDL generation.
6. For small designs (finite fields up to $\mathbb{F}_{2^{16}}$), optimizations can be handled by hardware synthesis tools without resorting to specialized rewriting or algebraic techniques.

The structural similarities from mathematical point of view (item 3) are explained in Table 1 in Sect. 3.2. Items 4 and 5 are related to the transition from

GAP objects to VHDL code, which was one of the biggest challenges in developing the toolkit. The problem in item 4 was tackled with a *template* used to connect the FSR modules in hardware and additional GAP objects, implemented as a part of the FSRtoVHDL package, e.g. an *inout* field to capture top-level input and output ports. The *template* must specify indeterminates, all parameters needed by FSR constructors and must include the actual FSR objects. Item 5 is partially addressed by the good design of FSR objects: all the information for VHDL implementation of a particular FSR is included in the GAP object in form of GAP attributes, e.g. the underlying finite field defines which VHDL data type to use for the signals.

The detailed structure of the toolkit is shown in Fig. 2(b). The objects LFSR, NLFSR and FILFUN are implemented in the FSR package. The FSRtoVHDL package contains VHDL generation functions, which generate hardware modules complying with typical hardware coding conventions. The *template* is used to provide information to the toolkit's function named *manager*. The *manager* generates a simple spreadsheet which is used by the designer to specify connections between source and target ports within the top-level module. Based on this information, the *manager* invokes a sequence of appropriate VHDL generation functions.

3.2   The GAP Package FSR: Feedback Shift Registers

The GAP package FSR implements feedback shift registers (Sect. 2.1). It allows creation, initialization and running of both LFSRs and NLFSRs, and can compute some of their properties, e.g. the internal state size of any FSR or the period of an LFSR. The FSR package also implements filtering functions, named FILFUNs. A FILFUN is an object of type FSR without feedback, shifting, or storing, whose functionality is defined by a multivariate polynomial. The justification for such a design decision is twofold: (i) filtering functions are similar to NLFSR feedback functions and (ii) FSRs with output filters are common, indicating that they will be used together. The structural similarities between the three FSR objects (item 3 in Sect. 3.1) are captured in Table 1. The differences between the FSRs and FILFUNs will be discussed shortly.

The two basic components of all three objects are *state*, holding the current state $(S_{n-1}, \ldots, S_0) \in \mathcal{F}^n$, and *basis* used for representation of the field elements. The constructors for the FSR objects are listed in Table 2, and their main functionality in Table 3. Table 3 differentiates between a regular and an external step and run. A stand-alone simple (N)LFSR object is self-contained: it is updated by the computed feedback value (regular step and run). Examples 2 and 3 show the regular run. The external *StepFSR* allows arbitrary filters to be added to the feedback of any (N)LFSR or e.g. to mask the output of the filtering function. The external step and run allow the FSRs to be used directly as building blocks of many ciphers.

While the LFSR and NLFSR differ only in feedback, the filters are a bit of an exception, which is indicated by the dashed line in Fig. 2(b). A filter alone does not require any feedback, shifting or stages, i.e. hardware registers: the component *state* is used to hold the current values needed to evaluate the filtering function. The component *state* is not updated, but rather loaded anew with each step: the method *RunFSR* takes a list of "initial" states as input as shown in Example 3, then calls *LoadStepFSR* for each list entry.

The FSR package also includes output formatting functions for testbench generation and drawing functions that can automatically generate tikz code. More detail can be found at https://github.com/nzidaric/fsr and its manual.

Example 2. The following example shows a regular run of an LFSR over $\mathbb{F}_{2^2}$, using the function call *RunFSR* with initial state ist and number of steps performed given as an argument. Stage S_0 is used to output the sequence elements.

```
―――――――――――――――― Example 2 ――――――――――
gap> K := GF(2);; x := X(K,"x");; f := x^2+x+Z(2)^0;;
gap> F := FieldExtension(K, f);; y := X(F, "y");;
gap> l := y^3+y^2+y+Z(2^2);;
gap> lfsr := LFSR(F, l);
< empty LFSR over GF(2^2)  given by FeedbackPoly = y^3+y^2+y+Z(2^2) >
gap> ist := [Z(2^2), Z(2)^0, 0*Z(2)];;
gap> RunFSR(lfsr, ist, 5);
[ 0*Z(2), Z(2)^0, Z(2^2), Z(2^2)^2, Z(2^2)^2, Z(2^2)^2 ]
```

Example 3. The following example shows a regular run of an FILFUN over $\mathbb{F}_2$, using the function call *RunFSR* with a sequence of inputs inputsequence.

```
―――――――――――――――― Example 3 ――――――――――
gap> f := x_0*x_1+x_2;; fil := FILFUN(K,f);
< FILFUN of length 3 over GF(2),
with the MultivarPoly = x_0*x_1+x_2>
gap> inputsequence := [[Z(2)^0, Z(2)^0, 0*Z(2)],[Z(2)^0, Z(2)^0,
Z(2)^0],[0*Z(2), Z(2)^0, 0*Z(2)]];;
gap> RunFSR(fil, inputsequence);
[ Z(2)^0, 0*Z(2), 0*Z(2) ]
```

3.3   The Top-Level Modules in GAP and VHDL

The toolkit supports numerous configurations of top-level modules, ranging from simple sequence generators using a single (N)LFSR to complex designs with multiple (N)LFSRs and filtering functions. The GAP package FSR was designed in such a way that each FSR object corresponds to a VHDL hardware module. The toolkit directly supports any FSR that can be modelled as shown in Table 2. A more detailed classification can be seen in Table 4. FSRs with less conventional feedback functions can be modelled using the FSR package by splitting the feedback function into a filter and a feedback which allows direct use of (N)LFSR objects and then connecting the filter using the external connection to the FSR.

Table 1. Structural similarities between LFSR, NLFSR and FILFUN objects

FSR object name	Multivariate polynomial $f(x)$		FSR (feedback, memory)	Output	
	Linear	Nonlinear		One or more state elm.	Computed value of $f(x)$
LFSR	✓		✓	✓	
NLFSR		✓	✓	✓	
FILFUN	✓	✓			✓

Table 2. Constructors for FSR objects

FSR constructor			FSRtoVHDL comments
Name	Mandatory	Optional	
LFSR	$\mathbb{F}_q$, $h(y)$ from Eq. (5)	basis, $(d_0, \ldots d_{t-1})$	$q = 2$ or $q = 2^m$
NLFSR	$\mathbb{F}_q$, $f(x_0, \ldots, x_{n-1})$ from Eq. (1), n	basis, $(d_0, \ldots d_{t-1})$	$q = 2$ or $q = 2^m$
FILFUN	$\mathbb{F}_q$, $f(x_0, \ldots, x_{t-1})$ from Eq. (1)	basis	

$h(y) : F_q \to \mathbb{F}_q$ $(d_0, \ldots d_{t-1})$ - output taps
$f(x_0, \ldots, x_{j-1}) : \mathbb{F}_q^j \to \mathbb{F}_q$, where $j = t, n$ n - length (number of stages)

Table 3. Main menthods for FSR objects

All FSR types: LFSR, NLFSR, FILFUN		
Method name	Options	Comments
LoadFSR	NA	Load the initial state
StepFSR	Regular step	FSR self-contained: compute value x
	External step	Adds an external elem. to the computed value: $x + \text{ext}$
	Compute the feedback/function value x, use x or $x + \text{ext}$ to:	
	○ Update S_{n-1} after shifting stages $S_{n-1}, \ldots, S_1$ for (N)LFSR	
	○ Output as the new element in case of FILFUN	
LoadStepFSR	Regular step	FSR self-contained: compute value $\hat{x}$
	External step	Adds an external elem. to the computed value: $x + \text{ext}$
	Combines methods *LoadFSR*, *StepFSR* to load the new values	
	For variables before evaluating the function	
	This method is used by *RunFSR* for FILFUN objects	
RunFSR	Regular run	With regular step
	External run	With external step
	Optional *LoadFSR* followed by sequence of *StepFSR* calls	

The recommended strategy for capturing top-level modules is a combination of a top-down and bottom-up approach. Before the implementation, whether in software or in hardware, the mathematical design can be transformed for the easiest representation as a collection of different FSRs. Usually, the implementation using the FSR package is straightforward: the objects from Table 2, for which the conditions $h(y) : \mathbb{F}_q \to \mathbb{F}_q$ hold for the LFSRs or conditions $f(x_0, \ldots, x_{j-1}) : \mathbb{F}_q^j \to \mathbb{F}_q$, where $j = t, n$, for the NLFSRs and FILFUNs, are very common and can be implemented directly. Example 4 shows a top-level design for which this process is more complicated. The FSR packages can also be used to create e.g. a 4-bit shift register (no feedback) in VHDL by creating an LFSR with a feedback x^4. The FILFUN objects can be used for arbitrary Boolean functions. The designs steps for implementation of the top-level module:

1. Represent the cryptographic module as a collection of different FSRs.
2. Identify possible modes of operation. For each mode and all FSRs define:
   - The number of steps performed
   - The FSR input for the *state* and the possible external input *ext*
3. Capture desired behaviour in GAP using methods *LoadFSR*, *StepFSR* and *LoadStepFSR* or using the *manager* function call.

Step 2 is very important: it ensures a successful transition to VHDL and a clean finite state machine (FSM), parametrized with an appropriate number of steps (clock cycles) and issuing correct control signals for the use of the external FSR inputs. Table 5 shows the modes of operation for the WG-7 cipher. All the steps performed in each mode should be exactly the same; discrepancies mean that a mode exists but was not captured and design step 2 must be repeated.

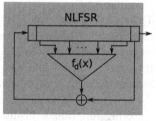

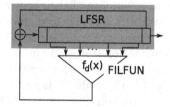

(a) original n-stage NLFSR   (b) n-stage LFSR with a filter

Fig. 3. Schematic of an NLFSR represented as an LFSR with a filter FILFUN

Example 4. Figure 3(a) represents the original schematic of an NLFSR from [15], generating a span-n sequence: the shift register itself is defined over $\mathbb{F}_2$, the t stages provide the coefficients for an element $x \in \mathbb{F}_{2^t}$, which is the input to the function $f_d : \mathbb{F}_{2^t} \to \mathbb{F}_2$. The FSR package (N)LFSR objects can have up to n output taps: in case of $t+1$ output tap positions, the method *StepFSR* returns the contents of $t+1$ stages, specified[1] by $(d_0, \ldots, d_t)$. Thus, the NLFSR from Fig. 3(a) can be implemented as an LFSR (shaded grey in Fig. 3(b)) with a FILFUN implementing function f_d. The LFSR is defined over $\mathbb{F}_2$ with feedback $x^n + 1$ and $t+1$ tap positions. It uses an external signal from the FILFUN for external step and run.

3.4   The FSRtoVHDL Package

The FSRtoVHDL package takes an FSR object, or a collection of FSRs objects, and a *template* specifying their relationship, as an input and generates the VHDL code for its hardware implementation. The capabilities of the FSRtoVHDL package are captured in Fig. 2(b): the package can write VHDL packages, basic building blocks, FSR modules, datapaths composed of FSR modules and can generate testvectors for these hardware modules.

Two VHDL packages are used to define the finite fields used in the design (`field_pkg.vhd`) and signals used by the FSRs (`fsr_pkg.vhd`): they contain all of the fields and FSRs used in the design. The finite fields and the FSRs are enumerated, which increases the readability of the generated code. The parameters defined in the `field_pkg.vhd` are used for basic building blocks, such as multipliers. The basic building blocks are implemented using naive methods[2] and for polynomial bases only. The VHDL modules that only need to be parametrized by the degree of $\mathcal{F}/\mathcal{K}$ are predefined, however, the blocks depending the field defining polynomial f are generated on the fly once f is known. Examples of the latter are reduction matrices and matrix-vector multipliers for multiplication by constants. FSRtoVHDL includes methods that generate corresponding matrices prior to generation of their corresponding VHDL modules.

The FSRs are classified into 6 cases based of their type, underlying finite field, number of variables t and kind of nonzero coefficients. Conditions, cases, implementation status and implementation comments are listed in Table 4. Each case has its own rules for generating the FSR architecture. A special case exists for the FILFUN filters: they can have one single input, i.e., use a univariate filtering function. For the small fields it is feasible to attempt a look-up table style implementation, called `const_array`[3]. The filtering function is evaluated for all field elements and stored as a constant VHDL array. The value the variable takes, represented w.r.t. the basis used by the FSR object, behaves as an address to this array.

---

[1] Output taps in Table 2.
[2] Which have a good performance for small fields.
[3] To differentiate it from the FPGA LUTs.

Table 4. FSRtoVHDL classification for FSR modules

Conditions				FSRtoVHDL		
FSR type	# vars t	Underlying finite field	$\forall i : c_i$ belongs to	Case	Implementation status	Comments
LFSR	NA	$\mathbb{F}_2$	$\mathbb{F}_2$	1	Fully	
		$\mathbb{F}_{2^m}$	$\mathbb{F}_2$			
			$\mathbb{F}_{2^m}$	2	fully	MV† for constants
NLFSR or FILFUN	$t > 0$	$\mathbb{F}_2$	$\mathbb{F}_2$	3	Fully	
	$t > 1$	$\mathbb{F}_{2^m}$	$\mathbb{F}_2$	4	Partially	
			$\mathbb{F}_{2^m}$	5	Partially	MV† for constants
	$t = 1$	$\mathbb{F}_{2^m}$	$\mathbb{F}_2$	6	Fully	const_array architecture
			$\mathbb{F}_{2^m}$			

NA - Not applicable   † - Matrix vector multiplier

4   Case Studies and ASIC Implementation Results

4.1   Case Study 1: The Datapath for WG Keystream Generators

WG-7 [16] and WG-8 [17] are members of the Welch-Gong (WG) family of bit-oriented stream ciphers. WG ciphers generate a keystream with proven randomness and cryptographic properties. They are composed of an LFSR over an extension field and a filtering function. The LFSR outputs an m-sequence, which is then filtered with the WG transformation over the same extension field.

Let m be an integer that is not a multiple of 3. The decimated[4] WG transformation from $\mathbb{F}_{2^m}$ to $\mathbb{F}_2$ consists of a WG permutation (Eq. (6)) and WG transformation (Eq. (7)) of $X \in \mathbb{F}_{2^m}$:

$$\mathsf{WGP}\text{-}m(X^d) = q(X^d + 1) + 1 \tag{6}$$

$$\mathsf{WGT}\text{-}m(X^d) = \mathrm{Tr}(\mathsf{WGP}(X^d)) \tag{7}$$

The polynomial $q(x) = x + x^{r_1} + x^{r_2} + x^{r_3} + x^{r_4}$ is a permutation polynomial from $\mathbb{F}_{2^m}$ to $\mathbb{F}_{2^m}$. Further details are omitted for brevity. Example 5 shows the WG-7 GAP code using Eq. (6).

---

[4] Decimation exponent $d > 1$ and $\gcd(d, 2^m - 1) = 1$.

```
 _____ Example 5: beginning of the WG-7 template _____
################    WG7 params    ################
f := x^7+x+Z(2)^0;;  F := FieldExtension(K, f);; ChooseField(F);
l := y^23+y^12+y^11+y^8+y^7+y^6+y^5+y^4+y^3+y^2+y+Z(2^7);;
exponents := [ 1, 33, 41, -23, 39 ];; d := 63;; # trace = x_0
################    WG7 FSRs    ################
lfsr := LFSR(F, 1, [Degree(1)-1]);
dwgpfun := One(F);
for j in [1..Length(exponents)] do
  r := exponents[j];
  if r<0 then r := r mod (Size(F)-1); fi;
  dwgpfun := dwgpfun + (x_0^d + One(F))^r;
od;
dwgpfil := FILFUN(F, dwgpfun);
```

Table 5 shows a filled out spreadsheet provided by the *manager*: the entries ×, [0], *load, init, run* and *always* are filled in by the designer. The values *load, init, run* are possible modes of operation (used by FSM), and the value *always* indicates an unconditional connection. The entry "[0] run" corresponds to the trace of an element of $\mathbb{F}_{2^7}$ represented in polynomial basis using the field defining polynomial $x^7 + x + 1$, i.e. the output is the WG transformation (Eq. (7))[5].

Table 5. WG-7 spreadsheet example

	i_data_1	fsr_1_o_fsr	fsr_2_o_fsr
o_data_1	×	×	[0] run
fsr_1_i_fsr	load	×	×
fsr_1_i_ext	×	×	init
fsr_2_i_fsr	×	always	×
fsr_2_i_ext	×	×	×

relationship to GAP code in **Example 5** above:

FSR object	VHDL module
lfsr	fsr_1
dwgpfil	fsr_2

ASIC implementation results for the WG-7 and WG-8 datapaths, created by FSRtoVHDL package, are listed in Table 6, and the FPGA implementation results for WG-8 in Table 7. Compared to a manual design, the FSRtoVHDL generated WG-7dp exhibits a slightly smaller area and comparable clock frequency, expect for the highest frequency circuit. All three WG-8 FSRtoVHDL circuits reached a significantly higher frequency, with a 2.5× speedup for the best optimality; they are expected to outperform the manual WG-8 implementation [17] after the FSM is added. The FSRtoVHDL FPGA implementation, however, reached only 65% of the frequency reached by the manual WG-8 implementation [17].

---

[5] $\mathbb{F}_{2^8}$ with defining polynomial $x^8 + x^4 + x^3 + x^2 + 1$: trace is bit 5, i.e. "[5] run".

Table 6. ASIC implementation results for WG-7 and WG-8 datapaths

Design used	Speed [GHz]	Area [GE]	Speed [GHz]	Area [GE]	Synthesis tools optimization goal
	FSRtoVHDL		Manual design		
WG-7dp	1.00	1320	0.91	1300	Smallest area
WG-7dp	1.43	1430	1.43	1740	Best optimality
WG-7dp	1.67	2260	2.00	2330	Highest frequency
	FSRtoVHDL		WG-8† [17]		
WG-8dp	0.83	1640			Smallest area
WG-8dp	1.25	1860	0.5	1786	Best optimality‡
WG-8dp	1.67	2950			Highest frequency

† including FSM   ‡ unknown synth. tools goal for †

Table 7. FPGA implementation results for WG-8 datapath

Design used	Speed [MHz]	Area [# slices]	Speed [MHz]	Area [# slices]	FPGA device used
	FSRtoVHDL		WG-8† [17]		
WG-8dp	124	74	190	137	xc3s1000–5fg320

† including FSM

4.2 Case Study 2: Grain V1

Grain v1 [18,19] is one of the three Profile 2 ciphers[6] included in the eSTREAM portfolio. The structure of Grain v1 is shown Fig. 4: it includes an 80-bit LFSR (with $f(x)$), an 80-bit NLFSR (with $g(x)$) and a filtering function $h(x)$, which takes the input bits from both the LFSR and the NLFSR. The result of this function is masked by a bit from the NLFSR to produce the keystream bit.

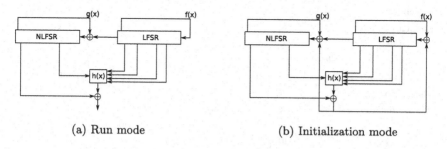

(a) Run mode   (b) Initialization mode

Fig. 4. Original schematic of Grain cipher [18]

---

[6] Stream ciphers for hardware applications with highly restricted resources.

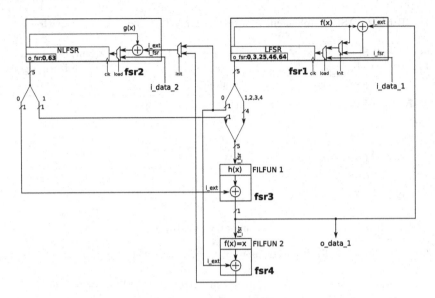

Fig. 5. Unified schematic of Grain v1 using FSR

The two modes of operation from Fig. 4 are combined into a single schematic in Fig. 5: the two original shift registers are presented as FSR package blocks with output taps and utilizing the external step. The filled out spreadsheet to define the datapath for Grain is shown in Table 8; this table provides the information that the *manager* needs to execute Grain and generate the VHDL for the Grain datapath (Fig. 5).

Table 8. Grain spreadsheet example

	i_data_1	i_data_2	fsr_1_o_fsr	fsr_2_o_fsr	fsr_3_o_fsr	fsr_4_o_fsr
o_data_1	×	×	×	×	×	run
fsr_1_i_fsr	load	×	×	×	×	×
fsr_1_i_ext	×	×	×	×	init	×
fsr_2_i_fsr	×	load	×	×	×	×
fsr_2_i_ext	×	×	[0] run	×	×	init
fsr_3_i_fsr	×	×	$[1, 2, 3, 4, -1]$	$[-1, -1, -1, -1, 1]$	×	×
fsr_3_i_ext	×	×	×	[0]	×	×
fsr_4_i_fsr	×	×	×	×	always	×
fsr_4_i_ext	×	×	[0]	×	×	×

The NLFSR (fsr_2 in Fig. 5 and Table 8) uses the external input during both initialization and running mode, hence an extra multiplexer is needed (row fsr_2_i_ext in Table 8). The filtering function $h(x)$, now represented as FIL-FUN1, takes bits 1,2,3,4 from fsr_1 and bit 1 from fsr_2: this is encoded with two rules[7] in row fsr_3_i_fsr in Table 8. The masking bit from the NLFSR is used as the external input for the filtering function $h(x)$, now represented as FILFUN1 (fsr_3 in Table 8). The extra XOR gate, used for the NLFSR external input during initialization, is represented as the FILFUN2 identity function and an external input. While it may seem excessive to represent a simple XOR with a FILFUN, this aids the transition to VHDL.

Tables 9 and 10 show the ASIC and FPGA implementation results for the Grain datapath. The ASIC implementations are compared to a manual design of the Grain datapath and of the Grain cipher used in [17], which is to the best of authors' knowledge the only post place-and-route CMOS 65 nm implementation of Grain. The same circuit is presented in both the best optimality and highest frequency row for the manual design of grain datapath. The FSRtoVHDL datapath implementation results are very satisfactory in comparison with the smallest area and best optimality manual circuits. The reference implementation [17] includes the FSM, which makes a direct comparison difficult. However, the smallest area FSRtoVHDL datapath is expected to reach the hardware cost of Grain from [17] after the FSM is added.

The FPGA Grain_dp results are compared with the Grain cipher implementation results from [20], which includes the FSM. The generated datapath area is approximately 65% of the area for the full cipher. The speedup reached by the FSRtoVHDL datapath design is probably not representative: a drop in frequency is common after an FSM is added.

Overall, the FSRtoVHDL generated hardware is comparable to the manual designs and gives a good starting point for further manual optimizations.

Table 9. ASIC implementation results for Grain datapath

Design used	Speed [GHz]	Area [GE]	Speed [GHz]	Area [GE]	Synthesis tools optimization goal
	FSRtoVHDL		Manual design		
Grain_dp	1.11	977	1.00	1020	Smallest area
Grain_dp	1.67	1080	1.67	1110	Best optimality
Grain_dp	2.00	1610	1.67	1110	Highest frequency
			Grain† [17]		
Grain	-	-	1.02	1126	Unknown

† including FSM

---

[7] Meaning of −1: this signal is defined in the other rule.

Table 10. FPGA implementation results for Grain datapath

Design used	Speed [MHz]	Area [# slices]	Speed [MHz]	Area [# slices]	FPGA device used
	FSRtoVHDL		Grain† [20]		
Grain_dp	228	28	196	44	xc3s50-5pq208

† including FSM

5   Conclusion

This work presents an automation toolkit for the rapid hardware design of cryptographic modules with filtering structures, composed of feedback shift registers and filtering functions. The toolkit consists of two packages, FSR and FSRtoVHDL, written in the GAP language. A great advantage of the FSR package is the generality of the FSR objects that can be modelled, and as such, they can be used directly as building blocks of many cryptographic modules. The FSR package is the core of the toolkit for generation of hardware modules. Because the FSRs can be executed, the toolkit is also able to generate of test vectors for the (hardware) simulations.

The toolkit is based on exploitation of structural similarities between LFSRs, NLFSRs and filters, from both a mathematical and a hardware perspective. For each FSR object a corresponding hardware module can be generated, and the FSR objects themselves contain all the information needed for their execution and hardware implementation. The FSR package can be used to implement arbitrary primitives in GAP, which can operate e.g. as random number generators.

The optimization of the generated hardware is left to the synthesis tools. The results of the synthesis tools, e.g. critical path analysis, can be used to further optimize the generated hardware, or even change a part of the design entirely. Two case studies were used to show that the toolkit generated datapaths are comparable with manual designs. Overall it provides a good estimate of the hardware cost for a cryptographic primitive and gives a good starting point for further manual hardware optimizations.

References

1. Robshaw, M.: New Stream Cipher Designs - The eSTREAM Project. Springer, Heidelberg (2008). https://doi.org/10.1007/978-3-540-68351-3
2. Wu, H.: ACORN: A Lightweight Authenticated Cipher (v1). http://competitions.cr.yp.to/round1/acornv1.pdf
3. CAESAR: Competition for Authenticated Encryption. https://competitions.cr.yp.to/caesar.html
4. ETSI/SAGE Specification version 1.1: Specification of the 3GPP Confidentiality and Integrity Algorithms UEA2 & UIA2. Document 2: SNOW 3G Specification
5. ETSI/SAGE Specification Version 1.6: Specification of the 3GPP Confidentiality and Integrity Algorithms 128-EEA3 & 128-EIA3. Document 2: ZUC Specification

6. The GAP Group: GAP - Groups, Algorithms, and Programming, Version 4.8.8 (2017). https://www.gap-system.org
7. Lidl, R., Niederreiter, H.: Finite fields. In: Encyclopedia of Mathematics and its Applications, vol. 20, Cambridge University Press, Cambridge (1997)
8. Golomb, S.W., Gong, G.: Signal Design for Good Correlation: For Wireless Communication, Cryptography, and Radar. Cambridge University Press, Cambridge (2005)
9. Chen, L., Gong, G.: Communication System Security. CRC Press, Boca Raton (2012)
10. SageMath. http://www.sagemath.org/
11. SageMath Package Cryptography. http://doc.sagemath.org/html/en/reference/cryptography/index.html
12. Symbolic Linear Feedback Shift Registers. http://library.wolfram.com/infocenter/MathSource/5717/
13. MathWorks. https://www.mathworks.com/
14. Coussy, P., Gajski, D.D., Meredith, M., Takach, A.: An introduction to high-level synthesis. IEEE Design Test Comput. **26**(4), 8–17 (2009). https://doi.org/10.1109/MDT.2009.69
15. Mandal, K., Gong, G.: Generating good span n sequences using orthogonal functions in nonlinear feedback shift registers. In: Koç, Ç.K. (ed.) Open Problems in Mathematics and Computational Science, pp. 127–162. Springer, Cham (2014). https://doi.org/10.1007/978-3-319-10683-0_7
16. Gong, G., Aagaard, M., Fan, X.: Resilience to distinguishing attacks on WG-7 cipher and their generalizations. Cryptogr. Commun. **5**(4), 277–289 (2013)
17. Yang G., Fan X., Aagaard M., Gong G.: Design space exploration of the lightweight stream cipher WG-8 for FPGAs and ASICs. In: WESS 2013, Article No. 8. ACM, New York (2013). https://doi.org/10.1145/2527317.2527325
18. Hell, M., Johansson, T., Meier, W.: Grain - a stream cipher for constrained environments. Int. J. Wirel. Mob. Comput. **2**(1), 86–93 (2007). https://doi.org/10.1504/IJWMC.2007.013798
19. Hell, M., Johansson, T., Maximov, A., Meier, W.: The grain family of stream ciphers. In: Robshaw, M., Billet, O. (eds.) New Stream Cipher Designs. LNCS, vol. 4986, pp. 179–190. Springer, Heidelberg (2008). https://doi.org/10.1007/978-3-540-68351-3_14
20. Hwang, D., Chaney, M., Karanam, S., Ton, N., Gaj, K.: Comparison of FPGA-targeted hardware implementations of eSTREAM stream cipher candidates. SASC **2008**, 151–162 (2008)

Invited Talk 3

Sequences with Low Correlation

Daniel J. Katz$^{(\boxtimes)}$

Department of Mathematics, California State University, Northridge,
CA 91330, USA
daniel.katz@csun.edu

Abstract. Pseudorandom sequences are used extensively in communications and remote sensing. Correlation provides one measure of pseudorandomness, and low correlation is an important factor determining the performance of digital sequences in applications. We consider the problem of constructing pairs (f, g) of sequences such that both f and g have low mean square autocorrelation and f and g have low mean square mutual crosscorrelation. We focus on aperiodic correlation of binary sequences, and review recent contributions along with some historical context.

Keywords: Crosscorrelation · Autocorrelation · Aperiodic
Merit factor · Sequence

1 Introduction

Sequences with low correlation play many roles in technology, including remote sensing, design of scientific instruments, operation of communications networks, and acoustic design. The monographs of Golomb, Gong, and Schroeder [14,15,41] give some sense of the broad sweep of their applications. Golomb [14, p. 25] used correlation as a measure of pseudorandomness, a concept of significance for cryptography that has been developed extensively by Mauduit and Sárkozy in [34] and further works. Here we give an overview of recent progress on the problem of designing binary sequence pairs where both sequences have low aperiodic autocorrelation and the two sequences of the pair have low mutual aperiodic crosscorrelation.

This paper is organized as follows: Sects. 2, 3, and 4 give the basic definitions (of sequences, correlation, and merit factors). Section 5 lists some constructions of sequence families with low mean square autocorrelation. Sections 6, 7, and 8 describe how the constructions are done, and provide more details about autocorrelation performance. Section 9 examines the question of low mean square crosscorrelation, and Sect. 10 discusses a combined measure (called the Pursley-Sarwate criterion) of autocorrelation and crosscorrelation performance of a sequence pair. Section 11 discusses families of sequence pairs with low Pursley-Sarwate criterion, and Sect. 12 concludes with open questions.

This paper is based upon work supported in part by the National Science Foundation under Grants DMS-1500856 and CCF-1815487.

L. Budaghyan and F. Rodríguez-Henríquez (Eds.): WAIFI 2018, LNCS 11321, pp. 149–172, 2018.
https://doi.org/10.1007/978-3-030-05153-2_8

2   Sequences

If ℓ and m are positive integers, an *additive m-ary sequence of length ℓ* is an ℓ-tuple of elements of the additive group of $\mathbb{Z}/m\mathbb{Z}$, that is,

$$a = (a_0, a_1, \ldots, a_{\ell-1}) \in (\mathbb{Z}/m\mathbb{Z})^\ell. \tag{1}$$

When $m = 2$ we have an *additive binary sequence*, that is, an element of $\mathbb{F}_2^\ell$.

A *multiplicative m-ary sequence of length ℓ* is an ℓ-tuple of mth roots of unity in $\mathbb{C}$, that is,

$$b = (b_0, b_1, \ldots, b_{\ell-1}) \in \mu_m^\ell, \tag{2}$$

where $\mu_m = \{e^{2\pi i j/m} : 0 \le j < m\}$ is the multiplicative group of mth roots of unity in $\mathbb{C}$. Most often we have $m = 2$, so $\mu_2 = \{1, -1\}$; this gives *multiplicative binary sequences*, which we shall just call *binary sequences*.

Consider the group homomorphism $\varepsilon \colon \mathbb{Z}/m\mathbb{Z} \to \mu_m$ with $\varepsilon(x) = e^{2\pi i x/m}$. If the sequences a and b of (1) and (2) are related by $b_k = \varphi(a_k)$ for every k, then we say that *b is the multiplicative version of a*, and equivalently, that *a is the additive version of b*.

For the purposes of this paper, it will be more convenient to consider sequences in their multiplicative guise. Furthermore, we shall identify the sequence $f = (f_0, \ldots, f_{\ell-1}) \in \mathbb{C}^\ell$ in multiplicative form with the polynomial $f(z) = f_0 + f_1 z + \cdots + f_{\ell-1} z^{\ell-1} \in \mathbb{C}[z]$, whose coefficients are the terms of the sequence f. This identification makes calculations easier and we shall see in Sect. 4 that it forms a bridge between the study of correlation and harmonic analysis that has proved fruitful in these studies.

3   Correlation

Correlation is a measure of the similarity between the various shifted versions of a pair of sequences. When the sequences of the pair are the same, we are comparing a sequence to shifted versions of itself, which is self-correlation, or autocorrelation. Truly random sequences should have low correlation with shifted versions of themselves (unless the shift is zero) and of each other, so we demand that our pseudorandom sequences also have low correlation.

Let us now define correlation precisely. For two sequences

$$f = (f_0, f_1, \ldots, f_{\ell-1}) \in \mathbb{C}^\ell$$
$$g = (g_0, g_1, \ldots, g_{\ell-1}) \in \mathbb{C}^\ell, \tag{3}$$

and $s \in \mathbb{Z}$, the *aperiodic crosscorrelation of f with g at shift s*, denoted $C_{f,g}(s)$, is defined by

$$C_{f,g}(s) = \sum_{j \in \mathbb{Z}} f_{j+s} \overline{g_j},$$

where we use the convention that $f_j = g_j = 0$ when $j \notin \{0, 1, \ldots, \ell - 1\}$, so that the above sum only has a finite number of nonzero entries. We index over

$\mathbb{Z}$ because the underlying translation operation involved in aperiodic crosscorrelation is a non-cyclic shift. We can view the aperiodic correlation of f with g at shift s as the inner product between the overlapping portions of f and g when g is shifted s places to the right relative to f, as shown in Fig. 1.

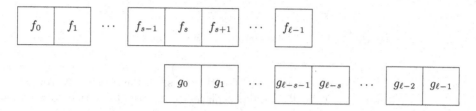

Fig. 1. Aperiodic correlation of f with g at shift $s > 0$

Let us identify f and g of (3) with the polynomials $f(z) = f_0 + f_1 z + \cdots + f_{\ell-1} z^{\ell-1}$ and $g(z) = g_0 + g_1 z + \cdots + g_{\ell-1} z^{\ell-1}$, respectively, as discussed at the end of Sect. 2. Furthermore, let us adopt the convention that for any Laurent polynomial

$$a(z) = \sum_{j \in \mathbb{Z}} a_j z^j,$$

in the ring $\mathbb{C}[z, z^{-1}]$ of Laurent polynomials over $\mathbb{C}$, the *conjugate of* $a(z)$ is defined to be

$$\overline{a(z)} = \sum_{j \in \mathbb{Z}} \overline{a_j} z^{-j}, \tag{4}$$

where $\overline{a_j}$ is the usual complex conjugate of a_j. Then it is not difficult to show that

$$f(z)\overline{g(z)} = \sum_{s \in \mathbb{Z}} C_{f,g}(s) z^s,$$

that is, the crosscorrelation of f with g at shift s is the coefficient of z^s in the product $f(z)\overline{g(z)}$. This interpretation allows us to discover quite easily the following basic symmetry of aperiodic correlation:

$$C_{f,g}(s) = \overline{C_{g,f}(-s)}. \tag{5}$$

There is also a periodic version of correlation that treats our sequences (3) as periodically repeating every ℓ terms. In this case, $\mathbb{Z}/\ell\mathbb{Z}$ is the natural set for indexing sequence terms and expressing shifts, reflecting the cyclic nature of the sequences and the shifting. Then for any $s \in \mathbb{Z}/\ell\mathbb{Z}$, the *periodic crosscorrelation of f with g at shift s*, denoted $\mathrm{PC}_{f,g}(s)$, is defined by

$$\mathrm{PC}_{f,g}(s) = \sum_{j \in \mathbb{Z}/\ell\mathbb{Z}} f_{j+s} \overline{g_j}.$$

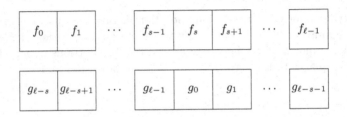

Fig. 2. Periodic correlation of f with g at shift s

We can view the periodic correlation of f with g at shift s as the inner product of f with g when g is cyclically shifted s places to the right relative to f, as shown in Fig. 2.

If $s \in \mathbb{Z}$, then we interpret $\mathrm{PC}_{f,g}(s)$ as $\mathrm{PC}_{f,g}(\sigma)$ where $\sigma \in \mathbb{Z}/\ell\mathbb{Z}$ is the congruence class of s modulo ℓ. In this case one can see that for any $s \in \mathbb{Z}$, we have

$$\mathrm{PC}_{f,g}(s) = \sum_{\substack{t \in \mathbb{Z} \\ t \equiv s \pmod{\ell}}} C_{f,g}(t),$$

where at most two of these terms can be nonzero, and in particular, if $0 \le s < \ell$, then

$$\mathrm{PC}_{f,g}(s) = C_{f,g}(s) + C_{f,g}(s - \ell). \tag{6}$$

One can obtain a polynomial interpretation of periodic crosscorrelation by regarding our sequences as lying in the ring $\mathbb{C}[z]/(z^\ell - 1)$ rather than $\mathbb{C}[z]$. The notion of a conjugate of a Laurent polynomial from (4) carries over naturally to $\mathbb{C}[z]/(z^\ell - 1)$: negative powers of z can be reinterpreted as positive powers since $z^\ell = 1$ in this ring, and the ideal $(z^\ell - 1)$ is closed under our conjugation since $\overline{z^\ell - 1} = -z^{-\ell}(z^\ell - 1)$. Then one can show that

$$f(z)\overline{g(z)} \equiv \sum_{s \in \mathbb{Z}/\ell\mathbb{Z}} \mathrm{PC}_{f,g}(s)z^s \pmod{z^\ell - 1},$$

and from this prove the symmetry

$$\mathrm{PC}_{f,g}(s) = \overline{\mathrm{PC}_{g,f}(-s)}.$$

Periodic correlation is more mathematically tractable than aperiodic correlation. For example, when we consider sequences derived from finite field characters (see Sects. 6 and 7), the periodic correlation values are complete character sums, while the aperiodic correlation values are incomplete character sums, which are much more difficult to handle. Equation (6) shows that the magnitude of any periodic correlation cannot be more than twice as large as the largest magnitude of any aperiodic correlation value. In consequence of this, Boehmer [1, p. 157] points out that having low periodic correlation at all shifts is a necessary but not sufficient condition for having low aperiodic correlation at all shifts. She then enunciates a design technique that has been used widely in attempts to design sequences with low aperiodic correlation: design sequences with low periodic correlation and hope that some of these will also have low aperiodic correlation.

Earlier we had mentioned autocorrelation, or correlation of a sequence with itself. If f is the sequence in (3) and $s \in \mathbb{Z}$, then the *aperiodic autocorrelation of f at shift s* is just the aperiodic crosscorrelation of f with itself at shift s, that is,

$$C_{f,f}(s) = \sum_{j \in \mathbb{Z}} f_{j+s} \overline{f_j}.$$

And for $s \in \mathbb{Z}/\ell\mathbb{Z}$, the *periodic autocorrelation of f at shift s* is just the periodic crosscorrelation of f with itself at shift s, that is,

$$PC_{f,f}(s) = \sum_{j \in \mathbb{Z}/\ell\mathbb{Z}} f_{j+s} \overline{f_j}.$$

Note that if the shift is zero in either the aperiodic or periodic case, then

$$C_{f,f}(0) = PC_{f,f}(0) = \sum_{j=0}^{\ell-1} |f_j|^2, \tag{7}$$

which is the squared Euclidean norm of f if it is regarded as a vector in $\mathbb{C}^\ell$. If all the terms of f are of unit magnitude, we say that f is a *unimodular sequence*. For example, all m-ary sequences are unimodular, since their terms are roots of unity. If f is unimodular, then

$$C_{f,f}(0) = PC_{f,f}(0) = \ell,$$

which is the length of the sequence, and is naturally as large as a correlation value involving unimodular sequences of length ℓ could possibly be. On the other hand, what we know about random walks suggests that typical correlation values for randomly selected binary sequences of length ℓ (with uniform probability distribution) should not have magnitudes much larger than $\sqrt{\ell}$.

4   Demerit Factors and Merit Factors

In a multi-user communications network, one can modulate the messages of the various users with different signature sequences. For efficient operation, it is desirable that the family of signature sequences used should have the following properties:

(i) Each sequence f should have low magnitude autocorrelation $|C_{f,f}(s)|$ at all nonzero shifts (all $s \neq 0$).
(ii) Each pair (f,g) of sequences should have low magnitude crosscorrelation $|C_{f,g}(s)|$ at all shifts s.

Notice that it is the magnitude of the correlation value that is typically considered, since often the argument is not discernible in our systems. In view of our comments on random sequences at the conclusion of the previous section, a typical correlation value can be considered small if it does not have magnitude much larger than the square root of the length of the sequences involved. Condition (4) helps the communications system maintain synchronization with the user represented by sequence f: the sharp difference between correlation of at shift 0 (when the sequence is aligned with a reference copy of itself) and at nonzero shifts (when it is not aligned) allows one to obtain very accurate timings. Condition (4) prevents the output from the user represented by sequence f from being confused with any output from the user represented by sequence g, regardless of any delays between these two signals.

In this paper, we consider the simplest possible case of this design problem for sequences with low autocorrelation and crosscorrelation, that is, we ask for pairs of sequences such that $|C_{f,f}(s)|$ and $|C_{g,g}(s)|$ are low for all $s \neq 0$ and $|C_{f,g}(s)|$ is low for all s. We could rate overall smallness of correlation in various ways. One method is to rate crosscorrelation performance for a sequence pair (f, g) by the worst case: the *peak crosscorrelation of f with g* is

$$\max_{s \in \mathbb{Z}} |C_{f,g}(s)|,$$

and we would want this to be small.

After studying this and some other common measures of smallness of crosscorrelation, Kärkkäinen [25, p. 149] expresses the view that one gets a better notion of likely performance from a mean square measure. Accordingly, for sequence pair (f, g), we define the *crosscorrelation demerit factor of f and g* as

$$\mathrm{CDF}(f, g) = \frac{\sum_{s \in \mathbb{Z}} |C_{f,g}(s)|^2}{|C_{f,f}(0)| \cdot |C_{g,g}(0)|},$$

which, in view of (7), is the sum of squared magnitudes of crosscorrelation values for the sequence pair we obtain from f and g if we scale each of them to have unit Euclidean magnitude. We should note that $\mathrm{CDF}(f, g) = \mathrm{CDF}(g, f)$ because of (5). Normally f and g are unimodular sequences of the same length ℓ, so the denominator of the CDF is simply ℓ^2. Since we want every term in the numerator to be as small as possible, a large CDF indicates poor performance. If one wants a measure that is larger for good sequence pairs, one defines the *crosscorrelation merit factor of f and g* to be

$$\mathrm{CMF}(f, g) = \frac{1}{\mathrm{CDF}(f, g)}.$$

We have analogous measures for autocorrelation. If f is a sequence, then the *autocorrelation demerit factor of f* is defined to be

$$\mathrm{ADF}(f) = \frac{\sum_{\substack{s \in \mathbb{Z} \\ s \neq 0}} |C_{f,f}(s)|^2}{|C_{f,f}(0)|^2} = \mathrm{CDF}(f, f) - 1, \qquad (8)$$

where one should note that we omit the autocorrelation at shift 0 in the numerator. This is because $C_{f,f}(0)$ is always large, and we want it to be large, so that it should not be construed as contributing to the demerit factor. And the *autocorrelation merit factor of f* is just the reciprocal of the demerit factor,

$$\text{AMF}(f) = \frac{1}{\text{ADF}(f)}.$$

Naturally we want to make the autocorrelation demerit factor small, or equivalently, to make the autocorrelation merit factor large.

Recall from the end of Sect. 2 that we always identify the sequence $f = (f_0, \ldots, f_{\ell-1}) \in \mathbb{C}^\ell$ with the polynomial $f(z) = f_0 + f_1 z + \cdots + f_{\ell-1} z^{\ell-1} \in \mathbb{C}[z]$. We shall now see how this point of view relates to merit factors. For any real number $r \geq 1$ and any function f defined on the complex unit circle, we define the L^r *norm of f on the complex unit circle* to be

$$\|f\|_r = \left(\frac{1}{2\pi} \int_0^{2\pi} |f(e^{i\theta})|^r d\theta \right)^{1/r},$$

provided that this integral exists (as it certainly will when f is a Laurent polynomial).

Then one can show that the crosscorrelation demerit factor is

$$\text{CDF}(f, g) = \frac{\|fg\|_2^2}{\|f\|_2^2 \|g\|_2^2}$$

and the autocorrelation demerit factor is

$$\text{ADF}(f) = \frac{\|f\|_4^4}{\|f\|_2^4} - 1 \tag{9}$$

This links the work of Littlewood (see [32] and [33, Problem 19]) on flatness of polynomials on the complex unit circle with the work of Golay [8,9] on merit factors.

Sarwate [38, Eqs. (13) and (38)] calculated expected values of demerit factors for randomly selected binary sequences (where each term is independent of the others and has equal probability of being $+1$ or -1). For a randomly selected sequence f of length ℓ, Sarwate calculated the expected value of the autocorrelation demerit factor to be

$$E[\text{ADF}(f)] = 1 - \frac{1}{\ell}. \tag{10}$$

For a randomly selected pair (f, g) of sequences of length ℓ, Sarwate calculated the expected value of the crosscorrelation demerit factor to be

$$E[\text{CDF}(f, g)] = 1. \tag{11}$$

So typical values of both autocorrelation and crosscorrelation demerit factors will be about 1 when the length ℓ is large, as it quite often is. For example, Gold sequences of length 1023 are used in the Global Positioning System (GPS), and code division multiple access communications (CMDA) protocols use even longer sequences. Thus we want constructions that produce families of low correlation sequence pairs of various lengths. Typically, our constructions produce families with unbounded lengths, and we rate a family by the *asymptotic demerit factors*, that is, the limit of the autocorrelation or crosscorrelation demerit factor as the length of the sequences tends to infinity. For sequences derived from finite field characters, it has been observed in many cases [2,27] that the limiting behavior of families is approached quite rapidly, so that even sequences of quite modest length (of the order of a hundred or more) already have demerit factors close to the limiting values.

5   High Asymptotic Autocorrelation Merit Factor

In this section we shall discuss constructions that give infinite families of binary sequences with high asymptotic autocorrelation merit factor. Recall (10), which says that randomly selected binary sequences of length ℓ have an average autocorrelation demerit factor of $1 - 1/\ell$, which is close to 1 for large ℓ. It is possible to obtain families where the asymptotic demerit factor is considerably lower. It is relatively rare to find such families, and to the author's best knowledge, the first one that was discovered derives from the Rudin-Shapiro polynomials, which shall be discussed further in Sect. 8. It was Littlewood who originally proved a result tantamount to showing that this family of polynomials has asymptotic demerit factor $1/3$ [33, pp. 27–28]. At the time, the concept of merit factor for correlation had not yet been defined: the formula for the autocorrelation merit factor would appear as a "factor" in a 1972 paper by Golay [8], who later called this the "merit factor" in another paper a few years later [9]. What Littlewood actually proved [33, p. 28] is a formula for the ratio of L^4 norm to L^2 norm of the Rudin-Shapiro polynomials, which via (9) is equivalent to finding the asymptotic autocorrelation demerit factor. The Rudin-Shapiro sequence family has one sequence f_n of length 2^n for each nonnegative integer n, and Littlewood's result shows that $\mathrm{ADF}(f_n) = (1 - (-1/2)^n)/3$, which tends to $1/3$ in the limit as n tends to infinity.

If we consider Littlewood's result as the first low asymptotic demerit factor record, then this record was broken by Høholdt and Jensen [17] with cyclically shifted Legendre sequences, and that record was again broken by Jedwab, Katz, and Schmidt [20, Theorem 1.1] with Legendre sequences that are cyclically shifted and appended (periodically extended). The Legendre sequences and their modifications shall be discussed in more detail in Sect. 7. We summarize these records in Table 1, which in addition to listing the asymptotic demerit factor also lists its reciprocal, the asymptotic merit factor, which is the way these results are usually presented in the literature. The asymptotic demerit factor of $0.157\ldots$ listed for the shifted and appended Legendre sequences is the smallest real root of the polynomial $27x^3 - 417x^2 + 249x - 29$. It should also be noted that, in addition to the records on Table 1, an important advance was the determination

Table 1. Records for high asymptotic autocorrelation merit factor

Sequence family	Asymptotic AMF	ADF	Proved by
Rudin-Shapiro	3	0.333...	Littlewood (1968) [33, pp. 28]
Legendre, shifted	6	0.166...	Høholdt-Jensen (1988) [17]
Legendre, shifted and appended	6.342...	0.157...	Jedwab-Katz-Schmidt (2013) [20, Theorem 1.1]

by Jensen and Høholdt [23, Sect. 5] of the asymptotic merit factor of a class of sequences known as maximal linear recursive sequences (m-sequences). These sequences shall be described in the next section, but they are of great interest because it is easy to generate large families of them for use in communications networks. Jensen and Høholdt showed that any infinite family of m-sequences has asymptotic autocorrelation demerit factor 1/3, which equals the performance of the Rudin-Shapiro polynomials.

6   Sequences from Additive Characters

In this section, we shall discuss sequences derived from additive characters of finite fields, of which the most fundamental are the *maximum length linear recursive shift register sequences*, which are also called *maximal linear recursive sequences*, or just *m-sequences*. Let $\mathbb{F}_q$ be a finite field of characteristic p and order $q = p^n$. An *additive character* is a homomorphism from the additive group $\mathbb{F}_q$ to the multiplicative group $\mathbb{C}^*$. We use Tr: $\mathbb{F}_q \to \mathbb{F}_p$ to denote the absolute trace from $\mathbb{F}_q$ to its prime field $\mathbb{F}_p$. Then for each $a \in \mathbb{F}_q$, the map $\varepsilon_a : \mathbb{F}_q \to \mathbb{C}^*$ with $\varepsilon_a(x) = \exp(2\pi i \operatorname{Tr}(ax)/p)$ is an additive character of $\mathbb{F}_q$, and $\{\varepsilon_a : a \in \mathbb{F}_q\}$ is the entire group of q additive characters from $\mathbb{F}_q$ into $\mathbb{C}^*$, with ε_0 being the trivial character (that maps every element of $\mathbb{F}_q$ to 1), while ε_1 is called the *canonical additive character*.

Let α be a primitive element of $\mathbb{F}_q$. Let us list the nonzero elements of $\mathbb{F}_q$ as powers of the primitive element α, that is, as

$$\alpha^0, \alpha^1, \alpha^2, \ldots, \alpha^{q-2},$$

and then apply a nontrivial additive character ψ to obtain a sequence

$$\left(\psi(\alpha^0), \psi(\alpha^1), \ldots, \psi(\alpha^{q-2})\right). \tag{12}$$

An *m-sequence* is any sequence obtained in this way. Any nontrivial additive character of a finite field of characteristic p has the complex pth roots of unity as its outputs, so the m-sequences produced from fields of characteristic p are p-ary sequences. We shall mainly be interested in binary m-sequences, which derive from fields of characteristic 2.

Changing the nontrivial character ψ in (12) simply causes a cyclic shift of the m-sequence, and each of the $q-1$ nontrivial additive characters of $\mathbb{F}_q$ produces a different cyclic shift, and the $q-1$ different cyclically shifted versions of our m-sequence are all distinct. If the character ψ we use is the canonical additive character, we call the m-sequence produced in (12) a *Galois sequence* or a *naturally shifted m-sequence*.

Changing the primitive element α in (12) to another primitive element $\beta = \alpha^d$ (where $\gcd(d, q-1) = 1$ to maintain primitivity) causes the m-sequence (12) to be decimated by d, that is, the new m-sequence based on β is what one obtains by selecting every dth element from the original sequence (starting at the beginning and proceeding cyclically modulo the length $q-1$). If β is a Galois conjugate of α over the prime field $\mathbb{F}_p$, that is, if d is a power of p modulo $q-1$, then one just gets back the original sequence (up to some cyclic shift); otherwise one gets a sequence that is distinct from every cyclic shift of the original sequence. Thus decimations d that are a power of p modulo $q-1$ are said to be *degenerate*. When the original sequence is a Galois sequence, decimation by a degenerate d yields back the original sequence exactly (not even cyclically shifted). Another type of decimation that will become useful later in Sect. 11 is a *reversing decimation*, which is any decimation d for which there is an integer k such that $d \equiv -p^k$ (mod $q-1$). If we decimate an m-sequence by such a d, one obtains the reverse of the original sequence (up to a cyclic shift).

We can now count the total number of m-sequences, based on our freedom to choose the character (cyclic shifting) and the primitive element modulo Galois conjugacy (decimation). If we organize m-sequences of length p^n-1 into classes of sequences modulo cyclic shifting (with p^n-1 sequences per class), the number of classes of m-sequences will be equal to the number of classes of primitive elements of $\mathbb{F}_q = \mathbb{F}_{p^n}$ modulo Galois conjugacy (with n Galois conjugates per class). Since the number of primitive elements in $\mathbb{F}_q = \mathbb{F}_{p^n}$ is $\varphi(p^n-1)$, where φ is Euler's φ-function, the total number of m-sequences of length p^n is $(p^n-1)\varphi(p^n-1)/n$. If α is a primitive element of our field $\mathbb{F}_q$, then α^{-1} will also be a primitive element and will not be a Galois conjugate of α unless $q \leq 4$ (in which case all primitive elements in $\mathbb{F}_q$ are Galois conjugates of each other). So $\varphi(p^n-1)/n > 1$ whenever $p^n > 4$, in which case it is possible to obtain at least two cyclically distinct m-sequences (related by a nondegenerate decimation) of length $p^n - 1$.

Our p-ary m-sequence (12) of length p^n-1 follows a linear recursion of depth n whose characteristic polynomial is the minimal polynomial of the primitive element α over the prime field $\mathbb{F}_p$. Because of this, our m-sequence of length $p^n - 1$ can be generated using a linear feedback shift register of length n. This efficient generation of very long sequences with rather small circuits makes m-sequences very popular in applications. Furthermore, from two m-sequences of length $p^n - 1$ related by a nondegenerate decimation d, one can construct a family of $p^n + 1$ Gold sequences of length $p^n - 1$. Gold's original construction [13, Sect. IV] uses carefully chosen decimations d to produce families where all the sequences have low periodic autocorrelation and all the pairs have low periodic crosscorrelation.

We now give an overview of findings on the asymptotic aperiodic autocorrelation merit factor of binary m-sequences and their relatives. As mentioned in the previous section, Jensen and Høholdt [23, Sect. 5] proved that m-sequences have asymptotic autocorrelation demerit factor 1/3. Jedwab and Schmidt [21, Theorems 11 and 12] applied some constructions described by Parker [35, Lemmas 3 and 4] to m-sequences to produce families of related sequences that also have asymptotic autocorrelation demerit factor 1/3. Parker gave two constructions, each of which takes a sequence as an input and gives a longer sequence as an output. The first construction, called the *negaperiodic construction*, doubles the length of the sequence, so we shall call it *Parker's doubling construction*. The second construction, called the *periodic construction*, quadruples the length of the sequence, so we shall call it *Parker's quadrupling construction*.

Another technique that was used to modify m-sequences is *appending*, which originates with studies by Kirilusha and Narayanaswamy [31]. We let $f(z) = \sum_{j=0}^{\ell-1} f_j z^j$ be a sequence of length ℓ (represented in polynomial form), and extend the definition of f_j so that $f_{j+\ell} = f_j$ for all $j \in \mathbb{Z}$. Then we can truncate or periodically extend f simply by changing the range of summation. For example if $m < \ell$, then $g(z) = \sum_{j=0}^{m-1} f_j z^j$ is a truncated version of f, while if $m > \ell$, then it is an periodically extended version of f. In these respective cases, we say that this new sequence g is f *truncated to m/ℓ times its usual length* or f *appended to m/ℓ times its usual length*. Jedwab, Katz, and Schmidt [19, Theorem 2.2] proved that if one applies this procedure to m-sequences, one can produce families with an asymptotic autocorrelation demerit factor of $0.299\ldots$, which is the smallest real root of the polynomial $3x^3 - 33x^2 + 33x - 7$. To achieve this, one should use m-sequences appended to about $1.115\ldots$ times their usual length, where $1.115\ldots$ is the middle root of $x^3 - 12x + 12$. Jedwab, Katz, and Schmidt also combined the appending procedure with Parker's constructions to produce further families with asymptotic autocorrelation demerit factor $0.299\ldots$.

The concept of an m-sequence can be generalized to a produce a larger family of sequences called the *Gordon-Mills-Welch sequences* [40]. The construction of Gordon-Mills-Welch sequences differs from that of m-sequences in that the character ψ used in the m-sequence construction (12) is replaced with a "twisted" version. Günther and Schmidt [16, p. 344–345] have recently shown that Gordon-Mills-Welch sequences attain the same asymptotic autocorrelation demerit factors that m-sequences do: 1/3 for natural length and $0.299\ldots$ if appended.

7   Sequences from Multiplicative Characters

Now we describe a sequence construction that is in some sense dual to the construction of m-sequences. The pseudorandom behavior of m-sequences can be traced to the fact that we form them by listing the nonzero elements of a finite field $\mathbb{F}_q$ in an order based on the *multiplicative structure* of the field (that is, as powers of a primitive element) and then apply an *additive character* to them (see (12) and the commentary preceding it). Our next construction is dual in the sense that we shall devise a listing of the elements of a finite field based on the *additive structure* of the field and then apply a *multiplicative character* to them.

To make a multiplicative character sequence, let us start with a finite field $\mathbb{F}_p$ of prime order p, and write its elements in an order based on the additive structure of the field. We can take 1 as our additive generator, and then the additive analogue of taking increasing powers of this element is to form sums of increasing numbers of 1, that is, we list the elements of $\mathbb{F}_p$ in the order

$$0, 1, 1+1 = 2, 1+1+1 = 3, \ldots, p-1. \tag{13}$$

Now let $\chi \colon \mathbb{F}_p \to \mathbb{C}^*$ be a multiplicative character, that is, a group homomorphism from $\mathbb{F}_p^*$ to $\mathbb{C}^*$, and make sure that χ is nontrivial, that is, does not map every element to 1. Normally one extends a multiplicative character χ by setting $\chi(0) = 0$. Then we apply our nontrivial multiplicative character χ to our list (13) of elements of $\mathbb{F}_p$ to obtain the sequence[1]

$$(\chi(0), \chi(1), \chi(2), \ldots, \chi(p-1)). \tag{14}$$

The multiplicative characters of $\mathbb{F}_p$ form a cyclic group of order $p-1$ under multiplication. If χ is a character whose order is m in this group, then all terms except $\chi(0) = 0$ are mth roots of unity. Normally, we replace the initial $\chi(0) = 0$ term with a complex number mth root of unity (typically one just uses 1) to get a true m-ary sequence.

If p is an odd prime, then the group of multiplicative characters of $\mathbb{F}_p$ always contains one and only one character of order 2, which is called the *quadratic character* or *Legendre symbol*. If we use this as our character χ in the construction above (and replace $\chi(0)$ with 1), then we obtain a binary sequence $h = (h_0, h_1, \ldots, h_{p-1})$, called a *Legendre sequence*, where

$$h_j = \begin{cases} +1 & \text{if } j \text{ is the square of some element in } \mathbb{F}_p, \\ -1 & \text{if } j \text{ is not the square of any element in } \mathbb{F}_p. \end{cases} \tag{15}$$

Since there is only one character of order 2 over each prime field $\mathbb{F}_p$ of odd order, this construction gives us only one binary sequence of length p for each odd prime p. Contrast this with the construction of m-sequences in Sect. 6, which often produces many sequences of the same length that are not related to each other by cyclic shifting.

One might ask why we only used prime fields in the construction of multiplicative character sequences, while we used arbitrary finite fields to construct m-sequences. The reason is that prime fields are the only finite fields that are

---

[1] This sequence was formed using the specific choice of 1 as the additive generator of $\mathbb{F}_p$. We could have replaced 1 with any other $a \in \mathbb{F}_p^*$ to form the list $0, a, 2a, \ldots, (p-1)a$ of elements of $\mathbb{F}_p$ instead of (13), and then apply χ to every term to get the sequence $(\chi(0), \chi(a), \chi(2a), \ldots, \chi((p-1)a))$. This would just give the sequence in (14) multiplied by the unimodular scalar $\chi(a)$. This scalar multiplciation has no effect on the magnitudes of correlation values.

cyclic groups under addition, so they are the only finite fields where one can generate a list of all the elements using a single additive generator. A finite field $\mathbb{F}_{p^n}$ of characteristic p and order p^n is an n-dimensional vector space over $\mathbb{F}_p$, so $\mathbb{F}_{p^n}$ can be generated by an $\mathbb{F}_p$-basis consisting of n elements. Using this n-dimensional description of $\mathbb{F}_{p^n}$, we can generalize our construction to create n-dimensional arrays whose entries are given by evaluations of multiplicative characters, and there are natural definitions of correlation for these arrays, with many results analogous to what we present about sequences in this paper, for example, see [26].

As noted above, the standard multiplicative character construction only gives one binary sequence of length p for each odd prime p. This is not very satisfactory if we are interested in finding pairs or larger families of binary sequences with low crosscorrelation. Boothby and Katz [2] discovered that one can often obtain sequences with good aperiodic autocorrelation and crosscorrelation properties using linear combinations of multiplicative characters. Among the sequences formed from linear combinations of multiplicative characters that Boothby and Katz studied are the *cyclotomic sequences*, whose periodic and aperiodic auto-correlation properties had been studied by Boehmer [1], and whose periodic autocorrelation and periodic crosscorrelation properties had been studied by Ding, Helleseth, and Lam [5,6].

We now describe the construction of cyclotomic sequences. Let m be an even positive integer and let p be a prime with $m \mid p - 1$. Then let $\mathbb{F}_p^{*m}$ be the set $\{a^m : a \in \mathbb{F}_p^*\}$ of mth powers, which is a subgroup of order $(p-1)/m$ in the group in $\mathbb{F}_p^*$. We form the quotient group $\mathbb{F}_p^*/\mathbb{F}_p^{*m}$ of order m, which consists of the m cosets of $\mathbb{F}_p^{*m}$ in $\mathbb{F}_p^*$. Partition $\mathbb{F}_p$ into two sets, A and B, as follows: A contains 0 along with the union of $m/2$ cosets of $\mathbb{F}_p^{*m}$, while B contains the union of the other $m/2$ cosets of $\mathbb{F}_p^{*m}$. Then we define a sequence $f = (f_0, f_1, \ldots, f_{p-1})$ where $f_j = 1$ if $j \in A$ and $f_j = -1$ if $j \in B$. The choices that we make when allocating cosets to A or B can influence the correlation behavior of the sequences.

Let us consider cyclotomic sequences in some of the simplest cases, that is, when m is small. When $m = 2$ and when we define A to be $\{0\} \cup \mathbb{F}_p^{*2}$, we recover the Legendre sequence h defined above (cf. (15)). When $m = 4$, we define two new sequences in this manner: let α be a primitive element of $\mathbb{F}_p$ and list the four cosets of $\mathbb{F}_p^{*4}$ as $R_j = \alpha^j \mathbb{F}_p^{*4}$ for $j \in \{0, 1, 2, 3\}$. Then define $f = (f_0, f_1, \ldots, f_{p-1})$ by

$$f_j = \begin{cases} +1 & \text{if } j \in R_0 \cup R_1 \cup \{0\} \\ -1 & \text{if } j \in R_2 \cup R_3, \end{cases} \tag{16}$$

and define $g = (g_0, g_1, \ldots, g_{p-1})$ by

$$g_j = \begin{cases} +1 & \text{if } j \in R_0 \cup R_3 \cup \{0\} \\ -1 & \text{if } j \in R_1 \cup R_2. \end{cases} \tag{17}$$

The Legendre sequence defined in (15) reappears in this formalism as $h = (h_0, h_1, \ldots, h_{p-1})$, where

$$h_j = \begin{cases} +1 & \text{if } j \in R_0 \cup R_2 \cup \{0\} \\ -1 & \text{if } j \in R_1 \cup R_3, \end{cases} \qquad (18)$$

because $R_0 \cup R_2 = \mathbb{F}_p^{*2}$.

As mentioned above, these sequences are all formed by applying linear combinations of multiplicative characters to the list (13). In the case of our Legendre sequence h, the linear combination is just the single quadratic character. For sequences f and g, we let θ be the multiplicative character of $\mathbb{F}_p$ of order 4 defined by $\theta(\alpha^j) = e^{\pi i j/2} = i^j$, where we recall that α is the primitive element of $\mathbb{F}_p$ used to define the cosets R_j. The other multiplicative character of $\mathbb{F}_p$ of order 4 is $\overline{\theta}$, which has $\overline{\theta}(\alpha^j) = (-i)^j$. To get the sequence f, one first applies the linear combination of characters

$$\lambda(x) = \frac{1-i}{2}\theta(x) + \frac{1+i}{2}\overline{\theta}(x)$$

to the elements of the list (13) to get the sequence

$$(\lambda(0) = 0, \lambda(1), \ldots, \lambda(p-1)),$$

and then one replaces $\lambda(0) = 0$ with 1. To get the sequence g, one uses the same procedure, but with

$$\mu(x) = \frac{1+i}{2}\theta(x) + \frac{1-i}{2}\overline{\theta}(x)$$

in place of λ.

Now that we have introduced our sequences, we need to discuss the modified versions of them that have been found to have good correlation properties. First of all, unlike the definition of m-sequences in Sect. 6, the above definitions of sequences derived from multiplicative characters do not embrace all cyclic shifts of a given sequence. Instead each sequence comes defined with a particular cyclic shift. We will want to cyclically shift our multiplicative character sequences, but then we call them *shifted multiplicative character sequences* to distinguish them from the originals. For example, Golay [12] reported Turyn's discovery that one can significantly increase the autocorrelation merit factor of Legendre sequences if one cyclically shifts them. We can also apply Parker's doubling and quadrupling constructions, as well as the appending technique described in Sect. 6 to sequences produced from the constructions mentioned above.

As mentioned in Sect. 5, Høholdt and Jensen [17] proved that appropriately cyclically shifted Legendre sequences achieve asymptotic autocorrelation demerit factor 1/6. One can also obtain the same asymptotic demerit factor with sequences formed from Legendre sequences with Parker's doubling construction, as proved by Xiong and Hall [45, Theorem 3.3], or with Parker's quadrupling construction (combined with appropriate shifting), as proved by Schmidt, Jedwab, and Parker [39, Theorem 8].

If one also allows appending (with or without Parker's constructions and with appropriate shifting) of Legendre sequences, then Jedwab, Katz, and Schmidt (in [20, Theorem 1.1] an [19, Theorem 2.1]) showed that one can achieve an asymptotic autocorrelation demerit factor of $0.157\ldots$, which is the smallest real root of the polynomial $27x^3 - 417x^2 + 249x - 29$. To achieve this, one appends the sequences to be about $1.057\ldots$ times their usual length, where $1.057\ldots$ is the middle root of $4x^3 - 30x + 27$.

We note that one can generalize the notion of Legendre sequences, which are based on quadratic characters of prime fields, to *Jacobi sequences*, which are based on quadratic characters of integer residue rings (which allows for composite lengths). Jacobi sequences and their modifications (using shifting, Parker's constructions, and appending) often behave similarly to Legendre sequences, and are able to achieve the same asymptotic autocorrelation demerit factor of $1/6$ in their natural lengths and $0.157\ldots$ when appended. Although there are some detailed conditions that must be respected to obtain this behavior, Jacobi sequences provide good autocorrelation at a wider variety of lengths than one could obtain with Legendre sequences alone. See the papers of Jensen, Jensen, and Høholdt [24, Theorem 2.4, Sects. IV–V]; Xiong and Hall [45, Sect. V], [46]; Jedwab and Schmidt [22]; and Jedwab et al. [19, Theorem 2.3, Corollary 2.4, and Sect. 6] for the principal results.

Boothby and Katz [2, Theorem 19] and Günther and Schmidt [16, p. 347] show that carefully selected families of the cyclotomic sequences can produce asymptotic demerit factor $1/6$ when suitably cyclically shifted, and if one also appends appropriately, this can be improved to the same $0.157\ldots$ value as for appended Legendre sequences. When designing a particular family of such sequences it is necessary to make judicious choices of which cyclotomic classes to assign $+1$ values and which cyclotomic classes to assign -1 values, otherwise the demerit factors will be bounded away from $0.157\ldots$. Boothby and Katz [2, Theorem 10] give conditions under which one can achieve limiting autocorrelation demerit factor $0.157\ldots$ for sequences derived from linear combinations of multiplicative characters, which includes the binary cyclotomic sequences as a proper subclass. Boothby and Katz's conditions are met by the cyclotomic sequences derived from quartic characters described at (16) and (17), and these sequences were used by both Boothby and Katz [2, Theorem 19] and Günther and Schmidt [16, p. 347] as examples showing that one can achieve asymptotic demerit factor $0.157\ldots$. Günther and Schmidt also give examples of sequence designs from six cyclotomic classes, and exhibit families that achieve limiting demerit factor $0.157\ldots$ and families that do not.

Even if one uses sequence designs from four or six cyclotomic classes that can achieve asymptotic autocorrelation demerit factor $0.157\ldots$, one must restrict one's sequence family to contain only sequences derived from fields $\mathbb{F}_p$ whose orders fulfill exacting number-theoretic conditions (see [2, Theorem 19] and [16, Corollaries 2.5 and 2.6]). As such, if one adopts one of these cyclotomic sequence designs, the sequences produced will, at most lengths, fall short of Legendre sequences in terms of autocorrelation performance. Accordingly Boothby and

Katz [2, p. 6162] point out that there is little reason to use these cyclotomic sequences in applications where one wants a single sequence with good autocorrelation performance; rather, the real interest of cyclotomic sequences is that there is more than one of them of a given length, so they can be used in applications where crosscorrelation is important. We shall discuss this further in Sect. 11.

Günther and Schmidt [16, pp. 344–345] also studied another family of pseudorandom sequences called the *Sidel'nikov sequences* [43]. These sequences are derived from quadratic characters of finite fields, but in a different way than Legendre sequences. Günther and Schmidt proved that Sidel'nikov sequences have the same asymptotic autocorrelation demerit factors as m-sequences: 1/3 in their natural length, and 0.299... for appropriately appended versions.

8   Rudin-Shapiro-Like Sequences

In his master's thesis [42, p. 42], Shapiro devised a construction of a family $f_0, f_1, f_2, \ldots$ of sequences, where f_n is a binary sequence of length 2^n. Shapiro's construction is easier to understand when one introduces a companion family of sequences, $g_0, g_1, g_2, \ldots$. Recall from Sect. 2 our identification of sequences with polynomials. The construction is the recursion with

$$
\begin{aligned}
f_0(z) &= g_0(z) = 1 \\
f_{n+1}(z) &= f_n(z) + z^{2^n} g_n(z) \\
g_{n+1}(z) &= f_n(z) - z^{2^n} g_n(z).
\end{aligned}
\tag{19}
$$

In terms of sequences, this says that f_{n+1} is the concatenation of f_n and g_n, while g_{n+1} is the concatenation of f_n and $-g_n$. Shapiro's sequences (polynomials) are what one gets when one retains $f_0, f_1, f_2, \ldots$ and discards the companion sequences. These sequences were rediscovered by Rudin [37, Eq. (1.5)] somewhat later, and are now known as *Rudin-Shapiro sequences* (or *Rudin-Shapiro polynomials*).

Around the same time as Shapiro, Golay [7] discovered an equivalent construction that produced what he called *complementary pairs* (now called *Golay complementary pairs* or just *Golay pairs*). These are pairs (f, g) of sequences of the same length with $C_{f,f}(s) + C_{g,g}(s) = 0$ for all $s \neq 0$, and Golay originally devised them for use in multislit spectrometry. We say that a Golay pair has length ℓ to mean that it consists of two sequences, each of length ℓ. If one pairs the Shapiro sequences with their companions, that is, if one considers $(f_0, g_0), (f_1, g_1), (f_2, g_2), \ldots$, then one obtains one infinite family of complementary pairs constructed by Golay. As mentioned in Sect. 5 above, Littlewood [33, pp. 27–28] performed a calculation tantamount to showing that $\mathrm{ADF}(f_n) = (1 - (-1/2)^n)/3$, which proves that the family $f_0, f_1, f_2, \ldots$ of Rudin-Shapiro sequences has asymptotic demerit factor 1/3.

Brillhart and Carlitz [4, Theorem 1] showed that the companion sequences in the construction (19) were related to the main sequences by

$$
g_n(z) = (-1)^n z^{2^n - 1} f_n(-1/z)
$$

for every n. For any polynomial $h(z) \in \mathbb{C}[z]$, we define the *reciprocal polynomial of* $h(z)$, denoted $h^*(z)$, to be $z^{\deg h} h(1/z)$, that is, the polynomial obtained from h by writing the coefficients in reverse order. Then the result of Brillhart and Carlitz becomes $g_n(z) = (-1)^{2^n+n-1} f_n^*(-z)$, so that we could restate the construction without the companion sequences:

$$f_0(z) = 1$$
$$f_{n+1}(z) = f_n(z) + (-1)^{2^n+n-1} z^{\deg f_n+1} f_n^*(-z). \tag{20}$$

It turns out that one gets similar asymptotic autocorrelation behavior no matter how the sign is chosen on the second term, as observed by Høholdt, Jensen, and Justesen [18, Theorem 2.3], so we may generalize construction (20) to

$$f_0(z) = 1$$
$$f_{n+1}(z) = f_n(z) + \sigma_n z^{\deg f_n+1} f_n^*(-z), \tag{21}$$

where $\sigma_0, \sigma_1, \ldots$ is any sequence of values in $\{+1, -1\}$, called the *sign sequence* for our construction. Høholdt, Jensen, and Justesen show [18, Theorem 2.3] that regardless of the choice of sign sequence, one still obtains $\mathrm{ADF}(f_n) = (1 - (-1/2)^n)/3$, so the asymptotic autocorrelation demerit factor is $1/3$.

Construction (21) was further generalized by Borwein and Mossinghoff [3, pp. 1159 and 1161], by allowing much more freedom at the start:

$$f_0(z) = \text{any polynomial with coefficients in } \{+1, -1\}$$
$$f_{n+1}(z) = f_n(z) + \sigma_n z^{\deg f_n+1} f_n^*(-z), \tag{22}$$

where again $\sigma_0, \sigma_1, \ldots$ is any sequence of values in $\{+1, -1\}$, called the *sign sequence* for our construction. We call f_0 the *seed* of the construction, and we call the family $f_0, f_1, f_2, \ldots$ of polynomials the *stem* generated by that seed and sign sequence. Following Borwein and Mossinghoff, we call families of sequences (polynomials) generated from construction (22) *Rudin-Shapiro-like sequences (polynomials)*.

Borwein and Mossinghoff found a precise formula [3, Theorem 1] for the autocorrelation demerit factor of Rudin-Shapiro-like polynomials produced by their construction (22). That is, they have a formula for computing $\mathrm{ADF}(f_n)$ for every n, and from this they compute the asymptotic autocorrelation merit factor, which depends on the seed but not the sign sequence. They show that the asymptotic autocorrelation demerit factor is always greater than or equal to $1/3$, but only achieves a value of $1/3$ for certain seeds, which we call *optimal seeds*. Borwein and Mossinghoff performed a computer search (informed by some of their theoretical results) over all binary sequences of length 40 or less, and found optimal seeds of lengths 1, 2, 4, 8, 16, 20, 32, and 40, and no optimal seeds of other lengths in this range. Later Katz et al. [28, Table 1] performed a larger search that extended to all seeds of length 52 or less, and found new optimal seeds at length 52 (but at no other length between 40 and 52). This data was explained by the following classification of optimal seeds [30, Theorem 1]: a seed of length greater than 1 is optimal if and only if it is the interleaving of a Golay complementary pair, where the *interleaving* of two sequences $a = (a_0, a_1, \ldots, a_{\ell-1})$ and $b = (b_0, b_1, \ldots, b_{\ell-1})$ of length ℓ is the the sequence $(a_0, b_0, a_1, b_1, \ldots, a_{\ell-1}, b_{\ell-1})$ of length 2ℓ. In polynomial terms, the interleaving is $a(z^2) + zb(z^2)$. This result, along with the known fact that the two seeds ($+1$ and -1) of length 1 are optimal, gives a full classification of the optimal seeds. A construction of Turyn [44, Corollary to Lemma 5] shows that there is a Golay complementary pair of length $2^a 10^b 26^c$ for every choice of nonnegative integers a, b, c. Thus there are infinitely many optimal seeds.

9   High Asymptotic Crosscorrelation Merit Factor

Consider the sequences

$$f_\ell = (+1, +1, +1, +1, \ldots, +1, +1)$$
$$g_\ell = (+1, -1, +1, -1, \ldots, +1, -1)$$

of even length ℓ. It is not difficult to calculate that

$$\mathrm{CDF}(f_\ell, g_\ell) = \frac{1}{\ell},$$

so that the asymptotic crosscorrelation demerit factor of the family of pairs $\{(f_\ell, g_\ell) : \ell \in 2\mathbb{Z}\}$ is zero (so asymptotic crosscorrelation merit factor is infinite). But it is also not difficult to calculate that

$$\mathrm{ADF}(f_\ell) = \mathrm{ADF}(g_\ell) = \frac{2\ell^2 + 1}{3\ell},$$

so that the families $\{f_\ell : \ell \in 2\mathbb{Z}\}$ and $\{g_\ell : \ell \in 2\mathbb{Z}\}$ both have infinite asymptotic autocorrelation demerit factor (so asymptotic autocorrelation merit factor is 0). Thus it is not interesting to seek families of sequence pairs with low asymptotic crosscorrelation demerit factor in isolation from the asymptotic autocorrelation demerit factor of the constituent sequences. What we really want to know is whether there is a way to make asymptotic autocorrelation and crosscorrelation demerit factors small at the same time. In the next section we explore a measure that will help us quantify this goal.

10   Pursley-Sarwate Criterion

Pursley and Sarwate [36, Eqs. (3) and (4)] proved that any pair (f, g) of binary sequences has

$$1 - \sqrt{\mathrm{ADF}(f)\,\mathrm{ADF}(g)} \le \mathrm{CDF}(f, g) \le 1 + \sqrt{\mathrm{ADF}(f)\,\mathrm{ADF}(g)}. \qquad (23)$$

Their proof is based on the Cauchy-Schwarz inequality. We define the *Pursley-Sarwate criterion* for a pair (f, g) of sequences to be

$$\mathrm{PSC}(f, g) = \sqrt{\mathrm{ADF}(f)\,\mathrm{ADF}(g)} + \mathrm{CDF}(f, g),$$

and then the bound (23) tells us that

$$\mathrm{PSC}(f,g) \geq 1. \tag{24}$$

We would like sequence pairs (f,g) with $\mathrm{ADF}(f)$, $\mathrm{ADF}(g)$, and $\mathrm{CDF}(f,g)$ as small as possible, but the bound (24) shows that we cannot make them all simultaneously close to zero. In view of Sarwate's expected values of demerit factors for randomly selected binary sequences in (10) and (11), we expect a typical randomly selected pair (f,g) of sequences to have $\mathrm{PSC}(f,g)$ of about 2. We would like to construct sequence pairs (f,g) with $\mathrm{PSC}(f,g)$ as close to 1 as possible. We often consider *asymptotic* PSC of families of sequence pairs, that is, the limiting value of PSC as the length of the sequences tends to infinity.

11 Pairs with Low Asymptotic Pursley-Sarwate Criterion

One should recall the sequence constructions described in Sects. 6, 7, and 8 above: we now consider pairs of such sequences that have low Pursley-Sarwate criterion. Table 2 lists some constructions that produce families of binary sequence pairs with low asymptotic PSC.

Table 2. Families of sequence pairs with low asymptotic PSC

Sequence pair (f,g) construction	Asymptotic Values		
	$\mathrm{ADF}(f) = \mathrm{ADF}(g)$	$\mathrm{CDF}(f,g)$	$\mathrm{PSC}(f,g)$
Katz (2016) [27, pp. 5240, 5247]			
m-sequences, typical	1/3	1	4/3
m-sequence, reversing	1/3	5/6	7/6
half Legendre	7/12	7/12	7/6
Boothby-Katz (2017) [2, pp. 6160–6161]			
quartic cyclotomics	in $[1/6, 5/6]$	in $[1/3, 1]$	7/6
Legendre + quartic	1/6	1	7/6
Katz-Lee-Trunov [28, Table 3]			
Rudin-Shapiro-like	1/3	77/100	331/300
Katz-Moore [29, Theorem 1.1]			
Golay pair	1/3	2/3	1

Let us provide some context and details for the table entries. If we fix a $d \in \mathbb{Z}$ with $|d|$ not a power of 2 and produce an infinite family of binary m-sequence pairs (f_n, g_n) where g_n is (up to cyclic shift) a decimation of f_n by d, then Katz [27, Theorem 1] showed that the asymptotic crosscorrelation demerit factor will tend to 1. Since we have seen in Sect. 6 that the autocorrelation demerit factor tends to 1/3, this produces a sequence family with asymptotic PSC of 4/3.

We call this a *typical m-sequence construction*. If we instead use $d = -2^k$ for some nonnegative integer k, then we produce a family of binary m-sequence pairs (f_n, g_n) where g_n is related to f_n by the reversing decimation, and then Katz [27, Theorem 2] showed that one can lower the asymptotic crosscorrelation demerit factor to $5/6$ by appropriately cyclically shifting the sequences. This results in families with asymptotic PSC of $7/6$. We call this a *reversing m-sequence construction*. (We never allow d to be a power of 2, since that will give degenerate decimations, and we will be correlating an m-sequence with cyclic shifts of itself.)

Another construction of Katz [27, p. 5247] takes a Legendre sequence (which has length equal to some odd prime p), cyclically shifts it in a certain way, discards the last term, and cuts the remaining sequence into a pair of two sequences of length $(p-1)/2$. In this way one can obtain a family of sequence pairs (f_n, g_n) with asymptotic $ADF(f_n)$, $ADF(g_n)$, and $CDF(f_n, g_n)$ all equal to $7/12$, and thus asymptotic PSC equal to $7/6$. We call this the *half Legendre construction*.

Boothby and Katz [2, Theorem 21] crosscorrelated the two cyclotomic sequences (16) and (17) derived from quartic characters, and also the cyclically shifted versions of these two sequences. As mentioned in Sect. 7, one obtains very low asymptotic autocorrelation demerit factor only for certain lengths, depending on a number-theoretic criterion. It turns out that the crosscorrelation demerit factor of our sequence pairs tends to decrease as their autocorrelation demerit factor increases. In fact, for any real number A with $1/6 \leq A \leq 5/6$, there is an infinite family of pairs (f_n, g_n) of these cyclically shifted cyclotomic sequences such that asymptotic $ADF(f_n)$ and $ADF(g_n)$ are A, asymptotic $CDF(f_n, g_n)$ is $7/6 - A$, and asymptotic PSC is $7/6$.

One can also crosscorrelate cyclically shifted Legendre sequences (see (15), or equivalently (18)) with cyclically shifted versions of either of our quartic cyclotomic sequences (see (16) or (17)). In this case Boothby and Katz [2, Theorem 20] show that one always obtains asymptotic CDF of 1, so one should choose the shifted Legendre sequences and shifted quartic cyclotomic sequences to have limiting ADF of $1/6$, and thus obtain limiting PSC of $7/6$.

It should be noted that all the constructions of Katz and Boothby-Katz discussed here employ sequences in their usual length, but their results allow for the possibility of truncating and appending the sequences. They showed [2, Eq. (7)] that modest appending can be used to produce families of sequence pairs with asymptotic PSC slightly lower than $7/6$.

Katz, Lee, and Trunov crosscorrelated pairs of Rudin-Shapiro-like polynomials [28, Theorem 2.4], by beginning with two different seeds, f_0 and g_0, and applying recursion (22) to produce two stems $f_0, f_1, \ldots$ and $g_0, g_1, \ldots$ (using the same sign sequence in recursion (22) to produce the two stems). They derive a precise formula for $CDF(f_n, g_n)$ for each n. This reduces to Borwein and Mossinghoff's precise formula [3, Theorem 1] for $ADF(f_n)$ when we set $g_n = f_n$ and subtract 1 (see (8)). From the formula of Katz, Lee, and Trunov one can compute $PSC(f_n, g_n)$ precisely and from this determine the limiting PSC. Katz, Lee, and Trunov [28, Table 3] found a pair of seeds, each of length 40, that yield stems with limiting ADF of 1/3 and limiting CDF of 77/100, for a limiting PSC of $331/300 = 1.10\overline{3}$.

Finally, Katz and Moore [29, Theorem 1.1] proved that a pair (f, g) of binary sequences has $\text{PSC}(f, g) = 1$ if and only if (f, g) is a Golay complementary pair. In Sect. 8, we noted that there are known to be Golay pairs of lengths $2^a 10^b 26^c$ for all nonnegative integers a, b, and c, so we have infinitely many binary sequence pairs with PSC exactly equal to 1. Thus we obtain an asymptotic PSC of 1 with the Golay pairs. We note that if (f, g) is a Golay pair, then $\text{ADF}(f) = \text{ADF}(g)$. The Golay pairs (f_n, g_n) produced by recursion (19) have Rudin-Shapiro sequences as the first sequence in each pair. Thus they have asymptotic $\text{ADF}(f_n)$ equal to 1/3 by the result of Littlewood described in Sects. 5 and 8. So asymptotic $\text{ADF}(g_n)$ is also 1/3 for these pairs, and thus asymptotic $\text{CDF}(f_n, g_n)$ is 2/3.

We note that the families of sequence pairs on Table 2 all have equal asymptotic autocorrelation demerit factors for the first and second elements of the pairs. While this must be the case for Golay complementary pairs, is not always observed in other constructions with low asymptotic PSC. For example, Katz, Lee, and Trunov [28, Table 2] exhibit constructions of families $(f_0, g_0), (f_1, g_1), \ldots$ of pairs of Rudin-Shapiro-like sequences with low asymptotic $\text{PSC}(f_n, g_n)$ where the asymptotic $\text{ADF}(f_n)$ is not equal to the asymptotic $\text{ADF}(g_n)$.

12   Open Questions

We present two open questions that arise naturally from the considerations above.

Question 1. What is the lowest asymptotic autocorrelation demerit factor for binary sequences?

Or equivalently, what is the highest asymptotic autocorrelation merit factor for binary sequences? Littlewood [33, pp. 28–29] made a conjecture (stated in terms of norms of polynomials) that there is a infinite family of binary sequences with autocorrelation demerit tending to zero, or equivalently, autocorrelation merit factor tending to infinity. Golay, on the other hand, conjectured that autocorrelation merit factor is bounded, and he proposed [10] that asymptotic merit factor can never exceed $2e^2 = 14.77\ldots$. Later [11] he revised his proposed upper bound on asymptotic merit factor to a value of about 12.32.

We saw in Sect. 9 a construction of sequence pairs with asymptotic crosscorrelation demerit factor of zero, but at the expense of poor autocorrelation performance. It would be interesting to know how low asymptotic crosscorrelation demerit factor can be made without having poor autocorrelation demerit factors.

Question 2. Among infinite families of binary sequence pairs (f, g) such that $\text{ADF}(f)$, $\text{ADF}(g)$, and $\text{CDF}(f, g)$ tend to limits as the length of the sequences tends to infinity, and such that the limiting values for $\text{ADF}(f)$ and $\text{ADF}(g)$ are not greater than 1, what is the lowest possible limiting value for $\text{CDF}(f, g)$?

In Sect. 11 we saw that the construction of Boothby-Katz [2, p. 6160] involving pairs of cyclotomic sequences derived from quartic characters furnishes families of sequence pairs (f, g) with asymptotic $\mathrm{CDF}(f, g)$ of $1/3$ and asymptotic $\mathrm{ADF}(f)$ and $\mathrm{ADF}(g)$ of $5/6$ (so limiting $\mathrm{PSC}(f, g)$ is $7/6$). If one wants to get even lower asymptotic CDF, then one can use appending. One would use Theorems 19 and 21 of [2] with the following parameters: one would let $\gamma = \pi/2$, let $\Lambda = 1.207\ldots$ be the middle root of $4x^3 - 36x^2 + 60x - 27$, and let $R = (1 - 2\Lambda)/4$. This would produce a family of pairs of quartic cyclotomic sequences that are cyclically shifted and then appended to $\Lambda = 1.207\ldots$ times their normal length. The limiting ADF of these sequences is 1 and the limiting CDF is $0.254\ldots$, the middle root of $729x^3 + 981x^2 - 1245x + 241$. Note that the PSC for this family is $1.254\ldots$, which is considerably worse than the $7/6$ that one obtains without appending, so a considerable sacrifice in autocorrelation performance is being made for this increase in crosscorrelation performance. One should note that when one appends like this, there will be a rather large value in the autocorrelation spectrum when the appended portion on one copy of the sequence comes into alignment with the initial portion of the other copy. The magnitude of this large autocorrelation value will be about equal to the length of the appended sequence times $1 - 1/\Lambda = 0.172\ldots$.

This paper has confined itself to analyzing the autocorrelation and crosscorrelation demerit factors of sequence pairs. In many applications one needs larger families of sequences with low mutual correlation, and it would be interesting to extend the concepts here to that more general setting.

Acknowledgement. The author thanks Yakov Sapozhnikov for his careful reading of this paper and his helpful suggestions.

References

1. Boehmer, A.M.: Binary pulse compression codes. IEEE Trans. Inform. Theory **13**(2), 156–167 (1967)
2. Boothby, K.T.R., Katz, D.J.: Low correlation sequences from linear combinations of characters. IEEE Trans. Inform. Theory **63**(10), 6158–6178 (2017)
3. Borwein, P., Mossinghoff, M.: Rudin-Shapiro-like polynomials in L_4. Math. Comput. **69**(231), 1157–1166 (2000)
4. Brillhart, J., Carlitz, L.: Note on the Shapiro polynomials. Proc. Amer. Math. Soc. **25**, 114–118 (1970)
5. Ding, C., Helleseth, T., Lam, K.Y.: Several classes of binary sequences with three-level autocorrelation. IEEE Trans. Inform. Theory **45**(7), 2606–2612 (1999)
6. Ding, C., Helleseth, T., Lam, K.Y.: Duadic sequences of prime lengths. Discrete Math. **218**(1–3), 33–49 (2000)
7. Golay, M.J.E.: Static multislit spectrometry and its application to the panoramic display of infrared spectra. J. Opt. Soc. Am. **41**(7), 468–472 (1951)
8. Golay, M.J.E.: A class of finite binary sequences with alternate autocorrelation values equal to zero. IEEE Trans. Inform. Theory **18**, 449–450 (1972)
9. Golay, M.J.E.: Hybrid low autocorrelation sequences. IEEE Trans. Inform. Theory **21**, 460–462 (1975)

10. Golay, M.J.E.: Sieves for low autocorrelation binary sequences. IEEE Trans. Inform. Theory **23**, 43–51 (1977)
11. Golay, M.J.E.: The merit factor of long low autocorrelation binary sequences. IEEE Trans. Inform. Theory **28**(3), 543–549 (1982)
12. Golay, M.J.E.: The merit factor of Legendre sequences. IEEE Trans. Inform. Theory **29**, 934–936 (1983)
13. Gold, R.: Optimal binary sequences for spread spectrum multiplexing. IEEE Trans. Inform. Theory **13**(4), 619–621 (1967)
14. Golomb, S.W.: Shift register sequences. With portions co-authored by Lloyd R. Welch, Richard M. Goldstein, and Alfred W. Hales, Holden-Day Inc., San Francisco, California-Cambridge-Amsterdam (1967)
15. Golomb, S.W., Gong, G.: Signal Design for Good Correlation. Cambridge University Press, Cambridge (2005)
16. Günther, C., Schmidt, K.U.: Merit factors of polynomials derived from difference sets. J. Combin. Theory Ser. A **145**, 340–363 (2017)
17. Høholdt, T., Jensen, H.E.: Determination of the merit factor of Legendre sequences. IEEE Trans. Inform. Theory **34**(1), 161–164 (1988)
18. Høholdt, T., Jensen, H.E., Justesen, J.: Aperiodic correlations and the merit factor of a class of binary sequences. IEEE Trans. Inform. Theory **31**(4), 549–552 (1985)
19. Jedwab, J., Katz, D.J., Schmidt, K.U.: Advances in the merit factor problem for binary sequences. J. Combin. Theory Ser. A **120**(4), 882–906 (2013)
20. Jedwab, J., Katz, D.J., Schmidt, K.U.: Littlewood polynomials with small L^4 norm. Adv. Math. **241**, 127–136 (2013)
21. Jedwab, J., Schmidt, K.U.: The merit factor of binary sequence families constructed from m-sequences. Finite fields: theory and applications. Contemp. Math. **518**, 265–278 (2010)
22. Jedwab, J., Schmidt, K.U.: The L_4 norm of Littlewood polynomials derived from the Jacobi symbol. Pacific J. Math. **257**(2), 395–418 (2012)
23. Jensen, H.E., Høholdt, T.: Binary sequences with good correlation properties. In: Huguet, L., Poli, A. (eds.) AAECC 1987. LNCS, vol. 356, pp. 306–320. Springer, Heidelberg (1989). https://doi.org/10.1007/3-540-51082-6_87
24. Jensen, J.M., Jensen, H.E., Høholdt, T.: The merit factor of binary sequences related to difference sets. IEEE Trans. Inform. Theory **37**((3 part 1)), 617–626 (1991)
25. Kärkkäinen, K.H.A.: Mean-square cross-correlation as a performance measure for department of spreading code families. In: IEEE Second International Symposium on Spread Spectrum Techniques and Applications, pp. 147–150 (1992)
26. Katz, D.J.: Asymptotic L^4 norm of polynomials derived from characters. Pacific J. Math. **263**(2), 373–398 (2013)
27. Katz, D.J.: Aperiodic crosscorrelation of sequences derived from characters. IEEE Trans. Inform. Theory **62**(9), 5237–5259 (2016)
28. Katz, D.J., Lee, S., Trunov, S.A.: Crosscorrelation of Rudin-Shapiro-like polynomials. Preprint, arXiv:1702.07697 (2017)
29. Katz, D.J., Moore, E.: Sequence pairs with lowest combined autocorrelation and crosscorrelation. Preprint, arXiv:1711.02229 (2017)
30. Katz, D.J., Lee, S., Trunov, S.A.: Rudin-Shapiro-like polynomials with maximum asymptotic merit factor. Preprint, arXiv:1711.02233 (2017)
31. Kirilusha, A., Narayanaswamy, G.: Construction of new asymptotic classes of binary sequences based on existing asymptotic classes. Summer Science Technical report, Department of Mathematics & Computer Science, University of Richmond, VA (1999)

32. Littlewood, J.E.: On polynomials $\sum^n \pm z^m$, $\sum^n e^{\alpha_m i} z^m$, $z = e^{\theta_i}$. J. London Math. Soc. **41**, 367–376 (1966)

33. Littlewood, J.E.: Some Problems in Real and Complex Analysis. D. C. Heath and Co., Raytheon Education Co., Lexington, Massachusetts (1968)

34. Mauduit, C., Sárközy, A.: On finite pseudorandom binary sequences. I. Measure of pseudorandomness, the Legendre symbol. Acta Arith. **82**(4), 365–377 (1997)

35. Parker, M.G.: Even length binary sequence families with low negaperiodic autocorrelation. In: Boztaş, S., Shparlinski, I.E. (eds.) AAECC 2001. LNCS, vol. 2227, pp. 200–209. Springer, Heidelberg (2001). https://doi.org/10.1007/3-540-45624-4_21

36. Pursley, M.B., Sarwate, D.V.: Bounds on aperiodic cross-correlation for binary sequences. Electron. Lett. **12**(12), 304–305 (1976)

37. Rudin, W.: Some theorems on Fourier coefficients. Proc. Amer. Math. Soc. **10**, 855–859 (1959)

38. Sarwate, D.V.: Mean-square correlation of shift-register sequences. IEE Proc. F-Commun. Radar Signal Process. **131**(2), 101–106 (1984)

39. Schmidt, K.U., Jedwab, J., Parker, M.G.: Two binary sequence families with large merit factor. Adv. Math. Commun. **3**(2), 135–156 (2009)

40. Scholtz, R.A., Welch, L.R.: GMW sequences. IEEE Trans. Inform. Theory **30**(3), 548–553 (1984)

41. Schroeder, M.R.: Number Theory in Science and Communication. With Applications in Cryptography, Physics, Digital Information, Computing, and Self-similarity. Springer Series in Information Sciences, vol. 7, 4th edn. Springer, Heidelberg (2006). https://doi.org/10.1007/b137861

42. Shapiro, H.S.: Extrenal problems for polynomials and power series. Master's thesis. Institute of Technology, Cambridge, Massachusetts (1951)

43. Sidel'nikov, V.M.: Some k-valued pseudo-random sequences and nearly equidistant codes. Problemy Peredači Informacii **5**(1), 16–22 (1969)

44. Turyn, R.J.: Hadamard matrices, Baumert-Hall units, four-symbol sequences, pulse compression, and surface wave encodings. J. Combin. Theory Ser. A **16**, 313–333 (1974)

45. Xiong, T., Hall, J.I.: Construction of even length binary sequences with asymptotic merit factor 6. IEEE Trans. Inform. Theory **54**(2), 931–935 (2008)

46. Xiong, T., Hall, J.I.: Modifications of modified Jacobi sequences. IEEE Trans. Inform. Theory **57**(1), 493–504 (2011)

Arithmetic and Applications of Finite Fields

Vector-Valued Modular Forms on Finite Upper Half Planes

Yoshinori Hamahata$^{(\boxtimes)}$

Department of Applied Mathematics, Okayama University of Science, Ridai-cho 1-1,
Okayama 700-0005, Japan
hamahata@xmath.ous.ac.jp

Abstract. Finite upper half planes are finite field analogs of the Poincaré upper half plane. Vector-valued modular forms on finite upper half planes are introduced, and then equivariant functions on these planes are defined. The existence of these functions is an application of vector-valued modular forms.

Keywords: Vector-valued modular form · Equivariant function
Finite upper half plane

1 Introduction

Let $SL(2, \mathbb{Z})$ be the classical modular group. This group acts on the Poincaré upper half plane $\mathfrak{H} = \{z \in \mathbb{C} \mid \text{Im}(\tau) > 0\}$ by the linear fractional transformation

$$\begin{pmatrix} a & b \\ c & d \end{pmatrix} z = \frac{az + b}{cz + d}.$$

Let $\rho : SL(2, \mathbb{Z}) \to GL(n, \mathbb{C})$ be an n-dimensional complex representation. A holomorphic map $F : \mathfrak{H} \to \mathbb{C}^n$ is called a *vector-valued modular form* of weight w (w any real number) and multiplier ρ if for any $\begin{pmatrix} a & b \\ c & d \end{pmatrix} \in SL(2, \mathbb{Z})$, we have

$$F\left(\frac{az + b}{cz + d}\right) = (cz + d)^w \rho\left(\begin{pmatrix} a & b \\ c & d \end{pmatrix}\right) F(z),$$

and if a cuspidal condition holds [2,11,12]. The classical vector-valued modular forms have been investigated as a generalization of scalar-valued modular forms. As pointed out by Selberg [16], these modular forms can be used in the study of modular forms for finite index subgroups of $SL(2, \mathbb{Z})$. The Jacobi forms developed by Eichler and Zagier [6] are related to these modular forms. In physics, they appear as the characters in rational conformal field theory [5,7].

Supported by JSPS KAKENHI Grant Number 15K04801.

L. Budaghyan and F. Rodríguez-Henríquez (Eds.): WAIFI 2018, LNCS 11321, pp. 175–187, 2018.
https://doi.org/10.1007/978-3-030-05153-2_9

A meromorphic function h on $\mathfrak{H}$ is called an *equivariant function* for $SL(2,\mathbb{Z})$ if it satisfies the condition

$$h\left(\frac{az+b}{cz+d}\right) = \frac{ah(z)+b}{ch(z)+d}$$

for $\begin{pmatrix} a & b \\ c & d \end{pmatrix}$ and $z \in \mathfrak{H}$. Such a function is related to modular forms [13–15].

In the mid-1980s, Terras introduced a finite upper half plane H_q that is defined over a finite field $\mathbb{F}_q$ as an analog of the Poincaré upper half plane $\mathfrak{H}$. Specifically, she and her coworkers investigated special functions on H_q in [1,3,18,19]. In [9], modular forms of a new type were studied on H_q. In the present paper, modular forms of other types are considered on H_q. Generalized and subsequently vector-valued modular forms are introduced. In particular, the definition of vector-valued modular forms is new. Moreover, when q is a prime number p, for a complex representation $\rho : GL(2,\mathbb{F}_p) \to GL(2,\mathbb{C})$, equivariant functions on H_p are defined. The existence of these functions is an application of vector-valued modular forms.

Notation. For a field F, let $F^\times = F \setminus \{0\}$.

2   Generalized Modular Forms

In this section, the generalized modular forms on finite upper half planes are introduced. For the classical modular forms, the reader is referred to [10].

2.1   Generalized Modular Forms

Let q be a power of an odd prime number p, and let $\mathbb{F}_q$ be the finite field with q elements. Let a non-square element $\delta \in \mathbb{F}_q$ be fixed, and let

$$H_q = \{z = x + y\sqrt{\delta} \mid x, y \in \mathbb{F}_q, y \neq 0\},$$

which is called a *finite upper half plane*. This plane is a finite field analog of the Poincaré upper half plane $\mathfrak{H}$. It should be noted that $\sqrt{\delta}$ plays the role of $i = \sqrt{-1}$ in $\mathfrak{H}$ and that H_q is a subset of $\mathbb{F}_q(\sqrt{\delta})$, which is analogous to the fact that $\mathfrak{H}$ is a subset of the field of complex numbers $\mathbb{C} = \mathbb{R}(i)$. For $z = x + y\sqrt{\delta} \in H_q$, let

$$x = \mathrm{Re}(z), \ y = \mathrm{Im}(z), \ \bar{z} = x - y\sqrt{\delta}, \ N(z) = z\bar{z} = x^2 - \delta y^2, \ \mathrm{Tr}(z) = z + \bar{z} = 2x.$$

Moreover, let $G_q = GL(2,\mathbb{F}_q)$ be the general linear group over $\mathbb{F}_q$. This group acts on H_q by the following linear fractional transformation: for $z \in H_q$ and $\gamma = \begin{pmatrix} a & b \\ c & d \end{pmatrix} \in G_q$, let

$$\gamma z = \frac{az+b}{cz+d}.$$

The fixed subgroup of $\sqrt{\delta}$ in G_q is

$$K_q = \left\{ \begin{pmatrix} a & b\delta \\ b & a \end{pmatrix} \mid a, b \in \mathbb{F}_q, a^2 - \delta b^2 \neq 0 \right\},$$

which is an analog of the orthogonal group $O(2)$. It is known that the action of G_q on H_q is transitive. Hence, H_q is expressed as $H_q = G_q/K_q$.

Let Γ be a subgroup of G_q. The map $m : \Gamma \times H_q \to \mathbb{C}^\times$ is called a *multiplier system* for Γ if

$$m(\gamma\gamma', z) = m(\gamma, \gamma'z)m(\gamma', z)$$

holds for all $\gamma, \gamma' \in \Gamma$ and $z \in H_q$. In the classical case, the definition of a multiplier system is wider, as in [4,8]. However, in this paper, the definition in [1,9,18,19] was used. Let $\mu : G_q \to \mathbb{C}^\times$ be a multiplicative character. For these Γ, m, and μ, a $\mathbb{C}$-valued function $f : H_q \to \mathbb{C}$ is called a *generalized modular form* for Γ, m, and μ if for any $\gamma \in \Gamma$, it holds that

$$f(\gamma z) = m(\gamma, z)\mu(\gamma)f(z).$$

The space of generalized modular forms of this type is denoted by $M(\Gamma, m, \mu)$. When μ is a trivial character, a generalized modular form, which is called a *modular form*, was studied in [1,9,18,19].

2.2   Special Cases

Herein, the generalized modular forms are discussed when Γ is the unipotent subgroup of G_q or K_q.

2.2.1 Case $\Gamma = N_p$. Let p be an odd prime number, and let

$$N_p = \left\{ \begin{pmatrix} 1 & u \\ 0 & 1 \end{pmatrix} \mid u \in \mathbb{F}_p \right\}.$$

Let $\chi : \mathbb{F}_p^\times \to \mathbb{C}^\times$ be a multiplicative character, and let $\psi : \mathbb{F}_p \to \mathbb{C}^\times$ be an additive character. Using these, a function on H_p is defined by

$$f(z; \chi, \psi) = \sum_{u \in \mathbb{F}_p} \chi \left(\mathrm{Im} \left(\frac{-1}{z + u} \right) \right) \psi(u). \tag{1}$$

This function, which is an analog of the classical K-Bessel function, was first defined in [3].

Theorem 1. (i) *For* $\gamma_u = \begin{pmatrix} 1 & u \\ 0 & 1 \end{pmatrix} \in N_p$, *we have*

$$f(\gamma_u z; \chi, \psi) = \mu(\gamma_u)f(z; \chi, \psi),$$

where $\mu : N_p \to \mathbb{C}^\times$ *is the character defined by* $\mu(\gamma_u) = \psi(-u)$. *That is,* $f(z; \chi, \psi) \in M(N_p, 1, \mu)$.

(ii) *If* χ *and* ψ *are non-trivial, then* $f(z; \chi, \psi)$ *is non-zero.*

Proof. See [3, Lemma 3].   □

2.2.2 Case $\Gamma = K_q$. Let $\pi : \mathbb{F}_q(\sqrt{\delta})^{\times} \to \mathbb{C}^{\times}$ be a multiplicative character. For $\gamma = \begin{pmatrix} a & b \\ c & d \end{pmatrix} \in K_q$, let $J_{\pi}(\gamma, z) = \pi(cz + d)$. Then, for $\gamma, \gamma' \in K_q$, it holds that

$$J_{\pi}(\gamma\gamma', z) = J_{\pi}(\gamma, \gamma'z)J_{\pi}(\gamma', z). \tag{2}$$

Hence, $J_{\pi} : K_q \times H_q \to \mathbb{C}^{\times}$ is a multiplier system for K_q. A map $m : K_q \times H_q \to \mathbb{C}^{\times}$ is defined by

$$m(k, z) = \frac{J_{\pi}(k, z)}{J_{\pi}(k, \sqrt{\delta})}. \tag{3}$$

It is easy to prove that m is a multiplier system for K_q. For a multiplicative character $\mu : K_q \to \mathbb{C}^{\times}$, let

$$E(z; \pi, \mu) = \frac{1}{|K_q|} \sum_{k \in K_q} \frac{J_{\pi}(k, \sqrt{\delta})}{J_{\pi}(k, z)} \cdot \mu(k)^{-1}, \tag{4}$$

where $|K_q|$ is the number of elements of K_q. $E(z; \pi, \mu)$ is called the *Eisenstein sum* for K_q, π, and μ. This is a finite field analog of the Eisenstein series on the Poincaré upper half plane.

Theorem 2. $E(z; \pi, \mu) \in M(K_q, m, \mu)$.

Proof. By (2), for $k' \in K_q$, it holds that

$$|K_q|E(k'z; \pi, \mu) = \sum_{k \in K_q} \frac{J_{\pi}(kk', \sqrt{\delta})}{J_{\pi}(k', \sqrt{\delta})} \cdot \frac{J_{\pi}(k', z)}{J_{\pi}(kk', z)} \cdot \mu(k)^{-1}$$

$$= m(k', z)\mu(k') \sum_{k \in K_q} \frac{J_{\pi}(kk', \sqrt{\delta})}{J_{\pi}(kk', z)} \cdot \mu(kk')^{-1},$$

which yields the result.   □

In general, it is difficult to determine the dimension of $M(\Gamma, m, \mu)$ over $\mathbb{C}$. Using Eisenstein sums, an easy example may be provided.

Example 1. Let $q = p = 3$, $\delta = -1$, and $i = \sqrt{-1}$. Then, $1 + \sqrt{\delta}$ is a generator of $\mathbb{F}_3(\sqrt{\delta})^{\times}$. The multiplicative character $\pi : \mathbb{F}_3(\sqrt{\delta})^{\times} \to \mathbb{C}^{\times}$ is defined by $\pi(1 + \sqrt{\delta}) = \exp(2\pi i/8)$. The group K_3 is a cyclic group generated by $g = \begin{pmatrix} 1 & -1 \\ 1 & 1 \end{pmatrix}$. The multiplicative character $\mu : K_3 \to \mathbb{C}^{\times}$ is defined by $\mu(g) = \exp(2\pi i/4)^{-1} = -i$. H_3 can be decomposed into a union of K_3-orbits as follows:

$$H_3 = \{\sqrt{\delta}\} \cup \{-\sqrt{\delta}\} \cup \{\pm 1 \pm \sqrt{\delta}\}.$$

Using π, a multiplier system $m_4 : K_3 \times H_3 \to \mathbb{C}^\times$ is defined by

$$m_4(g^j, z) = \frac{J_{\pi^4}(g^j, z)}{J_{\pi^4}(g^j, \sqrt{\delta})} \qquad (z \in H_3, j = 1, \ldots, 8).$$

Then, $m_4(g^j, \sqrt{\delta}) = m_4(g^j, -\sqrt{\delta}) = 1$ for $j = 1, \ldots, 8$. Each modular form $f \in M(K_3, m_4, \mu)$ is determined by the values $f(\sqrt{\delta})$, $f(-\sqrt{\delta})$, and $f(1 + \sqrt{\delta})$. For $f \in M(K_3, m_4, \mu)$, we have

$$f(\sqrt{\delta}) = m(g, \sqrt{\delta}) \mu(g) f(\sqrt{\delta}) = -i f(\sqrt{\delta}),$$

which implies that $f(\sqrt{\delta}) = 0$. Similarly, it follows that $f(-\sqrt{\delta}) = 0$. By definition, we have

$$E(1 + \sqrt{\delta}; \pi^4, \mu) = \frac{1}{8} \sum_{j=1}^{8} \frac{J_{\pi^4}(g^j, \sqrt{\delta})}{J_{\pi^4}(g^j, 1 + \sqrt{\delta})} \cdot \mu(g)^{-1} = \frac{1+i}{2}.$$

Consequently, $\dim_{\mathbb{C}} M(K_3, m_4, \mu) = 1$.

3 Vector-Valued Modular Forms

In this section, to extend the generalized modular forms defined in Sect. 2, vector-valued modular forms on finite upper half planes are introduced. For the classical vector-valued modular forms, the reader is referred to [2,11,12].

3.1 Vector-Valued Modular Forms

Let $\mathcal{F}_q$ be the set of all $\mathbb{C}$-valued functions on H_q, Γ be a subgroup of G_q, and $\rho : \Gamma \to GL_n(\mathbb{C})$ be an n-dimensional complex representation. A *vector-valued modular form* for Γ, the multiplier system $m : \Gamma \times H_q \to \mathbb{C}^\times$, and ρ is an element $F(z) = (f_1(z), \ldots, f_n(z))^t \in \mathcal{F}_q^n$ satisfying

$$F(\gamma z) = m(\gamma, z) \rho(\gamma) F(z)$$

for $\gamma \in \Gamma$ and $z \in H_q$. The space of all vector-valued modular forms of this type is denoted by $M(\Gamma, m, \rho)$.

The following is a basic result.

Theorem 3. *Let Γ be a subgroup of G_q.*

(i) For two complex representations ρ_1 and ρ_2 of Γ, there exists a linear isomorphism

$$M(\Gamma, m, \rho_1) \oplus M(\Gamma, m, \rho_2) \cong M(\Gamma, m, \rho_1 \oplus \rho_2).$$

(ii) Let $\rho : \Gamma \to GL(n, \mathbb{C})$ be an n-dimensional complex representation. For $F_i \in M(\Gamma, m_i, \rho)$ $(i = 1, \ldots, n)$, let $F = (F_1, \ldots, F_n)$. Then, $\det F \in M(\Gamma, m_1 \cdots m_n, \det \rho)$.

(iii) Let Γ be an abelian subgroup of G_q, and let $\rho : \Gamma \to GL(n, \mathbb{C})$ be a complex representation. For any vector-valued modular form $F(z) \in M(\Gamma, m, \rho)$, there exists $U \in GL(n, \mathbb{C})$ such that $U F(z)$ can be written as a direct sum of some generalized modular forms.

Proof. (i) is immediate from the definition of vector-valued modular forms.

(ii) For $\gamma \in \Gamma$,

$$F(\gamma z) = (m_1(\gamma, z)\rho(\gamma)F_1(z), \ldots, m_n(\gamma, z)\rho(\gamma)F_n(z))$$

$$= \rho(\gamma)F(z) \begin{pmatrix} m_1(\gamma, z) & & O \\ & \ddots & \\ O & & m_n(\gamma, z) \end{pmatrix},$$

which yields the result.

(iii) By assumption, there exist $U \in GL(n, \mathbb{C})$ and 1-dimensional representations $\mu_1, \ldots, \mu_n : \Gamma \to \mathbb{C}^\times$ such that for all $\gamma \in \Gamma$,

$$\rho(\gamma) = U^{-1} \begin{pmatrix} \mu_1(\gamma) & & O \\ & \ddots & \\ O & & \mu_n(\gamma) \end{pmatrix} U.$$

Let $UF(z) = (f_1(z), \ldots, f_n(z))^t$. Then, it holds that for all $\gamma \in \Gamma$,

$$\begin{pmatrix} f_1(\gamma z) \\ \vdots \\ f_n(\gamma z) \end{pmatrix} = \begin{pmatrix} m(\gamma, z)\mu_1(\gamma)f_1(z) \\ \vdots \\ m(\gamma, z)\mu_n(\gamma)f_n(z) \end{pmatrix}.$$

$\square$

For a subgroup Γ of G_q, let $m : \Gamma \times H_q \to \mathbb{C}^\times$ be a multiplier system. For $f \in \mathcal{F}_q$ and $\gamma \in \Gamma$, the map $\mathcal{F}_q \times \Gamma \to \mathcal{F}_q$, $(f, \gamma) \mapsto f(\gamma z)m(\gamma, z)^{-1}$ defines a right action of Γ on $\mathcal{F}_q$. Using this action, we have the following result, which is an analog of a result proved in [11].

Theorem 4. *Let $M \subset \mathcal{F}_q$ be a finite dimensional Γ-module with generators $f_1, \ldots, f_n$. Then, there exists an n-dimensional complex representation $\rho : \Gamma \to GL(n, \mathbb{C})$ such that $F(z) = (f_1(z), \ldots, f_n(z))^t \in M(\Gamma, m, \rho)$.*

Proof. Changing the order of $f_1, \ldots, f_n$, it may be assumed that $\{f_1, \ldots, f_d\}$ is a basis of M, and that $f_{d+1}, \ldots, f_n$ are written as linear combinations of $f_1, \ldots, f_d$. Let $G = (f_1, \ldots, f_d)^t$. For any $\gamma \in \Gamma$, there exists a unique element $a_{jk}(\gamma) \in \mathbb{C}$ such that

$$f_j(\gamma z)m(\gamma, z)^{-1} = \sum_{k=1}^{d} a_{jk}(\gamma)f_k(z) \quad (j = 1, \ldots, d).$$

Letting $a(\gamma) = (a_{jk}(\gamma))$, a d-dimensional complex representation $a : \Gamma \to GL(d, \mathbb{C})$ is obtained. Using this, we have $G(\gamma z)m(\gamma, z)^{-1} = a(\gamma)G(z)$ for $\gamma \in \Gamma$ and $z \in H_q$.

Let $e = n - d$. By assumption, there exists a matrix $Q \in \mathrm{Mat}_{e \times d}(\mathbb{C})$ such that $(f_{d+1}, \ldots, f_n)^t = QG$. Let $P = \begin{pmatrix} I_d \\ Q \end{pmatrix}$. Then, there exists a matrix $R \in$

$\mathrm{Mat}_{n \times n}(\mathbb{C})$ such that $RP = \begin{pmatrix} I_d \\ O \end{pmatrix}$. An n-dimensional complex representation $\rho : \Gamma \to GL(n, \mathbb{C})$ is defined by

$$\rho(\gamma) = R^{-1} \begin{pmatrix} a(\gamma) & O \\ O & I_e \end{pmatrix} R.$$

As

$$F = \begin{pmatrix} G \\ f_{d+1} \\ \vdots \\ f_n \end{pmatrix} = \begin{pmatrix} I_d \\ Q \end{pmatrix} G = PG,$$

it follows that for $\gamma \in \Gamma$,

$$F(\gamma z)m(\gamma, z)^{-1} = PG(\gamma z)m(\gamma, z)^{-1} = Pa(\gamma)G = R^{-1} \begin{pmatrix} a(\gamma) \\ O \end{pmatrix} G$$

$$= R^{-1} \begin{pmatrix} a(\gamma) & O \\ O & I_e \end{pmatrix} \begin{pmatrix} I_d \\ O \end{pmatrix} G = \rho(\gamma)PG = \rho(\gamma)F.$$

$\square$

Example 2. A vector-valued Maass Eisenstein series on H_q is introduced to generalize the scalar-valued Maass Eisenstein series defined in [17]. Let $\chi : \mathbb{F}_q^\times \to \mathbb{C}^\times$ be a multiplicative character. Let Γ be a subgroup of G_q, an n-dimensional complex representation $\rho : \Gamma \to GL(n, \mathbb{C})$ is considered, and a vector $\mathbf{a} \in \mathbb{C}^n$ is chosen. The *vector-valued Maass Eisenstein sum* for Γ, χ, ρ, and $\mathbf{a}$ is defined by

$$E_\Gamma(z; \chi, \rho, \mathbf{a}) = \sum_{\gamma \in \Gamma} \chi\left(\mathrm{Im}(\gamma z)\right) \rho(\gamma)^{-1} \mathbf{a}^t.$$

It is easily seen that for $\gamma' \in \Gamma$,

$$E_\Gamma(\gamma' z; \chi, \rho, \mathbf{a}) = \rho(\gamma')E_\Gamma(z; \chi, \rho, \mathbf{a}),$$

which implies that $E_\Gamma(z; \chi, \rho, \mathbf{a}) \in M(\Gamma, 1, \rho)$.

Let $n = 2$. When $F(z) := \sum_{\gamma \in \Gamma} \chi\left(\mathrm{Im}(\gamma z)\right) \rho(\gamma)^{-1}$ is not zero, there exists $E_\Gamma(z; \chi, \rho, \mathbf{a})$ whose lowest component is not zero. Indeed, if the $(2,1)$-entry of $F(z)$ is not zero, then $\mathbf{a}$ may be chosen to be $(1,0)$. If the $(2,2)$-entry of $F(z)$ is not zero, then $\mathbf{a}$ may be chosen to be $(0,1)$. If the $(1,1)$-entry of $F(z)$ is not zero, then by replacing ρ with $\rho' = \begin{pmatrix} 0 & -1 \\ 1 & 0 \end{pmatrix} \rho \begin{pmatrix} 0 & -1 \\ 1 & 0 \end{pmatrix}^{-1}$, $\mathbf{a}$ may be chosen to be $(0,1)$. If the $(1,2)$-entry of $F(z)$ is not zero, then by replacing ρ with $\rho' = \begin{pmatrix} 0 & -1 \\ 1 & 0 \end{pmatrix} \rho \begin{pmatrix} 0 & -1 \\ 1 & 0 \end{pmatrix}^{-1}$, $\mathbf{a}$ may be chosen to be $(1,0)$.

3.2 Special Cases

Herein, vector-valued modular forms are discussed when Γ is N_q or K_q.

3.2.1 Case $\Gamma = N_q$. To construct a vector-valued modular forms, a generalization of the Gauss sum is introduced. Let $\chi : \mathbb{F}_q^\times \to \mathbb{C}^\times$ be a multiplicative character, and let $\psi : \mathbb{F}_q \to GL(n, \mathbb{C})$ be a group homomorphism. Using these, the *matrix-valued Gauss sum* $G_q(\chi, \psi)$ is defined by

$$G_q(\chi, \psi) = \sum_{c \in \mathbb{F}_q^\times} \chi(c)\psi(c) \qquad (\in \mathrm{Mat}_{n \times n}(\mathbb{C})) \,.$$

The definition of this Gauss sum may not be new; however, the author is unfamiliar with the related references. The following proposition is easy to prove.

Proposition 1. *Let $\chi_0 : \mathbb{F}_q^\times \to \mathbb{C}^\times$ be a trivial multiplicative character, and let $\psi_0 : \mathbb{F}_q \to GL(n, \mathbb{C})$ be a trivial group homomorphism. Then,*
  (i)

$$G_q(\chi, \psi_n) = \begin{cases} (q-1)I_n & \text{if } \chi = \chi_0 \text{ and } \psi_n = \psi_0, \\ -I_n & \text{if } \chi = \chi_0 \text{ and } \psi_n \neq \psi_0, \\ O_n & \text{if } \chi \neq \chi_0 \text{ and } \psi_n = \psi_0. \end{cases}$$

  (ii) *If $\chi \neq \chi_0$ and $\psi_n \neq \psi_0$, then $\overline{G_q(\chi, \psi)}G_q(\chi, \psi) = qI_n$, where $\overline{G_q(\chi, \psi)}$ is the complex conjugate of $G_q(\chi, \psi)$.*

Henceforth, let $q = p$ and $n = 2$. For a multiplicative character $\chi : \mathbb{F}_p^\times \to \mathbb{C}^\times$, a group homomorphism $\psi : \mathbb{F}_p \to GL(2, \mathbb{C})$, and a vector $\mathbf{a} \in \mathbb{C}^2$, a function $F(z; \chi, \psi, \mathbf{a})$ on H_q is defined by

$$F(z; \chi, \psi, \mathbf{a}) = \sum_{u \in \mathbb{F}_p} \chi\left(\mathrm{Im}\left(\frac{-1}{z+u}\right)\right) \psi(u)\mathbf{a}^t,$$

which is a generalization of (1).

Theorem 5. (i) *For $\gamma_u = \begin{pmatrix} 1 & u \\ 0 & 1 \end{pmatrix} \in N_p$, we have*

$$F(\gamma_u z; \chi, \psi, \mathbf{a}) = \rho(\gamma_u)F(z; \chi, \psi, \mathbf{a}),$$

where $\rho : N_p \to GL(2, \mathbb{C})$ is a 2-dimensional complex representation defined by $\rho(\gamma_u) = \psi(-u)$. That is, $F(z; \chi, \psi, \mathbf{a}) \in M(N_p, 1, \rho)$.
  (ii) *If χ and ψ are non-trivial, then there exists a vector $\mathbf{a} \in \mathbb{C}^2$ such that $F(z; \chi, \psi, \mathbf{a})$ is non-zero.*

Proof. (i) When $z = x + y\sqrt{\delta}$, it holds that

$$\mathrm{Im}\left(\frac{-1}{z+u}\right) = \frac{y}{(x+u)^2 - \delta y^2},$$

which yields

$$F(z; \chi, \psi, \mathbf{a}) = \sum_{u \in \mathbb{F}_p} \chi(y) \overline{\chi((x+u)^2 - \delta y^2)} \psi(u) \mathbf{a}^t$$

$$= \chi(y) \psi(-x) \sum_{v \in \mathbb{F}_p} \overline{\chi(v^2 - \delta y^2)} \psi(v) \mathbf{a}^t.$$

Hence, it holds that

$$F(z + u; \chi, \psi, \mathbf{a}) = \psi(-u) F(z; \chi, \psi, \mathbf{a}).$$

From this, the result follows.

   (ii) Using (i), we obtain

$$\sum_{y \in \mathbb{F}_p^{\times}} \overline{\chi}(y) \psi(x) F(z; \chi, \psi, \mathbf{a}) = \sum_{\substack{y \in \mathbb{F}_p^{\times} \\ v \in \mathbb{F}_p}} \overline{\chi(v^2 - \delta y^2)} \psi(v) \mathbf{a}^t$$

$$= \sum_{\substack{w \in \mathbb{F}_{p^2} \\ \mathrm{Im}(w) \neq 0}} \overline{\chi}(N(w)) \psi\left(\frac{1}{2} \mathrm{Tr}(w)\right) \mathbf{a}^t$$

$$= \sum_{w \in \mathbb{F}_{p^2}} \overline{\chi}(N(w)) \psi\left(\frac{1}{2} \mathrm{Tr}(w)\right) \mathbf{a}^t - \sum_{u \in \mathbb{F}} \overline{\chi}(u^2) \psi(u) \mathbf{a}^t$$

$$= \left(G_{p^2}\left(\overline{\chi} \circ N, \psi \circ \frac{1}{2} \mathrm{Tr}\right) - G_p\left(\overline{\chi}^2, \psi\right)\right) \mathbf{a}^t.$$

From Proposition 1 (ii), the difference of the two Gauss sums in the last equation is not zero, and the result follows. $\qquad \square$

Remark 1. Let $X(\delta, a)$ be a finite upper half plane graph, which is a Ramanujan graph when $a \neq 0, 4\delta$, and let A_a be the adjacency operator of $X(\delta, a)$. Then, it is easily seen that for any $a \in \mathbb{F}_p$, $F(z; \chi, \psi, \mathbf{a})$ is an eigenfunction of A_a.

3.2.2 Case $\Gamma = K_q$. We use the notations in Sect. 2. For the multiplier system $m : K_q \times H_q \to \mathbb{C}^{\times}$ in (3), a complex representation $\rho : K_q \to GL(n, \mathbb{C})$, and a vector $\mathbf{b} \in \mathbb{C}^n$, let

$$E(z; \pi, \rho, \mathbf{b}) = \frac{1}{|K_q|} \sum_{k \in K_q} \frac{J_{\pi}(k, \sqrt{\delta})}{J_{\pi}(k, z)} \cdot \rho(k)^{-1} \mathbf{b}^t.$$

$E(z; \pi, \rho, \mathbf{b})$ is called the *vector-valued Eisenstein sum* for K_q, ρ, and $\mathbf{b}$. This sum, which is a finite field analog of the vector-valued Eisenstein series on H_q, is a generalization of the sum in (4).

Theorem 6. $E(z; \pi, \rho, \mathbf{b}) \in M(K_q, m, \rho).$

Proof. By (2), for $k' \in K_q$, it holds that

$$|K_q|E(k'z; \pi, \rho, \mathbf{b}) = \sum_{k \in K_q} \frac{J_\pi(kk', \sqrt{\delta})}{J_\pi(k', \sqrt{\delta})} \cdot \frac{J_\pi(k', z)}{J_\pi(kk', z)} \cdot \rho(k)^{-1}\mathbf{b}^t$$

$$= m(k', z)\rho(k') \sum_{k \in K_q} \frac{J_\pi(kk', \sqrt{\delta})}{J_\pi(kk', z)} \cdot \rho(kk')^{-1}\mathbf{b}^t,$$

and the result follows.   □

4   Equivariant Functions

4.1   Definitions

For a subgroup Γ of $G_q = GL(2, \mathbb{F}_q)$, let $\rho : \Gamma \to GL(2, \mathbb{C})$ be a 2-dimensional complex representation. The quotient $h(z) = f_1(z)/f_2(z)$ with $f_1, f_2 \in \mathcal{F}_q$ ($f_2(z) \neq 0$) is called a ρ-*equivariant function* with respect to Γ if

$$h(\gamma z) = \rho(\gamma) \cdot h(z)$$

holds for $\gamma \in \Gamma$ and $z \in H_q$. Here the action on both sides is given by linear fractional transformations.

If $(f_1(z), f_2(z))^t \in M(\Gamma, m, \rho)$ with $f_2(z) \neq 0$, then it is easily seen that the quotient $f_1(z)/f_2(z)$ is a ρ-equivariant function. The following question is now raised: *for a given representation ρ, does there exist a ρ-equivariant function?* In the classical case, the corresponding problem was solved in [13,14].

Let $n = 2$ in Example 2. If the lowest component of $E_\Gamma(z; \chi, \rho, \mathbf{a})$ is non-zero, then using this function, a ρ-equivariant function can be constructed.

4.2   Special Cases

Let p be an odd prime number. Herein, ρ-equivariant functions are discussed when Γ is N_p or K_3.

4.2.1   Case $\Gamma = N_p$. Using a generator t of $\mathbb{F}_p^\times$, a multiplicative character $\chi : \mathbb{F}_p^\times \to \mathbb{C}^\times$ is defined by $\chi(t) = e^{2\pi i/p}$. Moreover, an additive character $\psi : \mathbb{F}_p \to \mathbb{C}^\times$ is defined by $\psi(1) = e^{2\pi i/p}$. Then, we have the following.

Theorem 7. *For any complex representation $\rho : N_p \to GL(2, \mathbb{C})$, there exists a ρ-equivariant function.*

Proof. By Theorem 1 (ii), the function $f(z; \chi, \psi)$ in (1) is non-zero. As N_p is abelian, ρ is equivalent to the direct sum of 1-dimensional representations $\alpha, \beta : N_p \to \mathbb{C}^\times$. When

$$\rho(\gamma_u) = \begin{pmatrix} \alpha(\gamma_u) & 0 \\ 0 & \beta(\gamma_u) \end{pmatrix} \qquad (\gamma_u \in N_p),$$

there exists i $(0 \leq i \leq p-1)$ such that $\alpha = \beta \mu^i$. Then, the pair of functions $(h_1(z), h_2(z))^t \in \mathcal{F}_p^2$ is defined as

$$(h_1(z), h_2(z))^t = (f(z; \chi, \psi)^{i+1}, f(z; \chi, \psi))^t \quad if \ \alpha = \beta \mu^i \quad (i = 0, 1, \ldots, p-1).$$

Then, for $\gamma_u \in N_p$, $h_1(\gamma_u z)/h_2(\gamma_u z) = \rho(\gamma_u) \cdot h_1(z)/h_2(z)$.

When there exists a matrix $U \in GL(2, \mathbb{C})$ such that

$$\rho(\gamma_u) = U \begin{pmatrix} \alpha(\gamma_u) & 0 \\ 0 & \beta(\gamma_u) \end{pmatrix} U^{-1} \quad (\gamma_u \in N_p),$$

let $(f_1(z), f_2(z))^t = U(h_1(z), h_2(z))^t$. Then, for $\gamma_u \in N_p$, $f_1(\gamma_u z)/f_2(\gamma_u z) = \rho(\gamma_u) \cdot f_1(z)/f_2(z)$. □

4.2.2 Case $\Gamma = K_3$. We have the following result.

Proposition 2. *Let* $\rho : K_3 \to GL(2, \mathbb{C})$ *be a complex representation such that* $\rho\left(\begin{pmatrix} -1 & 0 \\ 0 & -1 \end{pmatrix}\right) = \begin{pmatrix} -1 & 0 \\ 0 & -1 \end{pmatrix}$ *or* $\begin{pmatrix} 1 & 0 \\ 0 & 1 \end{pmatrix}$. *Then, there exists a ρ-equivariant function.*

Proof. We use the notations in Example 1. By direct computation, $E(\sqrt{\delta}; \pi^4, 1) = 1, E(1 + \sqrt{\delta}; \pi^4, \mu) = (1 + i)/2$. Hence, we have non-zero modular forms $E(z; \pi^4, 1) \in M(K_3, m_4, 1)$ and $E(z; \pi^4, \mu) \in M(K_3, m_4, \mu)$. As K_3 is abelian, ρ is equivalent to the direct sum of 1-dimensional representations $\alpha, \beta : K_3 \to \mathbb{C}^\times$.

When

$$\rho(k) = \begin{pmatrix} \alpha(k) & 0 \\ 0 & \beta(k) \end{pmatrix} \quad (k \in K_3),$$

by assumption, there exists j $(0 \leq j \leq 3)$ such that $\alpha = \beta \mu^j$. Then, the pair of functions $(h_1(z), h_2(z))^t \in \mathcal{F}_3^2$ is defined as

$$(h_1(z), h_2(z))^t = \begin{cases} (E(z; \pi^4, 1), E(z; \pi^4, 1))^t & if \ \alpha = \beta, \\ (E(z; \pi^4, \mu)^j, E(z; \pi^4, 1)^j)^t & if \ \alpha = \beta \mu^j \ (j = 1, 2, 3). \end{cases}$$

Then, for $k \in K_3$, $h_1(kz)/h_2(kz) = \rho(k) \cdot h_1(z)/h_2(z)$.

When there exists a matrix $U \in GL(2, \mathbb{C})$ such that

$$\rho(k) = U \begin{pmatrix} \alpha(k) & 0 \\ 0 & \beta(k) \end{pmatrix} U^{-1} \quad (k \in K_3),$$

let $(f_1(z), f_2(z))^t = U(h_1(z), h_2(z))^t$. Then, for $k \in K_3$, $f_1(kz)/f_2(kz) = \rho(k) \cdot f_1(z)/f_2(z)$. □

5   Concluding Remarks

Vector-valued modular forms on finite upper half planes have been introduced and then applied to equivariant functions. It is interesting to consider the following questions:

- It is difficult to determine the dimension of the space of vector-valued modular forms. Can a formula for its dimension be established?
- There is a shortage of interesting examples of the modular forms discussed in this paper. Can Poincaré series be defined on finite upper half planes?
- In [9], Hilbert modular forms on $H_q \times \cdots \times H_q$, i.e., the product of H_q, were introduced. Can Siegel modular forms be defined on "finite Siegel upper half spaces"? This may possibly be accomplished by starting with the finite symplectic group factored out by a stabilizer as in the case of H_q in Sect. 2. It is known that classical vector-valued modular forms are related to Jacobi forms [6]. Can an analog of Jacobi forms be defined?

Acknowledgments. The author would like to thank the anonymous referees for careful reading and insightful comments that improved this paper.

References

1. Angel, J., Celinker, N., Poulos, S., Terras, A., Trimble, C., Velasquez, E.: Special functions on finite upper half planes. Contemp. Math. **138**, 1–26 (1992)
2. Bantay, P.: Vector-valued modular forms. Contemp. Math. **497**, 19–31 (2009)
3. Celinker, N., Poulos, S., Terras, A., Trimble, C., Velasquez, E.: Is there life on finite upper half planes? Contemp. Math. **143**, 65–88 (1993)
4. Cohen, H., Strömberg, F.: Modular Forms. AMS, Providence (2017)
5. Dong, C., Li, H., Mason, G.: Modular-invariance of trace functions in orbifold theory and generalized moonshine. Comm. Math. Phys. **214**, 1–56 (2000)
6. Eichler, M., Zagier, D.: The Theory of Jacobi Forms. Birkhäuser, Basel (1985)
7. Eholzer, W., Skoruppa, N.-P.: Modular invariance and uniqueness of conformal characters. Comm. Math. Phys. **174**, 117–136 (1995)
8. Freitag, E., Busam, R.: Complex Analysis, 2nd edn. Springer, Heidelberg (2009). https://doi.org/10.1007/978-3-540-93983-2
9. Hamahata, Y.: A note on modular forms on finite upper half planes. In: Carlet, C., Sunar, B. (eds.) WAIFI 2007. LNCS, vol. 4547, pp. 18–24. Springer, Heidelberg (2007). https://doi.org/10.1007/978-3-540-73074-3_3
10. Knopp, M., Mason, G.: Generalized modular forms. J. Number Theory **99**, 1–28 (2003)
11. Knopp, M., Mason, G.: On vector-valued modular forms and their Fourier coefficients. Acta Arithmetica **110**, 117–124 (2003)
12. Knopp, M., Mason, G.: Vector-valued modular forms and Poincaré series. Illinois J. Math. **48**, 1345–1366 (2004)
13. Saber, H., Sebbar, A.: Equivariant functions and integrals of elliptic functions. Geom. Dedicata **160**, 373–414 (2012)
14. Saber, H., Sebbar, A.: Equivariant functions and vector-valued modular forms. Int. J. Number Theory **10**, 949–954 (2014)

15. Saber, H., Sebbar, A.: On the existence of vector-valued automorphic forms. Kyushu J. Math. **71**, 271–285 (2017)
16. Selberg, A.: On the estimation of Fourier coefficients of modular forms. In: Proceedings of a Symposium in Pure Mathematics, vol. VIII, pp. 1–15. AMS (1965)
17. Shaheen, A., Terras, A.: Fourier expansions of complex-valued Eisenstein series on finite upper half planes. Int. J. Math. Sci. **2006**, 1–17 (2006)
18. Terras, A.: Fourier Analysis on Finite Groups and Applications. London Mathematical Society Student Texts, vol. 43. Cambridge University Pres, Cambridge (1999)
19. Terras, A.: Harmonic Analysis on Symmetric Spaces - Euclidean Space, the Sphere, and the Poincaré Upper Half-Plane, 2nd edn. Springer, New York (2013). https://doi.org/10.1007/978-1-4614-7972-7

Normal Basis Exhaustive Search: 10 Years Later

L. Moura[1], D. Panario[2], and D. Thomson[2]([✉])

[1] University of Ottawa, Ottawa, Canada
lucia@site.uottawa.ca
[2] Carleton University, Ottawa, Canada
{daniel,dthomson}@math.carleton.ca

Abstract. This paper concerns an exhaustive search for normal bases with minimum complexity in finite fields $\mathbb{F}_{2^n}$ over $\mathbb{F}_2$ for $n \leq 46$. This is a followup paper to [11], which appeared one decade ago in 2008 and completed the cases $n \leq 39$. We extend the results in [11] by taking advantage of a combination of algorithmic improvements, more efficient implementations and massive parallelism.

Keywords: Finite fields · Normal bases · NTL · Parallel computing

1 Introduction

For any finite field extension $\mathbb{F}_{2^n}$ over $\mathbb{F}_2$, upon choice of a basis, we can write any field element as an n-long binary vector; that is, $\mathbb{F}_{2^n} \cong \mathbb{F}_2^n$, and the basis makes the isomorphism explicit. When performing finite field arithmetic the choice of representation of elements is critical to the performance of various operations in the system.

For tiny fields, look-up tables can be employed for field arithmetic and for mid-sized fields *Zech logarithms* as a time-memory tradeoff on full lookup tables can be used, see [13, Sect. 2.1.7.5] for more details. For larger finite fields where lookup tables are no longer feasible, we exploit an explicit isomorphism between the field $\mathbb{F}_{2^n}$ and the vector space $\mathbb{F}_2^n$ by choice of a given basis. Most commonly, field extensions are given by *polynomial* or *power bases*; namely, $\mathbb{F}_{2^n} \cong \mathbb{F}_2[x]/(f)$, where $f \in \mathbb{F}_2[x]$ is an irreducible polynomial of degree n. Here, arithmetic is performed $(\bmod\ f)$, and is reasonably efficient in many cases, especially when a sparse choice of f is used. This paper deals with a different kind of basis, namely, *normal bases*.

Definition 1. *Let* $\mathcal{B} = \{\alpha, \alpha^2, \ldots, \alpha^{2^{n-1}}\}$. *If* $\mathcal{B}$ *is a basis of* $\mathbb{F}_{2^n}$ *over* $\mathbb{F}_2$ *(i.e., if its elements are linearly independent), then* $\mathcal{B}$ *is a* normal basis *of* $\mathbb{F}_{2^n}$ *over* $\mathbb{F}_2$ *and every* $\beta \in \mathcal{B}$ *is a* normal element *of* $\mathbb{F}_{2^n}$ *over* $\mathbb{F}_2$.

For any $\alpha \in \mathbb{F}_{2^n}$, the elements α^{2^i}, $0 \leq i \leq n-1$, are the (Galois) *conjugates* of α.

© Springer Nature Switzerland AG 2018
L. Budaghyan and F. Rodríguez-Henríquez (Eds.): WAIFI 2018, LNCS 11321, pp. 188–206, 2018.
https://doi.org/10.1007/978-3-030-05153-2_10

We highlight the usefulness of a normal basis representation of $\mathbb{F}_{2^n}$ over $\mathbb{F}_2$. Suppose $\gamma = \sum_{i=0}^{n-1} g_i \alpha^{2^i}$, then $\gamma^{2^j} = \sum_{i=0}^{n-1} g_{(i-j) \bmod n} \alpha^{2^i}$ for any integer j. Hence, squaring in normal basis representation is given by a cyclic bit shift of its underlying coefficient vector.

The multiplication of elements represented in normal basis also receives a simplification. Denote by t_{ijk} the *structure constants* of $\mathcal{B}$, given by the relations

$$\alpha^{2^i} \alpha^{2^j} = \sum_{k=0}^{n-1} t_{ijk} \alpha^{2^k}.$$

Since $\alpha^{2^i} \alpha^{2^j} = (\alpha \alpha^{2^{j-i}})^{2^i}$, we have $t_{ijk} = t_{0(j-i)(k-i)}$ for all i, j, k, where subscripts are taken modulo n. The complexity of multiplication of generic elements in $\mathbb{F}_{2^n}$ depends directly on the number of non-zero structure constants. By the above, therefore, the cost of multiplication is directly related to the number of nonzero t_{0jk}.

Definition 2. *Let $\mathcal{B} = \{\alpha, \alpha^2, \ldots, \alpha^{2^{n-1}}\}$ be a normal basis of $\mathbb{F}_{2^n}$ over $\mathbb{F}_2$ and let constants t_{ij} be given by the relations*

$$\alpha \alpha^{2^i} = \sum_{j=0}^{n-1} t_{ij} \alpha^{2^j} \quad 0 \leq i \leq n-1.$$

The complexity (or density) of $\mathcal{B}$ is given by $\mathcal{C}_{\mathcal{B}} = |\{t_{ij} \neq 0 \colon 0 \leq i, j \leq n-1\}|$. The matrix $T_{\mathcal{B}} = (t_{ij})$ is the multiplication table *of $\mathcal{B}$.*

By *low complexity normal bases*, we mean normal bases whose complexity is bounded by kn for a small integer constant k. Low complexity normal bases are desirable for efficient computations, particularly when the application requires a large number of squarings or exponentiations. For example, the National Institute for Standards in Technology prescribes low complexity normal bases for use in elliptic curve cryptography for curves over $\mathbb{F}_{2^n}$ [14], and they are also prescribed for use in decoding of Gabidulin codes in the rank metric [17]. The following proposition provides an achievable lower-bound on the complexity of normal bases.

Proposition 1. *[12] Let $\mathcal{B}$ be a normal basis, then $2n - 1 \leq \mathcal{C}_{\mathcal{B}} \leq n^2 - n$.*

Bases which meet the lower bound from Proposition 1 are *optimal normal bases*, and were characterized completely in [4]. Optimal normal bases exist only when $n + 1$ or $2n + 1$ are prime, and generalizations given by *Gauss periods* provide low complexity normal bases only when $kn + 1$ is prime for some k; see [13, Sect. 5.3]. Few other constructions of normal bases are known, nearly all of which construct normal bases in $\mathbb{F}_{2^n}$ from existing normal bases in either subfields or extensions. Hence, low-complexity normal bases are even required in these cases as a starting point. Heuristically, the complexity of a random normal basis is on the order of $n^2/2$, is tightly compacted about the mean, and up to half of all elements of $\mathbb{F}_{2^n}$ are normal, so random search is unlikely to yield low complexity normal bases; see [11].

Exhaustive searches for normal bases of $\mathbb{F}_{2^n}$ over $\mathbb{F}_2$ have been performed previously, for $n < 30$ in [12], for $n < 33$ in [8], and for $n < 40$ by the authors of this paper (with A. Masuda) in [11]. Restricting to a subset of normal bases, namely *self-dual* normal bases, allows for efficient search by the transitive action of orthogonal circulant matrices; this was completed for $n \leq 47$ in [8] and those results were verified and extended to odd characteristics in [1].

The purpose of this paper is to extend the results in [11] using a combination of algorithmic improvements, efficient implementations and increased availability of computational resources. In [11] we found the minimum complexity normal bases in $\mathbb{F}_{2^n}$ for all $n = 2, \ldots, 39$ using an implementation of Algorithm 1; see Sect. 3 for details. In this paper, we give an updated algorithm, Algorithm 2, see Sect. 4, and give a more efficient implementation, see Sect. 5. Using these improvements, we extend our search to include the cases $n = 40, \ldots, 46$. We give these results and some candidates for future work in Sect. 6.

2   Some Necessary Background and Notation

In this paper, we use field elements $\alpha \in \mathbb{F}_{2^n}$ interchangeably with their expansion as n-long binary vectors in some implicit underlying basis. When we write matrices as a single row of elements in $\mathbb{F}_{2^n}$ we mean their implicit expansion in the underlying $\mathbb{F}_2$-basis; for example, $P_\alpha = \left(\alpha \; \alpha^2 \cdots \alpha^{2^{n-1}} \right)$ should be interpreted as an $n \times n$ matrix with entries in $\mathbb{F}_2$ where the ith column is the expansion of α^{2^i}, $0 \leq i < n$, in the implicit underlying basis.

Throughout this paper, we denote by e_i the ith standard basis vector for $0 \leq i \leq n - 1$; that is $e_i = (e_{ij})$ with

$$e_{ij} = \begin{cases} 1 & j = i, \\ 0 & \text{otherwise.} \end{cases}$$

2.1   Complexity of Field Arithmetic

In Table 1, we give the computational complexity of the $\mathbb{F}_{2^n}$ arithmetic operations we use in this paper, given in number of $\mathbb{F}_2$ operations. In particular, we discuss the cost of $\mathbb{F}_{2^n}$ operations used by the NTL library [16] that we rely on for our implementation.

The quantity $M(n)$ gives the number of $\mathbb{F}_2$-operations for computing the multiplication of two elements of $\mathbb{F}_{2^n}$ over $\mathbb{F}_2$. Classically, this is done by polynomial convolution followed by division by an irreducible modulus of degree n with remainder. The NTL documentation states that $\mathbb{F}_{2^n}$ arithmetic is performed "using a combination of classical routines and Karatsuba." From our understanding of NTL internals, for $n < 64$, multiplication uses special 128-bit register arithmetic, when available, otherwise classically in an unrolled loop. For moderate sized $n < 512$, explicit unrolled Karatsuba multiplication is performed.

Since this is outside of our range of interest, we do not go into the details of Karatsuba multiplication costing here. Asymptotically faster methods are also available, but these methods are on efficient far outside of our range of interest. Division plus remainder exploits irreducible moduli of special forms: namely trinomial or pentanomial moduli, as applicable. We use the NTL default modulus for all n.

The NTL implementation of matrix inversion uses Gaussian reduction, so we give this cost with the observation that over $\mathbb{F}_2$ there is no row scaling required. Though asymptotically faster methods exist here too, the discussion in the preamble to [6, Chap. 12] indicates that these methods have crossover points also outside of our range of interest.

The cost $G(n)$ of a polynomial gcd in $\mathbb{F}_{2^n}[x]$ is only used as a lower bound to show that the computational complexity of our improved algorithm, Algorithm 2, is less than the original Algorithm 1. For this reason, we state the cost of $G(n)$ using fast arithmetic, regardless of crossover point, since this is enough to show the superiority of Algorithm 2.

Table 1. Upper bound on the cost of arithmetic operations in $\mathbb{F}_{2^n}$, given in bit operations [6].

Operation	Denoted	Cost
Addition in $\mathbb{F}_{2^n}$	$A(n)$	n
Multiplication (classical)	$M(n)$	$4n^2 - 5n + 1$
Matrix multiplication (classical) $C = AB$, $C, A, B \in \mathbb{F}_2^{n \times n}$	$\mathfrak{M}(n)$	$2n^3 - n^2$
Matrix inversion (Gaussian reduction) in $\mathbb{F}_2^{n \times n}$	$I(n)$	$\frac{2}{3}(n^3 + 3n^2 - 4n)$
$\gcd(f, x^n - 1), f \in \mathbb{F}_{2^n}[x]$ (fast methods)	$G(n)$	$\geq nM(n)\log n$

3   Original Algorithm From [11]

In this section we present Algorithm 1 and explain its various components. Algorithm 1 first appeared in [11], and is the main reference from which this work is based on. In Sect. 4 we analyze Algorithm 2, our updated algorithm from this work. We have a new implementation of Algorithm 2, so we leave a discussion of specific implementation details to Sect. 5.2.

3.1   Efficient Iteration Through $\mathbb{F}_{2^n}$

This section is devoted to justifying how explaining how lines 7 and 8 of Algorithm 1 admit efficient iteration through putative normal elements in $\mathbb{F}_{2^n}$.

For any element $\alpha \in \mathbb{F}_{2^n}$, calculating $\mathcal{B} = \{\alpha, \alpha^2, \dots, \alpha^{2^{n-1}}\}$ generically requires $n-1$ multiplications in $\mathbb{F}_{2^n}$ using repeated squaring. Suppose $\mathcal{B}$ is known,

---

Algorithm 1. Exhaustive search algorithm from [11]

1: **Input:** $n \in \mathbb{Z}_{>0}$
2: **Returns:** $\alpha \in \mathbb{F}_{2^n}$, a normal element with minimum complexity
3:
4: min_complexity $\leftarrow \infty$; min_element $\leftarrow 0$; $\alpha \leftarrow 0$; $\mathcal{B} = \{0, \ldots, 0\}$
5: Precompute $\{e_i^{2^j}\}, 0 \leq i, j \leq n - 1$
6: **for** $idx \leftarrow 0$ **to** $2^n - 1$ **do**
7:      $\gamma \leftarrow \Gamma(idx)$ ▷ See Section 3.1
8:      $\mathcal{B} \leftarrow \{\alpha + e_\gamma, \alpha^2 + e_\gamma^2, \ldots, \alpha^{2^{n-1}} + e_\gamma^{2^{n-1}}\}$
9:      **if** $\alpha \neq \min_<(\mathcal{B})$ **then**   ▷ See Section 3.2
10:         **continue**
11:      **if** $\gcd(\sum_{i=0}^{n-1} \alpha^{2^i} x^{n-1-i}, x^n - 1) \neq 1$ **then**   ▷ See Theorem 1
12:         **continue**
13:      cplex $\leftarrow \mathcal{C}_\mathcal{B}$ ▷ See Algorithm 3
14:      **if** cplex $<$ min_complexity **then**
15:         min_complexity $\leftarrow$ cplex; min_element $\leftarrow \alpha$
16: **return** min_complexity, min_element

---

and consider computing the putative basis $\mathcal{B}' = \{\alpha + \beta, \ldots, \alpha^{2^{n-1}} + \beta^{2^{n-1}}\}$. We show how to efficiently iterate through all of $\mathbb{F}_{2^n}$ after the precomputation of a small set of elements in $\mathbb{F}_{2^n}$.

Definition 3. *A Gray code is an ordering $\{g_0, g_1, \ldots, g_{2^n-1}\}$ of $\mathbb{F}_2^n$ such that $H(g_i, g_{i+1}) = 1$ for $i = 0, 1, \ldots, 2^n - 2$, where $H(a, b)$ is the Hamming distance of $a, b \in \mathbb{F}_2^n$.*

We use a Gray code for efficient iteration through $\mathbb{F}_2^n$ as follows. Any element $\alpha \in \mathbb{F}_{2^n}$ is isomorphic to its binary vector of coefficients in a fixed basis; say $\alpha \cong g_i$ for some g_i in a Gray code. The *successor* of α is the element $\beta \cong g_{i+1}$. For ease of notation we identify $\alpha, \beta \in \mathbb{F}_{2^n}$ with their coefficient vectors. In a Gray code, $\beta = \alpha + e_\gamma$ for some standard basis vector e_γ, $0 \leq \gamma \leq n - 1$. Then $\beta^{2^j} = \alpha^{2^j} + e_\gamma^{2^j}$, by the linearity of Frobenius. Hence, with a precomputation of the vectors $e_i^{2^j}$ for $0 \leq i, j \leq n - 1$, any β^{2^j} can be computed by a binary vector addition.

Given a Gray code $\{g_0, \ldots, g_{2^n-1}\}$ define $\Gamma \colon [0, 2^n - 1] \to [0, n - 1]$, where $\Gamma(i) = \gamma$ whenever $g_{i+1} = g_i + e_\gamma$. The function Γ is used to compute a Gray code successor in line 7–8 in Algorithm 1; a specific instantiation of Γ is given in Proposition 5.

3.2   Reducing Search Space Using a Canonical Element Check

In this section, we present Proposition 2 to justify line 9, which reduces the size of the search space in Algorithm 1 by a factor of n.

Proposition 2. *Let $\mathcal{B}_\alpha = \{\alpha, \alpha^q, \ldots, \alpha^{q^{n-1}}\}$ be a putative normal basis. For any $\beta \in \mathcal{B}_\alpha$, as an unordered multiset $\mathcal{B}_\beta = \{\beta, \beta^q, \ldots, \beta^{q^{n-1}}\} = \mathcal{B}_\alpha$ and hence $\mathcal{B}_\beta$ is a basis if and only if $\mathcal{B}_\alpha$ is. If both $\mathcal{B}_\alpha$ and $\mathcal{B}_\beta$ are bases, then $\mathcal{C}_{\mathcal{B}_\beta} = \mathcal{C}_{\mathcal{B}_\alpha}$.*

By Proposition 2, we need to check normality and compute complexity for only a canonical basis representative among its conjugates. Since elements of $\mathbb{F}_{2^n}$ are represented as n-long binary vectors under some implicit basis, so for any $\alpha, \beta \in \mathbb{F}_{2^n}$ we define the ordering $\leq$ by $\alpha < \beta$ if α has lexicographically smaller binary vector representation than β. Equality is well-defined under this ordering: α and β have equal order if and only if $\alpha = \beta$. Equality occurs in a putative basis $\mathcal{B}$ if and only if $\beta = \alpha^{2^i}$ for some i, $1 \leq i \leq n-1$, proving that $\mathcal{B}$ is not a basis. We denote by $\min_<(\mathcal{B})$ the unique minimal element of $\mathcal{B}$ under $\leq$, if it exists.

For implementation notes on computing lexicographically small elements, see Sect. 5.2.

3.3 Checking Normality

Line 11 of Algorithm 1 is used to filter non-normal elements of $\mathbb{F}_{2^n}$, based on Theorem 1.

Theorem 1. [13, Theorem 5.2.11] *Let* $\alpha \in \mathbb{F}_{2^n}$ *and let* $g_\alpha(x) = \alpha x^{n-1} + \alpha^2 x^{n-2} + \cdots + \alpha^{2^{n-2}} x + \alpha^{2^{n-1}}$. *Then* α *is normal over* $\mathbb{F}_2$ *if and only if* $\gcd(g_\alpha, x^n - 1) = 1$.

We combine all the ingredients in this section to justify the correctness of Algorithm 1.

Theorem 2. *Algorithm 1 computes the complexity of every normal basis of* $\mathbb{F}_{2^n}$ *over* $\mathbb{F}_2$.

Proof. Let $\mathcal{B} = \{\alpha, \alpha^2, \ldots, \alpha^{2^{n-1}}\} \subset \mathbb{F}_{2^n}$ be a normal basis of $\mathbb{F}_{2^n}$ over $\mathbb{F}_2$. If $\mathcal{B}$ is a basis, then $\mathcal{B}$ has a unique minimal element under $\leq$, so without loss of generality suppose $\alpha = \min_<(\mathcal{B})$.

By Proposition 5, α will be met for some index idx; hence, $\mathcal{B}$ will be computed in line 8 at index idx. By supposition, α passes line 9. Finally, $\gcd(\sum_{i=0}^{n-1} \alpha^{2^i} x^{n-1-i}, x^n - 1) = 1$ by Theorem 1, hence the complexity of $\mathcal{B}$ is computed in line 13. ∎

4 An Improved Algorithm for Exhaustive Search for Normal Bases

In this section, we present a number of algorithmic improvements to Algorithm 1 from [11]. Our improved algorithm is given in Algorithm 2. Section 4.1 justifies an additional cut-down of the search space, and we improve the overall run-time in Sect. 4.2 by deferring the normality check due to Theorem 1.

Algorithm 2. Updated exhaustive search for low complexity normal bases

---

1: **Input:** $n \in \mathbb{Z}_{>0}$
2: **Returns:** $\alpha \in \mathbb{F}_{2^n}$, a normal element with minimum complexity
3:
4: min_complexity $\leftarrow \infty$; min_element $\leftarrow 0$; $\alpha \leftarrow 0$; $\mathcal{B} = \{0, \ldots, 0\}$; trace $\leftarrow 0$
5: Precompute $\{e_i^{2^j}\}, 0 \leq i, j \leq n - 1$ and $\mathrm{Tr}(e_i), 0 \leq i \leq n - 1$
6: **for** $idx \leftarrow 0$ **to** $2^n - 1$ **do**
7:      $\gamma \leftarrow \Gamma(idx)$ ▷ See Section 3.1
8:      $\mathcal{B} \leftarrow \{\alpha + e_\gamma, \alpha^2 + e_\gamma^2, \ldots, \alpha^{2^{n-1}} + e_\gamma^{2^{n-1}}\}$
9:      trace $\leftarrow$ trace $\oplus \mathrm{Tr}(e_\gamma)$
10:     **if** trace $= 0$ **then**
11:         **continue**
12:     **if** $\alpha \neq \min_<(\mathcal{B})$ **then**   ▷ See Section 3.2
13:         **continue**
14:     cplex $\leftarrow \mathcal{C}_\mathcal{B}$ ▷ See Algorithm 3
15:     **if** cplex $<$ min_complexity **then**
16:         min_complexity $\leftarrow$ cplex; min_element $\leftarrow \alpha$
17: **return** min_complexity, min_element

4.1   Trace Precomputation

We use another pre-computation and cut-down due to the following observation, based on the fact that $\mathcal{B}$ is a normal basis if and only if its elements are linearly independent.

Definition 4. *Let* $\mathbb{F}_{2^n}$ *be the degree n extension of* $\mathbb{F}_2$*. Denote the* trace *of an element* $\alpha \in \mathbb{F}_{2^n}$ *by* $\mathrm{Tr}(\alpha) = \sum_{i=0}^{n-1} \alpha^{2^i}$.

Recall from Theorem 1 that an element α is normal if and only if $g_\alpha(x) = \sum_{i=0}^{n-1} \alpha^{2^i} x^{n-i}$ is coprime with $x^n - 1$. Since 1 is a root of $x^n - 1$ for all n, if 1 is a root of g_α, then α is not normal. We summarize the contrapositive in the next proposition.

Proposition 3. *If* α *is normal, then* $\mathrm{Tr}(\alpha) \neq 0$.

The trace is an additive map $\mathbb{F}_{2^n} \to \mathbb{F}_2$, and by linearity it is easy to see that $\mathrm{Tr}(\alpha^2) = \mathrm{Tr}(\alpha)$. Thus, the value of $\mathrm{Tr}(\alpha)$ is invariant across the set α^{2^j}. Hence, with an additional precomputation of $\mathrm{Tr}(e_i)$ for $0 \leq i < n$, when computing the successor $\alpha' \leftarrow \alpha + e_j$ for some j, we also compute $\mathrm{Tr}(\alpha') = \mathrm{Tr}(\alpha) + \mathrm{Tr}(e_j)$, with $\mathrm{Tr}(\alpha)$ and $\mathrm{Tr}(e_j)$ known. If $\mathrm{Tr}(\alpha + e_j) = 0$, then α' is not normal and we continue the outer loop. This can be seen in lines 9–11 of Algorithm 2.

Proposition 4 shows that with this small trace precomputation, requiring only a storage of n bits, we receive an immediate cut-down in the search space by a factor of 2.

Proposition 4. [10, Theorem 2.25] *Let* $\alpha \in \mathbb{F}_{2^n}$*. We have* $\mathrm{Tr}(\alpha) = 0$ *if and only if* $\beta^2 + \beta = \alpha$ *for some* $\beta \in \mathbb{F}_{2^n}$*. Hence, exactly* 2^{n-1} *elements of* $\mathbb{F}_{2^n}$ *have trace* 0.

4.2 Removing Explicit Normality Checking

A key difference between Algorithms 1 and 2 is the *removal* of line 11 from Algorithm 1, which ensures that only bases pass to the complexity calculation step given by Algorithm 3. In this section we show that, with a small change to Algorithm 3, the removal of this line clears a performance bottleneck from Algorithm 1.

We highlight the simple but important observation that $P_\alpha = \left(\alpha \; \alpha^2 \; \cdots \; \alpha^{2^{n-1}} \right)$ has full rank if and only if $\mathcal{B} = \{\alpha, \alpha^2, \ldots, \alpha^{2^{n-1}}\}$ is a normal basis of $\mathbb{F}_{2^n}$ over $\mathbb{F}_2$. Algorithm 3 calculates the complexity of $\mathcal{B}$, if it is normal, by computing the matrix $B_\alpha = \left(\alpha\alpha \; \alpha\alpha^2 \; \cdots \; \alpha\alpha^{2^{n-1}} \right)$ and then expressing it in the putative basis $\mathcal{B}$, if possible. The matrix P_α^{-1} is a "change of basis" matrix from the implicit underlying basis to $\mathcal{B}$. If $\mathcal{B}$ is not linearly independent (hence if the element α is not normal), then P_α is non-invertible. If P_α^{-1} is computed by Gaussian reduction, as soon as a non-pivot row is discovered, an inversion routine can return $0 = \det(P_\alpha)$. Hence, checking normality and matrix inversion can be performed simultaneously.

It is not obvious that deferring normality checking to the Gaussian reduction step is a net benefit. The remainder of this section is devoted to showing that this removal indeed is a benefit.

Theorem 3. [5] *The proportion $\psi(n)$ of normal bases of $\mathbb{F}_{2^n}$ over $\mathbb{F}_2$ satisfies*

$$\psi(n) = \frac{\Phi(x^n - 1)}{n2^n} \geq \frac{1}{ne^{0.83}(1 + \log_2(n))},$$

where Φ is Euler's phi function for polynomials over $\mathbb{F}_2$.

Since trace-0 elements are not normal, exactly $2n\psi(n)$ trace-1 elements of $\mathbb{F}_{2^n}$ are normal.

Let $G(n)$ be the cost of a gcd in $\mathbb{F}_{2^n}$, $M(n)$ be the cost of a field multiplication in $\mathbb{F}_{2^n}$ and $I(n)$ be the cost of an $n \times n$ matrix inversion over $\mathbb{F}_2$. The cost of first checking normality on $2^{n-1}/n$ elements and then calculating complexity on a lower-bound of the remaining $2\psi(n)$ elements is

$$\frac{2^{n-1}}{n} \left(G(n) + 2\psi(n) \left((n-1)M(n) + I(n) \right) \right). \tag{1}$$

On the other hand, using Gaussian reduction as a normality check takes time

$$\frac{2^{n-1}}{n} \left((n-1)M(n) + I(n) \right). \tag{2}$$

Clearly, if $G(n) > (1 - 2\psi(n))((n-1)M(n) + I(n))$, then Expression (1) is more expensive than Expression (2). By Table 1, $G(n) > nM(n)\log(n)$ for $n > 2$, so it is easy to verify that Expression (1) always dominates.

4.3   Computational Complexity of Algorithm 2

We focus on the main loop of Algorithm 2 and follow the notation from Table 1. Lines 7–11 are called 2^n times: lines 7 and 9 are constant time operations, so we ignore them, and line 8 has cost $nA(n)$. Indeed, the calculation of the Gray code index in line 7 is constant in an amortized sense, since $\sum_{i=1}^{n} i/2^i \leq 2$. Exactly 2^{n-1} elements pass to line 12 with cost $L(n)$, where $L(n)$ is the cost of checking canonicity (e.g., lex-first) on input of length n. Exactly $2^{n-1}/n$ elements enter Algorithm 3 in line 14. Line 7 of Algorithm 3 has cost $I(n)$ for each input; see Table 1. Exactly $\Phi(x^n - 1)/n$ elements pass beyond line 7 of Algorithm 3. The cost of the remaining lines is $(n-1)M(n) + \mathfrak{M}(n)$.

Theorem 4. *The cost $C(n)$ of the main loop of Algorithm 2, ignoring lower order terms, satisfies*

$$\frac{47}{6}n^2 > \frac{C(n)}{2^n} > \frac{11}{6}n^2 + 2.616\frac{n^2}{\log_2(n)}.$$

Proof. Let $C(n)$ be the cost of lines 6-16 of Algorithm 2 for $2 < n < 63$. Then we have

$$C(n) = 2^n nA(n) + 2^{n-1}L(n) + \frac{2^{n-1}}{n}I(n) + \frac{\Phi(x^n - 1)}{n}((n-1)M(n) + \mathfrak{M}(n)).$$

A lex-first calculation can be implemented in n^2 bit comparisons, so we use $L(n) = n^2$. We recall the costs from Table 1 using schoolbook methods: $A(n) = n$, $M(n) = 4n^2 - 5n + 1$ $I(n) = (2/3)(n^3 + 3n^2 - 4n)$ and $\mathfrak{M}(n) = 2n^3 - n^2$. By Theorem 3, $\frac{\Phi(x^n-1)}{n} \geq \frac{2^n}{ne^{0.83}(1+\log_2(n))}$ and we have the trivial upper bound $\Phi(x^n - 1) < 2^n$. Therefore,

$$\frac{47}{6}n^2 > \frac{C(n)}{2^n} > \frac{11}{6}n^2 + 2.616\frac{n^2}{\log_2(n)}.$$

∎

5   Implementation Details

In this section, we discuss some details of our implementation of Algorithm 2. We discuss our development methodology in Sect. 5.1 and give a demonstration of the importance of practical and efficient implementation in Sect. 5.2.

5.1   Sage Implementation for Testing

In this section, we discuss our methodology for determining which routines are most promising for implementation at scale. Our methodology is to use a platform that is agile and simple to test before spending a much larger effort to code an efficient C++ implementation.

For our experimental development, we use the Sage computer algebra system [15] running in JuPyter Notebooks. Sage is open-source and is naturally linked to many other mathematical libraries (like NTL). Sage is built on Python, so it is interpreted, object-oriented, not strongly typed, and memory is internally managed.

JuPyter notebooks provide a user interface in a web browser linked to various kernels in the back-end: in our case a Sage interpreter. Code is written in "cells", which is particularly convenient for testing experimental changes. Effectively, for our purpose, JuPyter notebooks provides an easy-to-use, graphical, interactive interface to Sage. The trade-off of this ease of development (say, over direct implementation in C++) is in overall performance and lack of control over internals that our highly-efficient implementations use.

In our Sage notebook, we have a Normal_Search object, which accepts functions as arguments: in particular, specific implementations of canonicity checking, normality checking and complexity computation (Algorithm 3). We find the most efficient implementation in each function class by profiling the main routine of each instance of Normal_Search instantiated with each set of distinct implementations in each function class.

Table 2. Runtime for complete exhaustive search of $\mathbb{F}_{2^n}$ over $\mathbb{F}_2$ with different normality checks: the first using Theorem 1, and the second the implicit normality check as performed in Algorithm 2. Runtime for single-threaded C++ implementation of Algorithm 2 is also given.

n	10	11	12	13	14	15	16	17	18	19
Theorem 1 (Hybrid)	210 ms	323 ms	604 ms	1.4 s	2.77 s	5.72 s	14.3 s	30.1 s	65 s	157 s
Implicit (Algorithm 2)	115 ms	166 ms	388 ms	773 ms	1.44 s	2.84 s	7.77 s	12.8 s	26.4 s	55.9 s
C++ (Algorithm 2)	3 ms	3 ms	4 ms	8 ms	13 ms	29 ms	52 ms	113 ms	188 ms	415 ms

We use simple timing and line-profiling inherited in the Sage notebook to compare implementations of Algorithms 1 and 2. In particular, we compare the use of a normality check in Theorem 1 against the implicit normality check in Algorithm 3. In Table 2, we present comparative timings for a Sage implementation of Algorithm 1 *with added trace precomputation* and Algorithm 2 (the only difference between these algorithms is therefore the presence or absence of normality checking due to Theorem 1). We observe that Theorem 1 gives slower runtimes; this led us to do the careful analysis in Sect. 4.2. The third row of Table 2 also provides the runtime of our high performance C++ implementation; see Sect. 5.2 for details.

We also used this comparative analysis to determine a more efficient implementation of canonicity checking between the one that appeared in [11] and this work. In Table 3, we give comparative timings for two different canonicity checks. We discuss this implementation in more detail in Sect. 5.2.

Our Sage JuPyter notebook appears under the GitHub project [18].

5.2   High Performance C++ Implementation of Algorithm 2

After determining candidates for efficient implementations in Sage, we code these in a C++ program using NTL for finite field arithmetic [16]. We first implemented a single-threaded version and profiled the code until the dominant subroutines are algorithmically necessary NTL internal calls; see Fig. 1, for example. We added two levels of parallelism: multi-node parallelism and multi-threading for intra-node parallelism. We give more details on our profiling in this section and on our parallel implementation in Sect. 5.3. As we see in Table 2, the absolute runtimes in our C++ implementation are 100-fold faster than in Sage, but importantly we notice that the relative timings of the two implementations are proportional.

Our reasoning for using NTL for arithmetic in our high-performance implementation included:

Familiarity: The authors have significant prior experience using NTL.

Comparability: Coding in NTL allowed us to compare our current code with the previous version from [11].

Performance: NTL is a mature library focusing on performance and contains a particularly fast implementation of $\mathbb{F}_{2^n}$ arithmetic.

Thread-safety: NTL has been thread-safe since version 7.0 (2014), though this feature has matured greatly as of version 9.8.0 (2016). For ease of intra-node parallelism, we make use of the BasicThreadPool library in NTL.

Exposure of Pointers: NTL exposes pointers to arrays and to the bit representatives of elements of $\mathbb{F}_{2^n}$, which is crucial for our implementation.

We do not make any attempt to write our own vectorized instructions, and we rely on compiler optimizations to unroll loops and assign AVX instructions, if available.

While we present our algorithms for clarity, their implementations may differ slightly from the presentation. Consider Algorithm 3, which accepts a putative normal basis and either returns its complexity or ∞ if it is not a basis. Here, we step through this algorithm in detail to show the differences that may occur between the implementation and the algorithmic description.

1. Each element α^{2^i} is represented as an n-vector over $\mathbb{F}_2$ using an implicit basis of $\mathbb{F}_{2^n}$ over $\mathbb{F}_2$.
2. The matrices P_α and B_α are initialized only once per thread, to save overhead.
3. The rows of the matrix $P_\alpha = \left(\alpha\, \alpha^2 \cdots \alpha^{2^{n-1}} \right)$ are pointers assigned to previously computed $\mathbb{F}_2$-vector representations of the α^{2^i}.
4. Since $\alpha\alpha^{2^i}$ with $i = 0$ is simply α^2, hence already stored, we save a multiplication in line 8.
5. In NTL, the matrix inverse function returns the determinant of a matrix, and if the matrix is invertible procedurally stores the inverse in a passed argument. Hence, lines 2–7 are computed using a single function call.
6. It is critical to performance to use the procedure versions of the matrix inverse in line 7 and matrix multiplication in line 9 to avoid pass-by-values. In practice, we store $P'_\alpha = P_\alpha^{-1}$ in the existing P_α matrix and the product $B_\alpha P'_\alpha$ in the B_α matrix.

Algorithm 3. Check normality and calculate the complexity of a putative normal basis.

1: **Input**: A putative normal basis $\mathcal{B} = \{\alpha, \alpha^2, \ldots, \alpha^{2^{n-1}}\}$
2: **Returns**: The complexity $C_\mathcal{B}$ of the normal basis $\mathcal{B}$ if $\mathcal{B}$ is a basis, otherwise ∞
3:
4: $P_\alpha \leftarrow \left(\alpha \; \alpha^2 \cdots \alpha^{2^{n-1}} \right)$
5: **if** $\det(P_\alpha) = 0$ **then**   ▷ New for Algorithm 2
6:     **return** ∞
7: $P'_\alpha \leftarrow P_\alpha^{-1}$
8: $B_\alpha \leftarrow \left(\alpha\alpha \; \alpha\alpha^2 \cdots \alpha\alpha^{2^{n-1}} \right)$
9: $T_\mathcal{B} = (t_{ij})_{i,j=0}^{n-1} \leftarrow B_\alpha P'_\alpha$ ▷ Multiplication table of $\mathcal{B}$
10: **return** $\sum_{ij} t_{ij}$ ▷ Complexity of $\mathcal{B}$, $C_\mathcal{B}$

---

7. NTL does not have a matrix weight function, so we compute the complexity of $\mathcal{B}$ from $T_\mathcal{B}$ row-by-row in line 10. We improved in performance over the NTL weight function by dereferencing pointers to underlying NTL objects and calling the compiler's builtin popcount intrinsic.

These low-level programming details are incredibly important to the efficiency of the *implementation* of the algorithm. We believe our presentation accurately represents the underlying algorithms of our search in a clear fashion for the reader.

Efficient Gray Code Successor. To realize an efficient iteration through $\mathbb{F}_{2^n}$, we use an efficient implementation of a Gray code successor algorithm, such as [9, Algorithm 2.3].

Proposition 5. *Denote by $H(g)$ the Hamming weight of $g \in \mathbb{F}_2^n$. Let $g_0 = (0, 0, \ldots, 0) \in \mathbb{F}_2^n$ and let $g_i = (g_{i0}, \ldots, g_{i(n-1)})$ for $i = 0, \ldots, 2^n - 1$. Define $S \colon \mathbb{Z}_{>0} \to \mathbb{Z}_{\geq 0}$, where $S(k)$ is the number of trailing 0s in the binary expansion of k. Let $g_{i+1} = g_i + e_{S(i+1)}$ for $i \geq 0$. Then $G = \{g_0, g_1, \ldots, g_{2^n-1}\}$ is a Gray code of length n.*

In some compilers (and in our code), the function S from Proposition 5 is given by a built-in intrinsic (i.e., the __builtin_ctzl intrinsic).

The Importance of Profiling: Efficient Canonical Element Checking. This section highlights the interplay between theory and practice: namely, the interplay between fast algorithms and computational complexity and the implementations of those algorithms. In particular, in Sect. 4.3 we describe the computational complexity of our new exhaustive search Algorithm 2 in comparison to Algorithm 1 from [11], and in this section we discuss evaluating the actual running time of the various components.

In our C++ development, we use the gperftools (formerly Google Performance Tools) packages for CPU profiling; detailed documentation is available [7].

The profiler aggregates data about the "call tree" of a program. Specifically it samples a running program at regular intervals, and records the function that is running at interrupt time along with all the functions and its call stack. By analyzing the call tree, we can determine in a very practical sense the CPU-time bottlenecks of the program. An example of a current call tree (given by kcachegrind run on the output of gperftools) appears in Fig. 1.

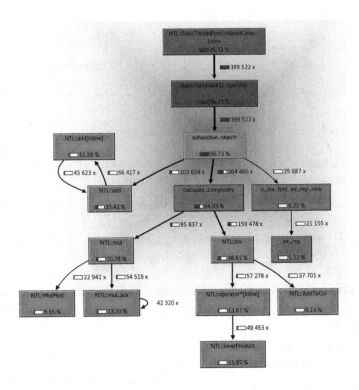

Fig. 1. Call tree for single-node C++/NTL implementation of Algorithm 2 for $\mathbb{F}_{2^{32}}$ over $\mathbb{F}_2$.

From the profiling output, we analyze bottlenecks in our implementation; for example, inefficiencies due to some objects being passed by value rather than by reference. The most surprising bottleneck came in our canonical element routine, where we found that approximately one-third of the total runtime was spent in this routine. While this routine is run $2^n/n$ times for each degree n, improving the runtime of this relatively simple subroutine was *as important* than algorithmic improvements to the overall performance of the program.

In [11], we implemented a naive bit-by-bit lexicographical checker. After profiling, we experimented with various canonical element checkers, the most performant of which uses an underlying integer representative of an element of $\mathbb{F}_{2^n}$. Explicitly, if $\alpha = \sum_{j=0}^{n-1} b_j \zeta_j$ is the expansion of α in an implicit basis

$\{\zeta_0, \ldots, \zeta_{n-1}\}$, then its integer representation is $\sum_{j=0}^{n-1} b_j 2^j$. We notice on average an approximate 25% improvement on overall runtime due to this new implementation in both Sage development and C++ production versions. We give a comparison of the runtimes of Algorithm 2 (in Sage) using these two routines in Table 3.

Table 3. Runtime for Sage implementation of the exhaustive search of $\mathbb{F}_{2^n}$ over $\mathbb{F}_2$ with two lex-first checks: integer representative and bit-by-bit comparison.

n	10	11	12	13	14	15	16	17	18	19
Integer rep	111 ms	168 ms	354 ms	748 ms	1.45 s	3.07 s	6.32 s	12.9 s	26.1 s	55.3 s
Bit-by-bit	130 ms	250 ms	456 ms	951 ms	1.87 s	3.89 s	7.71 s	16.7 s	32.2 s	72 s

As indicated in Fig. 1, after our optimizations the dominant functions in the call graph are NTL internal functions, indicating that we are reaching the limit of the efficiency of implementing the current algorithm.

5.3 Parallelism of Algorithm 2

Algorithm 2 is an example of an *embarrassingly parallel* application; that is, the search function requires no communication between workers, so with an appropriate implementation we can avoid overhead in parallelism.

Due to the embarrassingly parallel nature of the search, we do not attempt any thread boosting of the underlying arithmetic. For example, in the Gray code successor, we could update $(\alpha + e_j)^{2^i} = \alpha^{2^i} + e_j^{2^i}$ for multiple i in parallel. However, this would reduce the number of available threads and would require a barrier after this computation to ensure data coherence. Hence, we only implement parallelism by partitioning the search space according to Gray code rank.

Our test system was a single 2-core Core i7-4500 CPU with 1.80 GHz clock speed. We ran the code on a variety of available systems; in all cases the systems were Intel Core x86_64 architectures with AVX instructions. The code was compiled with the GNU C++ compiler version 7.x and NTL version 10.5.0 compiled with thread safety (NTL version 11.0.0 was released after the pre-proceedings version of this paper).

The combination of iteration through elements in Gray code order with a canonicity check forces an imbalance in the computation, since the relatively expensive Algorithm 3 computation is performed for elements with specific integer representations. To mitigate this imbalance, we partition the search space into a large number of "tasks". Each task represents a contiguous region of Gray code indices; in practice, the number of tasks is set to be much larger than the number of available compute nodes, and compute nodes are assigned to tasks asynchronously.

When a compute node is assigned a task, it initializes a pool of n_c threads, where n_c is the number of cores on the node. The task is partitioned into n_c equally-sized sub-tasks executed synchronously in parallel one per thread. While asynchronous execution at the thread level may be preferable, in practice we assign a sufficient number of tasks that any overhead is acceptable.

We have implemented parallelism in a variety of ways. For intra-node parallelism, we originally had a direct `pthreads` implementation, but for simplicity we eventually removed this in favour of using NTL's built in intrinsics. For inter-node parallelism, an earlier version of the code used MPI to distribute tasks to nodes. This is an acceptable strategy if a single executable is required, but specifying asynchronous computation, dealing with load-balancing and checkpointing prove to be complicated. Our published program accepts as input the task index as a command-line argument and is run independently one-per-node. This has the benefit of being able to rely on `gnu` compilers and not rely on platform-specific (and often less performant) compilers. In practice we assign a number of tasks so that all but the most expensive tasks finish within 5 (or so) min. This paradigm requires some additional job-scheduling by the user, but this can be a feature when borrowing time from highly-shared computing platforms.

We observe that this computation is an excellent candidate for GPU processing, since minimal data communication is required between the CPU and GPU. However, we do not have a pool of GPUs readily available for use, nor do we have the expertise to attempt a port of NTL, or other underlying field arithmetic library, to CUDA/GPUs. We leave this for future work.

6   Results and Future Work

We conclude this paper by presenting the main result of the search, an update to the minimum complexity of a normal basis of $\mathbb{F}_{2^n}$ over $\mathbb{F}_2$, $n = 40, \ldots, 46$. We also present a number of avenues for future work. Our reference code can be accessed through GitHub [18]. See the documentation on that page for instructions on building and running the code. We have also included options in our code for outputting all bases or all multiplication tables. Instructions and some compressed output is available through the GitHub repository.

In most cases, the basis with the minimum complexity in $\mathbb{F}_{2^n}$ over $\mathbb{F}_2$ is unique. Two exceptions occur when $n = 18$ and $n = 19$. The $n = 18$ case is very special: it satisfies the conditions for both types of so-called "optimal normal bases" (bases with minimum possible complexity). Whether the $n = 19$ case is causal or a combinatorial surprise is unknown to us.

6.1   Running Time of Algorithm 2

Table 4 provides the runtime for our high performance implementation of Algorithm 2 for $30 \leq n \leq 39$.

Our largest computation was for $n = 46$, which totaled $20,801$ CPU-hours on Intel Broadwell CPUs at $2.3\,\mathrm{GHz}$. In core-wall-time, this totaled $24,816\,\mathrm{h}$,

Table 4. CPU time for high performance implementation of Algorithm 2 on Intel Broadwell at 2.3 GHz.

n	30	31	32	33	34	35	36	37	38	39
CPU time ($\log_2(\text{sec})$)	9.35	10.58	11.60	12.54	13.75	14.59	15.60	16.87	17.84	18.84

indicating a near 20% overhead. This was a noticeable result of the synchronous execution of threads within the most compute-intensive tasks. For our eventual computation of $47 \leq n \leq 49$, we can mitigate this either by implementing asynchronous thread execution within a task, or reducing the size of the most compute-intensive tasks to mitigate any overhead.

By Theorem 4, Algorithm 2 scales exponentially in n, plus a quadratic factor. This can be seen for example, with some variability, in Tables 2 and 3. Based on the running time for $n = 46$ and assuming stability and homogeneity on production systems, we estimate approximately 287,000 CPU hours, or approximately 33 CPU years, to complete the searches for $47 \leq n \leq 49$. This is well within the resource allocation budgets of, for example, annual Compute Canada awards; see [2], for more information.

6.2 Changes to Table of Minimal Complexity Normal Bases

In [11], we presented tables of the minimum known complexity of a normal basis of any extension $\mathbb{F}_{2^n}$ over $\mathbb{F}_2$. These tables were repeated in [13, Sect. 2.2]. Our results here update the exhaustive tables in [11,13]. Also in [11,13], there appears a table of the minimum known complexity of $\mathbb{F}_{2^n}$ over $\mathbb{F}_2$ for $n \geq 40$ using all known methods and constructions. All non-explicit constructions take known normal bases from either subfields or extensions and use them to build normal bases in extensions and subfields, respectively.

In Table 5 we present the results of the search for $40 \leq n \leq 46$. Each number i in the set in the second column represents the term x^i appearing in the minimal polynomial of the basis. A Sage code snippet to check the minimal polynomial for $n = 45$ follows in Fig. 2. For $40 \leq n \leq 46$, we find no elements having smaller complexity than those that appear in [13, Table 2.2.10]. These complexities are

```
print minpoly45
F = GF(2)
Fx = PolynomialRing(GF(2), 'x')
modulus = 1 + Fx.gen() + Fx.gen()^3 + Fx.gen()^4 + Fx.gen()^(45) #The NTL default modulus for n=45
K = GF(2**n, name='x', modulus=modulus) #GF(2^n) = GF(2)[x]/(modulus)
congs = minpoly45.roots(K, multiplicities=False) #The roots of minpoly45 are a set of conjugates
B = matrix([ (congs[i]*congs[0])._vector_() for i in range(n) ]) #Construct the matrix B {cong[0]}
P = matrix([ congs[i]._vector_() for i in range(n) ]) #Construct the matrix P_{\cong[0]}
Bpinv = B*P.inverse() #Construct BP^{-1}

print len(Bpinv.nonzero_positions()) #The complexity of the basis is the number of nonzero positions in the matrix

x^45 + x^44 + x^42 + x^41 + x^40 + x^39 + x^38 + x^37 + x^33 + x^30 + x^23 + x^20 + x^18 + x^16 + x^14 + x^13 + x^1
2 + x^11 + x^9 + x^6 + x^2 + x + 1
153
```

Fig. 2. Code snippet to check the complexity of a normal basis from its minimal polynomial.

confirmed as the minimum complexities of any normal basis of $\mathbb{F}_{2^n}$ over $\mathbb{F}_2$, but no additional updates to [13, Table 2.2.10] result from bootstrapping methods.

Table 5. Minimum complexity normal basis generator of $\mathbb{F}_{2^n}$ over $\mathbb{F}_2$.

n	Nonzero terms of minimal polynomial	Complexity
40	$\{40, 39, 37, 34, 31, 26, 24, 23, 21, 19, 18, 16, 9, 5, 0\}$	189
41	$\{41, 40, 38, 37, 36, 33, 32, 22, 21, 20, 17, 16, 9, 8, 6, 5, 4, 1, 0\}$	81
42	$\{42, 41, 40, 38, 36, 35, 31, 30, 26, 23, 22, 20, 19, 18, 15, 12, 3, 2, 0\}$	135
43	$\{43, 42, 40, 37, 35, 33, 31, 30, 29, 28, 27, 25, 24, 22, 20, 18, 14, 12, 11, 9, 8, 7, 5, 3, 0\}$	165
44	$\{44, 43, 42, 40, 35, 33, 30, 28, 27, 26, 25, 24, 23, 21, 19, 18, 17, 12, 11, 10, 9, 5, 3, 2, 0\}$	147
45	$\{45, 44, 42, 41, 40, 39, 38, 37, 33, 30, 23, 20, 18, 16, 14, 13, 12, 11, 9, 6, 2, 1, 0\}$	153
46	$\{46, 45, 44, 42, 40, 39, 36, 32, 29, 28, 26, 23, 20, 15, 13, 10, 7, 4, 0\}$	135

Our eventual goal is to complete the search for $47 \leq n \leq 49$. The most likely candidate to find a previously unknown minimum complexity basis is in the $n = 48$ case, since the best known complexity here is 425: easily the largest minimum complexity for any $n < 57$. We do not expect to extend the table beyond $n = 50$, since there are optimal normal bases for $n = 50, 51, 52, 53$, and for $n = 54$ there is a normal basis with complexity $4n - 7$ by construction, so we expect that this basis is minimal. Searching $n \geq 55$ is out of reach for reasonable computational resources.

6.3   Choice of Basis Representation

All NTL field arithmetic is represented using a polynomial (or power) basis. Suppose instead that we choose a normal basis representation. For any $\alpha \in \mathbb{F}_{2^n}$, computing the set $\mathcal{B} = \{\alpha, \alpha^2, \ldots, \alpha^{2^{n-1}}\}$ under normal basis representation can be done by a series of cyclic bit shifts, so we no longer need to iterate through $\mathbb{F}_{2^n}$ in Gray code order.

In normal basis representation, we also keep an efficient cut-down from the trace computation. Recall that $\text{Tr}(\alpha) = \alpha + \alpha^2 + \cdots + \alpha^{2^{n-1}}$, where α^{2^i} is the i-fold shift of the underlying bit vector of α. Hence $\text{Tr}(\alpha) = \sum_{j=0}^{n-1} \alpha_j \pmod{2}$, where α_j is the jth bit of α in normal basis representation. The weight of a vector can be taken by modern computer architectures in a single popcount operation, hence the trace computation can be considered essentially free.

Regardless of basis, calculating normality (either by gcd as in Theorem 1 or by matrix inversion) and complexity requires computing field multiplications. Moreover, the cost of multiplication depends precisely on the complexity of the underlying basis. *A priori*, we expect a random basis of $\mathbb{F}_{2^n}$ over $\mathbb{F}_2$ to have complexity about $n^2/2$, see [11], hence this multiplication will be expensive *until after we find a low-complexity normal element*.

Other (non-normal) bases may be appropriate as well, but any other choice of basis would require hand-rolling a new finite field arithmetic library. Hence, we leave this for future work.

6.4   A Best-Case Scenario

Any algorithm that goes exhaustively through every normal element computing their bases must have cost at least

$$(n-1)M(n)\frac{\Phi(x^n - 1)}{n},$$

where $\Phi(x^n - 1)/n$ is the number of normal basis of $\mathbb{F}_{2^n}$ over $\mathbb{F}_2$. This expression assumes that we compute the representation of $B_i = \alpha\alpha^{2^i}$ for $i = 1, \ldots, n-1$, while looping only over normal elements and also having B_i represented immediately in the normal basis. It is unclear that the latter part is even possible. However, it is possible in spirit to loop over only normal elements using the *primary decomposition*, seen in [3], for example. We state the main result as follows.

Proposition 6. *Let $q = p^e$ for some prime p and positive integer e and let $\gcd(n, p) = 1$ so that $x^n - 1 = f_1 f_2 \cdots f_k$, where the f_i are distinct irreducible factors over $\mathbb{F}_q$. If $f_i(x) = \sum_{j=0}^n f_{ij}x^j$, let $F_i(x) = \sum_{j=0}^n f_{ij}x^{q^j}$. Each F_i is* a linearized polynomial, *hence a linear operator over $\mathbb{F}_q$, and its roots form a vector subspace of $\mathbb{F}_{q^n}$ over $\mathbb{F}_q$ that is stable under qth powers. Let $V_i = \ker(F_i)$, then $\alpha = \sum_{1 \le i \le k} \alpha_i$ with $\alpha_i \in V_i$ is a normal element of $\mathbb{F}_{q^n}$ over $\mathbb{F}_q$ if and only if $\alpha_i \ne 0$ for all $1 \le i \le k$.*

Proposition 6 can be modified to remove the restriction on the field degree, though the choices of α_i become more complicated. In order to iterate through only normal elements, we can iterate through the nonzero elements of each of the V_i vector subspaces of $\mathbb{F}_{q^n}$. It is unclear how to do this efficiently, so we leave this for future work.

6.5   The Case $n = 64$

We are still unaware of any low complexity normal bases of $\mathbb{F}_{2^n}$ over $\mathbb{F}_2$ when n is a power of 2, except for those obtained by exhaustive search ($n = 2^\ell, \ell \le 5$). When n is a power of 2, we realize a dramatic speedup due to the following specific version of Proposition 3.

Proposition 7. [13, Corollary 5.2.9] *An element $\alpha \in \mathbb{F}_{2^{64}}$ is normal if and only if $\mathrm{Tr}(\alpha) \ne 0$.*

By Proposition 7, precisely 2^{63} elements of $\mathbb{F}_{2^{64}}$ are normal, so with a trace pre-computation all elements passing into Algorithm 3 are normal. Hence, with the canonicity check, we compute the complexity of exactly 2^{57} elements. Each computation requires 63 multiplications (each estimated at 64^2 bit operations) and a 64×64 matrix inversion over $\mathbb{F}_2$ (estimated at 64^3 bit operations), so in total we estimate at least 2^{76} bit operations to calculate all of their complexities. This seems out of reach for reasonable modern computational power. This cost may be lowered by a few bits using an early-abort strategy, but any such exhaustive search is likely to be untenable for the foreseeable future. Of course, any algorithm that inspects every element of $\mathbb{F}_{2^{64}}$, as ours does, has 2^{64} as a lower bound on the running time.

Acknowledgement. We would like to thank the three reviewers for their helpful suggestions, which greatly improved the presentation of this paper.

References

1. Arnault, F., Pickett, E.J., Vinatier, S.: Construction of self-dual normal bases and their complexity. Finite Fields Appl. **18**, 458–472 (2012)
2. Compute Canada: 2018 Final Allocations for Publication. https://www.computecanada.ca/research-portal/accessing-resources/resource-allocation-competitions/rac-2018-results/. Accessed 17 May 2018
3. Blake, I.F., Gao, S., Mullin, R.C.: Specific irreducible polynomials with linearly independent roots over finite fields. Linear Algebr. Appl. **253**, 227–249 (1997)
4. Gao, S., Lenstra, H.W.: Optimal normal bases. Des. Codes Cryptogr. **2**, 315–323 (1992)
5. Gao, S., Panario, D.: Density of normal elements. Finite Fields Appl. **3**, 141–150 (1997)
6. von zur Gathen, J., Gerhard, J.: Modern Computer Algebra. Cambridge University Press, Cambridge (2013)
7. gperftools: Authors unlisted © Google Inc. (2005). www.github.io/gperftools. Accessed 27 Mar 2018
8. Jungnickel, D.: Finite Fields: Structure and Arithmetics, Bibliographisches Institut, Mannheim GE (1993)
9. Kreher, D.L., Stinson, D.R.: Combinatorial Algorithms: Generation Enumeration and Search. CRC Press, Boca Raton (1999)
10. Lidl, R., Niederreiter, H.: Finite Fields. Cambridge University Press, Oxford (1997)
11. Masuda, A., Moura, L., Panario, D., Thomson, D.: Low complexity normal elements over finite fields of characteristic two. IEEE Trans. Comput. **57**, 990–1001 (2008)
12. Mullin, R.C., Onyszchuk, I.M., Vanstone, S.A., Wilson, R.M.: Optimal normal bases in $GF(p^n)$, Discrete Appl. Math. **22**, 149–161 (1988/1989)
13. Mullen, G.L., Panario, D.: Handbook of Finite Fields. CRC Press, Boca Raton (2013)
14. National Institute for Standards in Technology: Digital Signature Standard FIPS PUB 186-4 (2013)
15. SageMath: the Sage Mathematics Software System (Version 7.1.0). The Sage Developers (2018). http://www.sagemath.org
16. Shoup, V.: NTL: number theory library. www.shoup.net/ntl. Accessed 3 Mar 2018
17. Silva, D., Kschischang, F.: Fast encoding and decoding of Gabidulin codes. IEEE Int. Symp. Inf. Theory **2009**, 2858–2862 (2009)
18. Thomson, D.: Exhaustive search for normal basis, Version 1.0.0. www.github.com/dgthoms/exh_normal. Accessed 26 May 2018

On Symmetry and Differential Properties of Generalized Boolean Functions

Thor Martinsen[1], Wilfried Meidl[2], Alexander Pott[3],
and Pantelimon Stănică[1(✉)]

[1] Department of Applied Mathematics, Naval Postgraduate School,
Monterey, CA 93943–5216, USA
{tmartins,pstanica}@nps.edu
[2] Johann Radon Institute for Computational and Applied Mathematics,
Austrian Academy of Sciences, Altenbergerstrasse 69, 4040 Linz, Austria
meidlwilfried@gmail.com
[3] Institute of Algebra and Geometry, Faculty of Mathematics,
Otto von Guericke University Magdeburg, Universitätsplatz 2,
39106 Magdeburg, Germany
alexander.pott@ovgu.de

Abstract. In this paper we investigate various differential properties of generalized Boolean functions defined on $\mathbb{F}_2^n$ with values in $\mathbb{Z}_{2^k}$, $k \geq 2$. We characterize linear structures for the generalized Boolean functions in terms of their binary expansion components, and find all symmetric generalized bent functions. Next, we show that there are no symmetric balanced functions defined on $\mathbb{F}_2^n$ with values in a group of order 2^k, $k \geq 2$, a contrast to the classical case for $k = 1$, commonly known as the bisection of binomial coefficients. Further, we characterize the avalanche features of a generalized Boolean function in terms of differentials. Lastly, we show that a partially gbent function is plateaued.

Keywords: Generalized Boolean functions · Linear structures
Generalized bent · Semibent · Partially bent · Avalanche features
Plateaued

1 Introduction

In [27], Schmidt found a connection between words in multi-carrier code-division multiple access (MC-CDMA) systems and generalized bent functions from $\mathbb{F}_2^n$ to $\mathbb{Z}_4$, as well as functions from $\mathbb{F}_2^n$ to $\mathbb{Z}_q$ were considered from the viewpoint of cyclic codes over rings. Shortly thereafter, generalized Boolean functions became an active area of research [17,21,23,24,27–29]. Many authors [8,9,18] have investigated linear structures of Boolean functions. However, thus far, little has been written about linear structures, symmetry, balancedness and avalanche features of generalized Boolean functions.

Let $\mathbb{V}_n$ be an n-dimensional vector space over the two-element field $\mathbb{F}_2$ and for an integer q, let $\mathbb{Z}_q$ be the ring of integers modulo q. By '+' and '−' we

L. Budaghyan and F. Rodríguez-Henríquez (Eds.): WAIFI 2018, LNCS 11321, pp. 207–223, 2018.
https://doi.org/10.1007/978-3-030-05153-2_11

respectively denote addition and subtraction modulo q, whereas '$\oplus$' denotes the addition over $\mathbb{V}_n$. We call a function from $\mathbb{V}_n$ to $\mathbb{Z}_q$ ($q \geq 2$) a *generalized Boolean function* on n variables and denote the set of all generalized Boolean functions by $\mathcal{GB}_n^q$ and when $q = 2$, by $\mathcal{B}_n$. If $q = 2^k$ for some $k \geq 1$ we can associate to any $f \in \mathcal{GB}_n^q$ a unique sequence of Boolean functions $a_i \in \mathcal{B}_n$ ($i = 0, 1, \ldots, k-1$) such that

$$f(\mathbf{x}) = a_0(\mathbf{x}) + 2a_1(\mathbf{x}) + \cdots + 2^{k-1}a_{k-1}(\mathbf{x}), \text{ for all } \mathbf{x} \in \mathbb{V}_n.$$

It has been observed, see [14, 22, 23, 30], that the Boolean functions a_i, and furthermore all Boolean functions of the form $a_{k-1} \oplus c_{k-2}a_{k-2} \oplus \cdots \oplus c_0a_0$, $c_i \in \mathbb{F}_2$, $0 \leq i \leq k-2$, play an important role in the analysis of properties of functions $f \in \mathcal{GB}_n^q$. In accordance with the terminology for vectorial bent functions, we call those Boolean functions *components* of f or *component functions* of f.

If $\mathbb{V}_n$ is $\mathbb{F}_2^n$, then the (*Hamming*) *weight* of $\mathbf{x} = (x_1, \ldots, x_n) \in \mathbb{V}_n$ is denoted by $wt(\mathbf{x})$ and equals $\sum_{i=1}^n x_i$ (the Hamming weight of a function is the weight of its truth table, that is, its output vector). The cardinality of a set S is denoted by $|S|$.

For a generalized Boolean function $f : \mathbb{V}_n \to \mathbb{Z}_q$ we define the *generalized Walsh-Hadamard transform* to be the complex valued function

$$\mathcal{H}_f^{(q)}(\mathbf{u}) = \sum_{\mathbf{x} \in \mathbb{V}_n} \zeta_q^{f(\mathbf{x})}(-1)^{\mathbf{u}\cdot\mathbf{x}},$$

where $\zeta_q = e^{\frac{2\pi i}{q}}$ and $\mathbf{u} \cdot \mathbf{x}$ denotes a (nondegenerate) inner product on $\mathbb{V}_n$ (for easy writing, we sometimes use ζ, $\mathcal{H}_f$, instead of ζ_q, respectively, $\mathcal{H}_f^{(q)}$, when q is fixed). If $\mathbb{V}_n = \mathbb{F}_2^n$, the vector space of the n-tuples over $\mathbb{F}_2$, then for $\mathbf{u} \cdot \mathbf{x}$ we use the conventional dot product. For $q = 2$, we obtain the usual *Walsh-Hadamard transform*

$$\mathcal{W}_f(\mathbf{u}) = \sum_{\mathbf{x} \in \mathbb{V}_n} (-1)^{f(\mathbf{x})}(-1)^{\mathbf{u}\cdot\mathbf{x}}.$$

The sum

$$\mathcal{C}_{f,g}(\mathbf{z}) = \sum_{\mathbf{x} \in \mathbb{V}_n} \zeta^{f(\mathbf{x})-g(\mathbf{x}\oplus\mathbf{z})}$$

is the *crosscorrelation* of f and g at $\mathbf{z} \in \mathbb{V}_n$. The *autocorrelation* of $f \in \mathcal{B}_n$ at $\mathbf{u} \in \mathbb{V}_n$ is $\mathcal{C}_{f,f}(\mathbf{u})$ above, which we denote by $\mathcal{C}_f(\mathbf{u})$. Recall [29] that if $f, g \in \mathcal{GB}_n^q$, then

$$\sum_{\mathbf{u} \in \mathbb{V}_n} \mathcal{C}_{f,g}(\mathbf{u})(-1)^{\mathbf{u}\cdot\mathbf{x}} = \mathcal{H}_f(\mathbf{x})\overline{\mathcal{H}_g(\mathbf{x})},$$

$$\mathcal{C}_{f,g}(\mathbf{u}) = 2^{-n} \sum_{\mathbf{x} \in \mathbb{V}_n} \mathcal{H}_f(\mathbf{x})\overline{\mathcal{H}_g(\mathbf{x})}(-1)^{\mathbf{u}\cdot\mathbf{x}}.$$

Taking the particular case $f = g$ we obtain

$$\mathcal{C}_f(\mathbf{u}) = 2^{-n} \sum_{\mathbf{x} \in \mathbb{V}_n} |\mathcal{H}_f(\mathbf{x})|^2(-1)^{\mathbf{u}\cdot\mathbf{x}}.$$

A function $f : \mathbb{V}_n \to \mathbb{Z}_q$ is called *generalized bent* (*gbent*) if $|\mathcal{H}_f(\mathbf{u})| = 2^{n/2}$ for all $\mathbf{u} \in \mathbb{V}_n$. We recall that a Boolean function f for which $|\mathcal{W}_f(\mathbf{u})| = 2^{n/2}$ for all $\mathbf{u} \in \mathbb{V}_n$ is a *bent* function, which only exists for even n. Further recall that $f \in \mathcal{B}_n$ is called *plateaued* if $|\mathcal{W}_f(\mathbf{u})| \in \{0, 2^{(n+s)/2}\}$ for all $\mathbf{u} \in \mathbb{V}_n$ for a fixed integer s depending on f (we then also call f *s-plateaued*). If $s = 1$ (n must then be odd), or $s = 2$ (n must then be even), we call f *semibent*.

Given a Boolean function f, the derivative of f with respect to a vector $\mathbf{a}$, denoted by $D_\mathbf{a} f$, is the Boolean function defined by

$$D_\mathbf{a} f(\mathbf{x}) = f(\mathbf{x} \oplus \mathbf{a}) \oplus f(\mathbf{x}), \text{ for all } \mathbf{x} \in \mathbb{V}_n.$$

For more on Boolean functions, the reader can consult the following excellent references [1–3, 7, 25, 31].

2 Derivatives and Linear Structures in the Generalized Boolean Functions' Context

Given a generalized Boolean function $f : \mathbb{V}_n \to \mathbb{Z}_q$, we define the *derivative* $D_\mathbf{a}^{(q)} f$ of f with respect to a vector $\mathbf{a} \in \mathbb{V}_n$ to be the generalized Boolean function $D_\mathbf{a}^{(q)} f : \mathbb{V}_n \to \mathbb{Z}_q$

$$D_\mathbf{a}^{(q)} f(\mathbf{x}) = f(\mathbf{x} \oplus \mathbf{a}) - f(\mathbf{x}), \text{ for all } \mathbf{x} \in \mathbb{V}_n.$$

When there is no danger of confusion, we write $D_\mathbf{a} f$ in lieu of $D_\mathbf{a}^{(q)} f$.

We say that $\mathbf{a} \in \mathbb{V}_n$ is a *linear structure* of a generalized Boolean function $f \in \mathcal{GB}_n^q$ if the derivative of f with respect to $\mathbf{a}$ is constant, that is, $f(\mathbf{x} \oplus \mathbf{a}) - f(\mathbf{x}) = c \in \mathbb{Z}_q$ constant, for all $\mathbf{x} \in \mathbb{V}_n$. Observe that if $\mathbf{a}_1, \mathbf{a}_2$ are linear structures for f, then there are constants c_1, c_2 such that $f(\mathbf{x} \oplus \mathbf{a}_1 \oplus \mathbf{a}_2) - f(\mathbf{x} \oplus \mathbf{a}_2) = c_1$, $f(\mathbf{x} \oplus \mathbf{a}_2) - f(\mathbf{x}) = c_2$, for all $\mathbf{x}$, which by summing renders $f(\mathbf{x} \oplus \mathbf{a}_1 \oplus \mathbf{a}_2) - f(\mathbf{x}) = c_1 + c_2$ for all $\mathbf{x}$. Thus, we see that the set of all linear structures (including $\mathbf{0}$) forms a vector subspace in $\mathbb{V}_n$, which we will denote by $LS_q(f)$ (and when q is fixed, we may write $LS(f)$). From here on, we let $q = 2^k$.

We begin with a characterization for the linear structures of a generalized Boolean function in terms of the generalized Walsh-Hadamard transform. Let $S_f = \{\mathbf{x} \in \mathbb{V}_n \,|\, \mathcal{H}_f(\mathbf{x}) \neq 0\}$ (the generalized Walsh-Hadamard support) and for a vector $\mathbf{a}$, let $\mathbf{a}^\perp$ be the orthogonal complement of $\mathbf{a}$, that is, $\mathbf{a}^\perp = \{\mathbf{x} \in \mathbb{V}_n \,|\, \mathbf{a} \cdot \mathbf{x} = 0\}$. This is a terminology widely used in linear algebra, although there may be a nontrivial intersection between a subspace and its orthogonal complement. From Parseval's identity, we immediately infer that $S_f \neq \emptyset$.

We might be tempted to conjecture that linear structures for the components transfer to linear structures for the generalized Boolean function, but that is not true, as we argue next: for example, let $n \geq 3, k \geq 2$, and $f(\mathbf{x}) = a_0(\mathbf{x})$ in $\mathcal{GB}_n^{2^k}$, where $a_0(x_1, \ldots, x_n) = x_1 \cdots x_{n-2}(x_{n-1} \oplus x_n)$ in $\mathcal{B}_n$. We observe that $(0, \ldots, 1, 1) \in LS_2(a_0)$, since $f(x_1, \ldots, x_{n-1} \oplus 1, x_n \oplus 1) = f(x_1, \ldots, x_{n-1}, x_n)$ over $\mathbb{F}_2$. However, $f(x_1, \ldots, x_{n-1} \oplus 1, x_n \oplus 1) = f(x_1, \ldots, x_{n-1}, x_n) + 2x_1 \cdots x_{n-2}$ over $\mathbb{Z}_{2^k}$, thus, $D_\mathbf{a} f(\mathbf{x}) = 2x_1 \cdots x_{n-2}$, and therefore $(0, \ldots, 0, 1, 1) \notin LS_{2^k}(f)$.

In reality, the next result settles the "score", by completely characterizing linear structures for the generalized Boolean functions in terms of their components.

Theorem 1. *Let* $f \in \mathcal{GB}_n^{2^k}$, *with* $f(\mathbf{x}) = \sum_{i=0}^{k-1} 2^i a_i(\mathbf{x})$, $a_i \in \mathcal{B}_n$. *The following are equivalent:*

(i) *The vector* $\mathbf{a}$ *is a linear structure for* f.
(ii) *The vector* $\mathbf{a}$ *satisfies* $\zeta^{f(\mathbf{a})-f(\mathbf{0})} = (-1)^{\mathbf{a} \cdot \mathbf{w}}$, *for all* $\mathbf{w} \in S_f$.
(iii) *The vector* $\mathbf{a}$ *is a linear structure for* a_i, $i \geq 0$, *such that* $a_i(\mathbf{a}) = a_i(\mathbf{0}), 0 \leq i < k - 1$. .

Proof. We first show $(i) \Leftrightarrow (ii)$. Let $g(\mathbf{x}) := f(\mathbf{x} \oplus \mathbf{a}) - c$, for some constant $c \in \mathbb{Z}_{2^k}$, $\mathbf{a} \in \mathbb{V}_n$. Then

$$\mathcal{H}_g(\mathbf{w}) = \sum_{\mathbf{x} \in \mathbb{V}_n} \zeta^{g(\mathbf{x})} (-1)^{\mathbf{x} \cdot \mathbf{w}}$$

$$= \sum_{\mathbf{x} \in \mathbb{V}_n} \zeta^{f(\mathbf{x} \oplus \mathbf{a}) - c} (-1)^{\mathbf{x} \cdot \mathbf{w}}$$

$$\overset{\mathbf{y} := \mathbf{x} \oplus \mathbf{a}}{=} \zeta^{-c} (-1)^{\mathbf{a} \cdot \mathbf{w}} \sum_{\mathbf{y} \in \mathbb{V}_n} \zeta^{f(\mathbf{y})} (-1)^{\mathbf{y} \cdot \mathbf{w}}$$

$$= \zeta^{-c} (-1)^{\mathbf{a} \cdot \mathbf{w}} \mathcal{H}_f(\mathbf{w}).$$

Now, if $\mathbf{a}$ is a linear structure, then (with the above notation) $g(\mathbf{x}) = f(\mathbf{x})$ (where $c = f(\mathbf{a}) - f(\mathbf{0})$), hence $\mathcal{H}_g(\mathbf{w}) = \mathcal{H}_f(\mathbf{w})$. Thus, $\zeta^{-c}(-1)^{\mathbf{a} \cdot \mathbf{w}} \mathcal{H}_f(\mathbf{w}) = \mathcal{H}_f(\mathbf{w})$, which renders $\mathcal{H}_f(\mathbf{w}) (1 - \zeta^{-c}(-1)^{\mathbf{a} \cdot \mathbf{w}}) = 0$. Therefore, taking any $\mathbf{w} \in S_f \neq \emptyset$, we get that $\zeta^c = (-1)^{\mathbf{a} \cdot \mathbf{w}}$, and since ζ is a primitive 2^k-root of unity, then necessarily, $c = 0$ or 2^{k-1}, depending on whether $\mathbf{w} \in \mathbf{a}^\perp$, or not. The converse is also true.

Next we show $(i) \Leftrightarrow (iii)$. If $f \in \mathcal{GB}_n^{2^k}$ with $f(\mathbf{x}) = \sum_{i=0}^{k-1} 2^i a_i(\mathbf{x})$, $a_0 \in \mathcal{B}_n$, then it is easy to see (by reducing modulo 2) that if $\mathbf{a}$ is a linear structure for f, and consequently, $D_{\mathbf{a}}^{(2^k)} f(\mathbf{x}) = \sum_{i=0}^{k-1} 2^i (a_i(\mathbf{x} \oplus \mathbf{a}) - a_i(\mathbf{x})) = c \in \mathbb{Z}_{2^k}$, then $\mathbf{a}$ is a linear structure for a_0 whose derivative is $D_{\mathbf{a}}^{(2)} f(\mathbf{x}) = a_0(\mathbf{x} \oplus \mathbf{a}) \oplus a_0(\mathbf{x}) = c$ (mod 2).

Let $f \in \mathcal{GB}_n^{2^k}$ and write $f(\mathbf{x}) = a_0(\mathbf{x}) + 2 f_1(\mathbf{x})$, where $a_0 \in \mathcal{B}_n$, $f_1 \in \mathcal{GB}_n^{2^{k-1}}$. Assume that $\mathbf{a}$ is a linear structure for f and so, $D_{\mathbf{a}} f(\mathbf{x}) = c \in \mathbb{Z}_q$ (independent of $\mathbf{x}$). We compute $D_{\mathbf{a}} f$, and obtain (using (ii))

$$D_{\mathbf{a}}(f)(\mathbf{x}) = (a_0(\mathbf{x} \oplus \mathbf{a}) - a_0(\mathbf{x})) + 2 (f_1(\mathbf{x} \oplus \mathbf{a}) - f_1(\mathbf{x})) = c \in \{0, 2^{k-1}\}. \quad (1)$$

Thus, $a_0(\mathbf{x}) = a_0(\mathbf{x} \oplus \mathbf{a})$, and from Eq. (1), we infer that $f_1(\mathbf{x} \oplus \mathbf{a}) - f_1(\mathbf{x}) = \frac{c}{2} \in \{0, 2^{k-2}\}$ in $\mathbb{Z}_{2^{k-1}}$. Therefore, $\mathbf{a}$ is a linear structure for f_1, in addition to being a linear structure for a_0. Inductively, we infer (by the uniqueness of the binary representation) that for all $\mathbf{x}$, $a_i(\mathbf{x} \oplus \mathbf{a}) - a_i(\mathbf{x}) = 0$, for $0 \leq i < k - 2$.

If $a_{k-1}(\mathbf{x} \oplus \mathbf{a}) - a_{k-1}(\mathbf{x}) = 0$, then $f(\mathbf{x} \oplus \mathbf{a}) - f(\mathbf{x}) = 0$, and if $a_{k-1}(\mathbf{x} \oplus \mathbf{a}) - a_{k-1}(\mathbf{x}) = \pm 1$, then $f(\mathbf{x} \oplus \mathbf{a}) - f(\mathbf{x}) = 2^{k-1}$. Certainly, the reciprocal is true and the claim is shown. $\square$

Corollary 1. *Let* $f \in \mathcal{GB}_n^{2^k}$. *If* $\mathbf{a}$ *is a linear structure for* f, *then either* $S_f \subseteq \mathbf{a}^{\perp}$, *or* $S_f \subseteq \overline{\mathbf{a}^{\perp}}$ *(the set complement of* $\mathbf{a}^{\perp}$*); also, if* $\mathbf{a}$ *is a linear structure for* f, *then* $f(\mathbf{a}) - f(\mathbf{0}) \in \{0, 2^{k-1}\}$.

Remark 1. It is immediate that if f is a generalized bent Boolean function then f has no linear structure.

Next, we will use the method of Lechner [19] and Lai [18] to simplify the algebraic normal form of a function admitting linear structures. The result is similar and we give here the proof for the reader's convenience. We shall use below the observation that if $\mathbf{a}$ is a linear structure for f, then $f(\mathbf{x} \oplus \mathbf{a}) - f(\mathbf{x}) = c$, where $c = f(\mathbf{a}) - f(\mathbf{0})$.

Proposition 1. *Let* $f \in \mathcal{GB}_n^{2^k}$ *and* $1 \le \dim LS_{2^k}(f) = r$. *Then, there exists an invertible* $n \times n$ *matrix* A *such that*

$$f((x_1, \ldots, x_n) \cdot A) = \sum_{i=1}^{r} \alpha_i x_i + g(x_{r+1}, \ldots, x_n),$$

where $\alpha_i \in \mathbb{Z}_{2^k}$ *and* $g \in \mathcal{GB}_{n-r}^{2^k}$ *is a generalized Boolean function with no linear structure.*

Proof. Since $\dim LS_{2^k}(f) = r$, we let $\{\mathbf{a}_1, \ldots, \mathbf{a}_r\}$ be a basis for $LS_{2^k}(f)$, which can be completed to a basis for $\mathbb{F}_2^n$, say, $\{\mathbf{a}_1, \ldots, \mathbf{a}_r, \mathbf{a}_{r+1}, \ldots, \mathbf{a}_n\}$. We now define the matrix A to be the matrix corresponding to the change of basis from the canonical basis $\{e_1 = (1, 0, \ldots, 0), \ldots, e_n = (0, \ldots, 0, 1)\}$ to the basis $\{\mathbf{a}_1, \ldots, \mathbf{a}_n\}$, that is, $\mathbf{a}_i = e_i A$, $1 \le i \le n$. Note that if $x \in \mathbb{F}_2$, then $f(x\mathbf{a}) - f(\mathbf{0}) = x(f(\mathbf{a}) - f(\mathbf{0}))$. Further, using the fact that $\mathbf{a}_i$ (hence $\mathbf{a}_i x_i$, as well), $1 \le i \le r$, are linear structures for f, we obtain

$$f((x_1, \ldots, x_n) \cdot A) = f\left(\left(\sum_{i=1}^{n} e_i x_i\right) \cdot A\right)$$

$$= f(\mathbf{a}_1 x_1 \oplus \cdots \oplus \mathbf{a}_r x_r \oplus \cdots \oplus \mathbf{a}_n x_n)$$

$$= f(\mathbf{a}_2 x_2 \oplus \cdots \oplus \mathbf{a}_r x_r \oplus \cdots + \mathbf{a}_n x_n) + f(x_1 \mathbf{a}_1) - f(\mathbf{0})$$

$$= f(\mathbf{a}_2 x_2 \oplus \cdots \oplus \mathbf{a}_r x_r \oplus \cdots + \mathbf{a}_n x_n) + x_1(f(\mathbf{a}_1) - f(\mathbf{0}))$$

$$= f(\mathbf{a}_3 x_3 \oplus \cdots + \mathbf{a}_n x_n) + x_1(f(\mathbf{a}_1) - f(\mathbf{0})) + x_2(f(\mathbf{a}_2) - f(\mathbf{0}))$$

$$\cdots\cdots\cdots$$

$$= \sum_{i=1}^{r} \alpha_i x_i + f(x_{r+1}\mathbf{a}_{r+1} \oplus \cdots \oplus x_n \mathbf{a}_n),$$

where $\alpha_i := f(\mathbf{a}_i) - f(\mathbf{0})$, so $g(x_{r+1}, \ldots, x_n) := f(x_{r+1}\mathbf{a}_{r+1} \oplus \cdots \oplus x_n \mathbf{a}_n)$.

To show the last claim, observe that if $(b_{r+1}, \ldots, b_n)$ is a linear structure for g, then $\mathbf{b} = (0, \ldots, 0, b_{r+1}, \ldots, b_n) \cdot A$ is a linear structure for f (and $\mathbf{b}$ is independent of $\mathbf{a}_1, \ldots, \mathbf{a}_r$), since A is invertible and

$$f(\mathbf{x} \cdot A \oplus \mathbf{b}) = \sum_{i=1}^{r} \alpha_i x_i + g((x_{r+1}, \ldots, x_n) \oplus (b_{r+1}, \ldots, b_n))$$

$$= \sum_{i=1}^{r} \alpha_i x_i + g(x_{r+1}, \ldots, x_n) + g(b_{r+1}, \ldots, b_n) - g(\mathbf{0})$$

$$= f(\mathbf{x} \cdot A) + f(\mathbf{b}) - f(\mathbf{0}), \quad \text{for all } \mathbf{x}.$$

This contradicts the fact that $\dim LS_{2^k}(f) = r$, and the theorem is shown. $\square$

3   Symmetric Generalized Boolean Functions

In this section $\mathbb{V}_n = \mathbb{F}_2^n$, the vector space of n-tuples over $\mathbb{F}_2$. Savicky [26] (see also [13]) showed that for each even n, the only symmetric bent functions are the quadratic symmetric functions $S_{c,d}(\mathbf{x}) = s_2(\mathbf{x}) \oplus c s_1(\mathbf{x}) \oplus d$, $c, d \in \mathbb{F}_2$, where s_1, s_2 are the elementary symmetric polynomials of degree 1, respectively 2. In this section we show that for any (even or odd) n, the only symmetric generalized bent Boolean function in $\mathcal{GB}_n^{2^k}$, $k \geq 2$, is essentially the quaternary function $s_1(\mathbf{x}) + 2s_2(\mathbf{x})$. In the second part, we show that there is no balanced symmetric generalized Boolean function in $\mathcal{GB}_n^{2^k}$, $k > 1$.

We shall be using the following result on generalized Boolean bent functions, which for n even first appeared in [23, Theorem 18]. For odd n we may refer to [22, 30].

Proposition 2. Let $f(\mathbf{x})$ be a gbent function in $\mathcal{GB}_n^{2^k}$, $k > 1$, (uniquely) given as

$$f(\mathbf{x}) = a_0(\mathbf{x}) + 2a_1(\mathbf{x}) + \cdots + 2^{k-2}a_{k-2}(\mathbf{x}) + 2^{k-1}a_{k-1}(\mathbf{x}), \qquad (2)$$

$a_i \in \mathcal{B}_n$, $0 \leq i \leq k - 1$.

(i) If n is even, then all Boolean functions of the form

$$g_{\mathbf{c}}(\mathbf{x}) = c_0 a_0(\mathbf{x}) \oplus c_1 a_1(\mathbf{x}) \oplus \cdots \oplus c_{k-2} a_{k-2}(\mathbf{x}) \oplus a_{k-1}(\mathbf{x}),$$

$\mathbf{c} = (c_0, c_1, \ldots, c_{k-2}) \in \mathbb{F}_2^{k-1}$, are bent functions. In particular, $a_0(\mathbf{x}) + 2a_1(\mathbf{x})$ is a quaternary gbent function if and only if a_1 and $a_1 \oplus a_0$ are bent.

(ii) If n is odd, then all Boolean functions of the form

$$g_{\mathbf{c}}(\mathbf{x}) = c_0 a_0(\mathbf{x}) \oplus c_1 a_1(\mathbf{x}) \oplus \cdots \oplus c_{k-2} a_{k-2}(\mathbf{x}) \oplus a_{k-1}(\mathbf{x}),$$

$\mathbf{c} = (c_0, c_1, \ldots, c_{k-2}) \in \mathbb{F}_2^{k-1}$, are semibent functions. Moreover for every $\mathbf{u} \in \mathbb{V}_n$ we either have $\mathcal{W}_{g_{\mathbf{c}}}(\mathbf{u}) \neq 0$ for all $\mathbf{c} = (c_0, c_1, \ldots, c_{k-2})$ with $c_{k-2} = 0$ or for all $\mathbf{c}$ with $c_{k-2} = 1$ (but not for both). In particular, $a_0(\mathbf{x}) + 2a_1(\mathbf{x})$ is a quaternary gbent function if and only if a_1 and $a_1 \oplus a_0$ are semibent such that $\mathcal{W}_{a_1}(\mathbf{u}) = 0$ if and only if $\mathcal{W}_{a_1 \oplus a_0}(\mathbf{u}) \neq 0$.

By Proposition 2, for a gbent function f in $\mathcal{GB}_n^{2^k}$ given as in (2), $a_{k-1} + \langle a_0, \ldots, a_{k-2} \rangle$ is an affine space of bent, respectively, semibent functions. As pointed out in [14], every gbent function for which the coordinate functions $\{a_0, \ldots, a_{k-2}\}$ are linearly dependent, can be reduced to a gbent function in $\mathcal{GB}_n^{2^{k'}}$, for some $k' < k$ and linearly independent coordinate functions. Conversely we can see f as a naturally lifted version of the gbent function in $\mathcal{GB}_n^{2^{k'}}$ with a very restricted value set in $\mathbb{Z}_{2^k}$. Hence for the classification of (symmetric) generalized bent functions, it is essential to consider only functions for which $\{a_0, \ldots, a_{k-2}\}$ are linearly independent.

A vectorial function $f : \mathbb{F}_2^n \to \mathbb{F}_2^k$ given as $f(\mathbf{x}) = (a_0(\mathbf{x}), a_1(\mathbf{x}), \ldots, a_{k-1}(\mathbf{x}))$ is symmetric if and only if every coordinate function a_i, $0 \leq i < k$, is symmetric. A similar statement applies to generalized Boolean functions (we omit the proof).

Lemma 1. *Let $f : \mathbb{F}_2^n \to \mathbb{Z}_{2^k}$, $k \geq 2$, and $f(\mathbf{x}) = \sum_{i=0}^{k-1} 2^i a_i(\mathbf{x})$, $a_i \in \mathcal{B}_n$. Then f is symmetric if and only if all components a_i are symmetric, $0 \leq i < k$.*

Since for odd n we require symmetric semibent functions, we will use the methods of [26] to also investigate semibent functions. The standard examples of semibent functions are partially bent functions with a one-dimensional linear space. Recall that a partially bent function f is defined as a function for which for all $\mathbf{a} \in \mathbb{F}_2^n$, the derivative $D_\mathbf{a} f$ is either balanced or constant. All quadratic functions are partially bent, but by a construction in [32] there exist semibent functions, which are not partially bent.

We start our analysis by observing that $s_2(\mathbf{x}) = \binom{wt(\mathbf{x})}{2} \bmod 2$ and $(s_2 \oplus s_1)(\mathbf{x}) = \binom{wt(\mathbf{x})}{2} + wt(\mathbf{x}) \bmod 2$, hence

$$s_2(\mathbf{x}) = \begin{cases} 0 & : \quad wt(\mathbf{x}) \equiv 0, 1 \bmod 4 \\ 1 & : \quad wt(\mathbf{x}) \equiv 2, 3 \bmod 4 \end{cases}$$

$$(s_2 \oplus s_1)(\mathbf{x}) = \begin{cases} 0 & : \quad wt(\mathbf{x}) \equiv 0, 3 \bmod 4 \\ 1 & : \quad wt(\mathbf{x}) \equiv 1, 2 \bmod 4 \end{cases} \tag{3}$$

Before we show that $S_{c,d} = s_2 \oplus cs_1 \oplus d$, $c, d \in \mathbb{F}_2$, is semibent, when n is odd, we recall that the Walsh transform of a symmetric function $f : \mathbb{F}_2^n \to \mathbb{F}_2$ is (see [26, Eq. (1)]),

$$\mathcal{W}_f(\mathbf{u}) = \sum_{k=0}^n (-1)^{c_k} \sum_{wt(\mathbf{x})=k} (-1)^{\mathbf{u} \cdot \mathbf{x}} = \sum_{k=0}^n (-1)^{c_k} P_k(wt(\mathbf{u}), n), \tag{4}$$

where $c_k = f(\mathbf{x})$ if $wt(\mathbf{x}) = k$, and P_k is the Krawtchouk polynomial [20]. In particular, if $wt(\mathbf{u}_1) = wt(\mathbf{u}_2)$, then $\mathcal{W}_f(\mathbf{u}_1) = \mathcal{W}_f(\mathbf{u}_2)$. We furthermore will use the generating function of P_k, which is given by (see [26, Eq. (2)]),

$$(1 - z)^{wt(\mathbf{u})}(1 + z)^{n - wt(\mathbf{u})} = \sum_{k=0}^n P_k(wt(\mathbf{u}), n) z^k. \tag{5}$$

Proposition 3. *Let n be odd, and $S_{c,d}(\mathbf{x}) = s_2(\mathbf{x}) \oplus cs_1(\mathbf{x}) \oplus d$, $c, d \in \mathbb{F}_2$.*

(i) *The symmetric function $S_{c,d}$ is a semibent function with linear space $LS(S_{c,d}) = \{\mathbf{0}, \mathbf{1}\}$.*

(ii) *A symmetric semibent function $f : \mathbb{F}_2^n \to \mathbb{F}_2$ cannot have the vectors of weight $(n-1)/2$ and the vectors of weight $(n+1)/2$ in the support of its Walsh transform.*

(iii) *The functions $S_{c,d}$ are the only symmetric semibent functions, which have a vector of weight $(n-1)/2$ or a vector of weight $(n+1)/2$ in the support of their Walsh transform.*

Proof. We first show (i). Since s_2 is quadratic, it is a partially bent function. We have to show that $LS(s_2)$ has dimension 1. Observe that

$$D_\mathbf{a}(s_2) = s_2(\mathbf{x} \oplus \mathbf{a}) \oplus s_2(\mathbf{x}) = \bigoplus_{i=1}^{n} a_i \bigoplus_{\substack{j=1 \\ j \neq i}}^{n} x_j \oplus C.$$

Inserting the unit vectors $\mathbf{e}_i$, we infer that $D_\mathbf{a}(s_2)$ is constant if and only if $\mathbf{a} = \mathbf{0}$ or (since n is odd) $\mathbf{a} = \mathbf{1}$. This shows for all $c, d \in \mathbb{F}_2$ that $S_{c,d}$ is semibent with linear space $\{\mathbf{0}, \mathbf{1}\}$.

We next show both (ii) and (iii). We determine the Walsh transform of a symmetric function f at a vector $\mathbf{u}_1$ of weight $wt(\mathbf{u}_1) = (n-1)/2$. With (5), we straightforwardly see that for $k = 2l$ and $k = 2l + 1$, $0 \leq l \leq (n-1)/2$, we have

$$\sum_{wt(\mathbf{x})=k} (-1)^{\mathbf{u}_1 \cdot \mathbf{x}} = (-1)^l \binom{\frac{n-1}{2}}{l}.$$

Consequently, by (4),

$$\mathcal{W}_f(\mathbf{u}_1) = \sum_{l=0}^{(n-1)/2} (-1)^{c_{2l}}(-1)^l \binom{\frac{n-1}{2}}{l} + \sum_{l=0}^{(n-1)/2} (-1)^{c_{2l+1}}(-1)^l \binom{\frac{n-1}{2}}{l}$$

$$= \sum_{l=0}^{(n-1)/2} (-1)^l \binom{\frac{n-1}{2}}{l} \left((-1)^{c_{2l}} + (-1)^{c_{2l+1}}\right).$$

Similarly for a vector $\mathbf{u}_2$ of weight $wt(\mathbf{u}_2) = (n+1)/2$ we obtain

$$\mathcal{W}_f(\mathbf{u}_2) = \sum_{l=0}^{(n-1)/2} (-1)^l \binom{\frac{n-1}{2}}{l} \left((-1)^{c_{2l}} - (-1)^{c_{2l+1}}\right).$$

Suppose that $\mathcal{W}_f(\mathbf{u}_1) = \pm 2^{(n+1)/2}$. Then we must have $(-1)^l((-1)^{c_{2l}} + (-1)^{c_{2l+1}}) = 2$ for all $0 \leq l \leq (n-1)/2$ (then $\mathcal{W}_f(\mathbf{u}_1) = 2^{(n+1)/2}$), or $(-1)^l((-1)^{c_{2l}} + (-1)^{c_{2l+1}}) = -2$ for all $0 \leq l \leq (n-1)/2$ (then $\mathcal{W}_f(\mathbf{u}_1) = -2^{(n+1)/2}$). In the first case, we have $c_{2l} = c_{2l+1} = 0$ if l is even, and

$c_{2l} = c_{2l+1} = 1$ if l is odd. (It is the other way around in the second case.) It immediately follows then that $\mathcal{W}_f(\mathbf{u}_2) = 0$. Moreover, with (3) we see that $\mathcal{W}_f(\mathbf{u}_1) = \pm 2^{(n+1)/2}$ implies that $f = s_2$ or $f = s_2 \oplus 1$.

If on the other hand $\mathcal{W}_f(\mathbf{u}_2) = \pm 2^{(n+1)/2}$, with the same reasoning we see that $\mathcal{W}_f(\mathbf{u}_1) = 0$, and $f = s_2 \oplus s_1$ or $f = s_2 \oplus s_1 \oplus 1$. □

Since for a symmetric function $f : \mathbb{F}_2^n \to \mathbb{F}_2$ we have $\mathcal{W}_f(\mathbf{u}_1) = \mathcal{W}_f(\mathbf{u}_2)$ if $wt(\mathbf{u}_1) = wt(\mathbf{u}_2)$ and the support of the Walsh transform of a semibent function has cardinality 2^{n-1}, a symmetric semibent function induces a bisection of the binomial coefficients, i.e., a subset S of $\{0, \ldots, n\}$ such that $\sum_{j \in S} \binom{n}{j} = \sum_{j \notin S} \binom{n}{j} = 2^{n-1}$. For odd n, the trivial bisections are the sets S that contain exactly one of $\binom{n}{j}$ and $\binom{n}{n-j}$ for all $0 \le j \le (n-1)/2$. Bisections of polynomial coefficients is a quite frequently studied problem [6,11,12,16]. It is not known for what values of n, a nontrivial bisection exists.

We expect that the functions $S_{c,d}$ are, unconditionally, the only symmetric semibent functions. We have the following partial result.

Corollary 2. *Let n be odd. The semibent function $S_{c,d} : \mathbb{F}_2^n \to \mathbb{F}_2$ is the only symmetric partially bent semibent function. If there is no nontrivial bisection of the binomial coefficients $\binom{n}{j}$, then $S_{c,d}$ is the only symmetric semibent function from $\mathbb{F}_2^n$ to $\mathbb{F}_2$.*

Proof. Suppose that f is a symmetric partially bent semibent function with linear space $\{\mathbf{0}, \mathbf{v}\}$. Then the support of $\mathcal{W}_f$ is $\{\mathbf{0}, \mathbf{v}\}^\perp$ or its coset. Observe that if $\mathbf{v} \ne \mathbf{1}$, then there always exist two vectors $\mathbf{u}_1, \mathbf{u}_2$ of the same weight, only one of which is in $\{\mathbf{0}, \mathbf{v}\}^\perp$. This contradicts the symmetry of f. Hence $LS(f) = \{\mathbf{0}, \mathbf{1}\}$, and the support of $\mathcal{W}_f$ consists either of the vectors of even weight or of the vectors of odd weight. With Proposition 3, $f = S_{c,d}$.

If n permits only the trivial bisection of the binomial coefficients, then every symmetric semibent function f has either the vectors of weight $(n-1)/2$ or the vectors of weight $(n+1)/2$ in the support of its Walsh transform. With Proposition 3, $f = S_{c,d}$. □

Since there is only one symmetric bent Boolean function up to addition of an affine function, there is no symmetric vectorial bent function for $k > 1$. As we show next, there is essentially only one example of a symmetric generalized bent Boolean function.

Theorem 2. *There are no symmetric vectorial bent functions $f : \mathbb{F}_2^n \to \mathbb{F}_2^k$ for $k > 1$. The only symmetric generalized bent Boolean functions $f \in \mathcal{GB}_n^{2^k}$, $k > 1$, are the quaternary functions $f(\mathbf{x}) = (s_1(\mathbf{x}) \oplus e) + 2S_{c,d}(\mathbf{x})$ for some $c, d, e \in \mathbb{F}_2$ (and their natural lifts to functions in $\mathcal{GB}_n^{2^k}$ with only four values in their value set in $\mathbb{Z}_{2^k}$).*

Proof. Let $f(\mathbf{x}) = \sum_{i=0}^{k-1} 2^i a_i(\mathbf{x})$, $a_i \in \mathcal{B}_n$, be a symmetric generalized bent Boolean function. Hence all $a_i \in \mathcal{B}_n$ are symmetric. If n is even, then all components $g_{\mathbf{c}}(\mathbf{x}) = c_0 a_0(\mathbf{x}) \oplus c_1 a_1(\mathbf{x}) \oplus \cdots \oplus c_{k-2} a_{k-2}(\mathbf{x}) \oplus a_{k-1}(\mathbf{x})$ are symmetric bent functions, i.e. $g_\mathbf{c}(\mathbf{x}) \in \{S_{c,d}, c, d \in \mathbb{F}_2\}$. Consequently, we are left with the 1-dimensional space of bent functions $S_{c,d}(\mathbf{x}) \oplus \langle s_1(\mathbf{x}) \oplus e\rangle$, $c, d, e \in \mathbb{F}_2$. Note that by Proposition 2(i), all quaternary functions $(s_1(\mathbf{x}) \oplus e) + 2S_{c,d}(\mathbf{x})$, $c, d, e \in \mathbb{F}_2$ are in fact generalized bent.

If n is odd, then for any $\mathbf{c} = (c_0, \dots, c_{k-3}, 0)$, $\mathbf{d} = (d_0, \dots, d_{k-3}, 1)$, $c_i, d_i \in \mathbb{F}_2$, the components

$$g_\mathbf{c} = a_{k-1} \oplus \bigoplus_{i=0}^{k-3} c_i a_i \quad \text{and} \quad g_\mathbf{d} = a_{k-1} \oplus a_{k-2} \oplus \bigoplus_{i=0}^{k-3} d_i a_i$$

are symmetric semibent functions, with the additional property that for any $\mathbf{u} \in \mathbb{F}_2^n$ we have $\mathcal{W}_{g_\mathbf{c}}(\mathbf{u}) = 0$ if and only $\mathcal{W}_{g_\mathbf{d}}(\mathbf{u}) \neq 0$. By Proposition 3(ii), $\mathcal{W}_{g_\mathbf{c}}(\mathbf{u}) \neq 0$ if $wt(\mathbf{u}) = (n-1)/2$ and $\mathcal{W}_{g_\mathbf{d}}(\mathbf{u}) \neq 0$ if $wt(\mathbf{u}) = (n+1)/2$, or vice versa. By Proposition 3(iii) then for all such $\mathbf{c}, \mathbf{d} \in \mathbb{F}_2^{k-1}$ we have $g_\mathbf{c}(\mathbf{x}) = s_2 \oplus d$, $d \in \mathbb{F}_2$, and $g_\mathbf{d} = s_2 \oplus s_1 \oplus e$, $e \in \mathbb{F}_2$, or vice versa. Therefore, the only candidate is $f(\mathbf{x}) = (s_1(\mathbf{x}) \oplus e) + 2S_{c,d}(\mathbf{x})$ for some $c, d, e \in \mathbb{F}_2$. By Proposition 2(ii) it remains to show that the supports of $\mathcal{W}_{s_2}$ and $\mathcal{W}_{s_2 \oplus s_1}$ are disjoint. With Proposition 3, the support of $\mathcal{W}_{s_2}$, respectively, the support of $\mathcal{W}_{s_2 \oplus s_1}$ is $\{\mathbf{0}, \mathbf{1}\}^\perp$ or its coset. Furthermore, exactly one of $\mathcal{W}_{s_2}, \mathcal{W}_{s_2 \oplus s_1}$ has the vectors of weight $(n-1)/2$ in its support, which completes the proof. $\qquad\square$

Remark 2. The possible lifts are described explicitly by $\tilde{f}(\mathbf{x}) = A + Bs_1(\mathbf{x}) + 2^{k-1}s_2(\mathbf{x})$ for constants $0 \leq A, B \leq 2^k - 1$ if n is even, and if n is odd by $\tilde{f}(\mathbf{x}) = C + 2^{k-2}s_1(\mathbf{x}) + 2^{k-1}S_{c,d}(\mathbf{x})$ for some constant $0 \leq C \leq 2^{k-1} - 1$ and $c, d \in \{0, 1\}$. Certainly those functions reduce to and are completely described with the quaternary gbent function $f : \mathbb{F}_2^n \to \mathbb{Z}_4$ given as $f(\mathbf{x}) = s_1(\mathbf{x}) + 2s_2(\mathbf{x})$.

We next impose balancedness to the symmetry of a generalized Boolean function, $f : \mathbb{F}_2^n \to \mathbb{Z}_{2^k}$ (in fact, our result is true for any function from $\mathbb{F}_2^n$ into a group of order 2^k). We adopt the classical approach by embedding the problem into one of multisection (not necessarily, bisection) of binomial coefficients. The connection is rather simple: since the function is symmetric, its value is independent of the weight of the input. Thus, the symmetric function f has constant value c_j for every weight j vector (of count $\binom{n}{j}$). Imposing balancedness, it means that each such c_j will occur the same number of times, that is, we can split the set of all binomial coefficients into 2^k sets, whose sums are equal, namely 2^{n-k}.

As mentioned above, if $k = 1$, this is an older and quite studied problem [6, 11, 12, 15, 16]. While there always exist trivial bisections, it is not known for what values of n, a nontrivial bisection exists. Many papers have been written, which employ heavy computations to find values of n, for which we can nontrivially bisect the binomial coefficients set $\left\{\binom{n}{j}\right\}_{0 \leq j \leq n}$. Thus, it is a natural question to ask whether a splitting of binomial coefficients of size other than two does exist. It was conjectured in [21] that no such 2^k-section for $k > 1$ existed. While this question originates from our attempt to investigate balanced and symmetric generalized Boolean functions, it has an interest of its own. As for the bisection, we say that we have a 2^k-*section* of a set of integers A (whose cardinality is divisible by 2^k) if there is a partition of cardinality 2^k of the set A such that the sum on each partition set is $\frac{1}{2^k} \sum_{x \in A} x$, $1 \leq j \leq 2^k$.

Theorem 3. *There is no symmetric balanced function from $\mathbb{F}_2^n$, $n \geq 1$, to any group of order 2^k, if $k \geq 2$. In particular, for $k \geq 2$, there are no 2^k-sections of binomial coefficients $\left\{ \binom{n}{j} \right\}_{0 \leq j \leq n}$.*

Proof. The result is easy to show for $1 \leq n \leq 10$, so we assume that $n \geq 10$. Freiman [10] considered the system of equations

$$a_{11}x_1 + a_{12}x_2 + \cdots + a_{1m}x_m = b_1$$
$$a_{21}x_1 + a_{22}x_2 + \cdots + a_{2m}x_m = b_2,$$

where $(0,0) \neq (a_{1j}, a_{2,j}) \in \mathbb{Z}^2$, $(b_1, b_2) \in \mathbb{Z}^2$, and he showed that the number of solutions $x_j \in \{0,1\}$ of the above system is exactly

$$J_{b_1, b_2} = 2^m \int_G \int e^{-2\pi i (xb_1 + yb_2)} \prod_{j=1}^m \frac{1}{2} \left(1 + e^{2\pi i (xa_{1j} + ya_{2j})} \right) dx \, dy,$$

where $G = \{ (x,y) \mid x, y \in \mathbb{R}, |x| \leq \frac{1}{2}, |y| \leq \frac{1}{2} \}$.

Let $n \geq 10$ be fixed, and we assume that there is a 2^k-section, $k \geq 2$ (we take k largest with this property). We consider such a 2^k-section and partition the binomial coefficients $\binom{n}{j}$ in 2^k (disjoint) sets $A_i, 1 \leq i \leq 2^k$ such that $\sum_{j \in A_i} \binom{n}{j} = 2^{n-k}$, $1 \leq i \leq 2^k$. Since we took k largest with this property (certainly, $k < n$), one of the sets, without loss of generality, say A_1, cannot be bisected further. We next consider the system

$$\sum_{j \in \cup_{i=2}^{2^k} A_i} x_j \binom{n}{j} + \sum_{j \in A_1} x_j \cdot 0 = (2^k - 1)2^{n-k}$$

$$\sum_{j \in \cup_{i=2}^{2^k} A_i} x_j \cdot 0 + \sum_{j \in A_1} x_j \binom{n}{j} = 2^{n-k},$$

and by Freiman's result there are exactly

$$J_{(2^k-1) \, 2^{n-k}, 2^{n-k}} = 2^{n+1} \int_{-1/2}^{1/2} \int_{-1/2}^{1/2} e^{-2\pi 2^{n-k}((2^k-1)x+y)}$$

$$\cdot \prod_{j \in \cup_{i=2}^{2^k} A_i} \frac{1}{2} \left(1 + e^{2\pi i x \binom{n}{j}} \right) \prod_{j \in A_1} \frac{1}{2} \left(1 + e^{2\pi i y \binom{n}{j}} \right) dx \, dy$$

$$= 2^{n+1} \int_{-1/2}^{1/2} e^{-(2^k-1)\pi 2^{n-k+1} x} \prod_{j \in \cup_{i=2}^{2^k} A_i} \frac{1}{2} \left(1 + e^{2\pi i x \binom{n}{j}} \right)$$

$$\cdot \int_{-1/2}^{1/2} e^{-\pi 2^{n-k+1}y} \prod_{j\in A_1} \frac{1}{2}\left(1 + e^{2\pi i y\binom{n}{j}}\right)$$

$$= 2^{n+1} \int_{-1/2}^{1/2} \prod_{j\in\cup_{i=2}^{2^k} A_i} \cos\left(\pi x\binom{n}{j}\right) \int_{-1/2}^{1/2} \prod_{j\in A_1} \cos\left(\pi x\binom{n}{j}\right),$$

solutions of that system. By our assumption, $J_{(2^k-1)\,2^{n-k},2^{n-k}} \geq 1$. We let $\langle,\rangle$ be the regular Euclidean scalar product, and observe that

$$\prod_{j\in A_1} \cos\left(\pi i x\binom{n}{j}\right) = \frac{1}{2^{|A_1|-1}} \sum_{\theta\in\{-1,1\}^{|A_1|-1}} \cos\left(\pi i x\left\langle(1,\theta),(\binom{n}{j})_{j\in A_1}\right\rangle\right).$$

Observe that $\left\langle(1,\theta),(\binom{n}{j})_{j\in A_1}\right\rangle \equiv \sum_{j\in A_1}\binom{n}{j} = 2^{n-k} \equiv 0 \pmod 2$, for all

$\theta \in \{-1,1\}^{|A_1|-1}$. Moreover, the scalar product $\left\langle(1,\theta),(\binom{n}{j})_{j\in A_1}\right\rangle \neq 0$, since we assumed that A_1 cannot be bisected further. Therefore, the integral

$$\int_{-1/2}^{1/2} \prod_{j\in A_1} \cos\left(\pi x\binom{n}{j}\right)$$

$$= \frac{1}{2^{|A_1|-1}} \int_{-1/2}^{1/2} \sum_{\theta\in\{-1,1\}^{|A_1|-1}} \cos\left(\pi x\left\langle(1,\theta),(\binom{n}{j})_{j\in A_1}\right\rangle\right)$$

$$= \frac{1}{2^{|A_1|-1}} \sum_{\theta\in\{-1,1\}^{|A_1|-1}} \int_{-1/2}^{1/2} \cos\left(\pi x\left\langle(1,\theta),(\binom{n}{j})_{j\in A_1}\right\rangle\right)$$

$$= \frac{1}{2^{|A_1|-1}\pi\left\langle(1,\theta),((\binom{n}{j}))_{j\in A_1}\right\rangle} \sum_{\theta\in\{-1,1\}^{|A_1|-1}} \sin\left(\pi x\left\langle(1,\theta),(\binom{n}{j})_{j\in A_1}\right\rangle\right)\Big|_{-1/2}^{1/2}$$

$$= 0,$$

since $\left\langle(1,\theta),(\binom{n}{j})_{j\in A_1}\right\rangle \equiv 0 \pmod 2$, which shows that our assumption that, for $k \geq 2$, there are 2^k-sections of binomial coefficients is false. The theorem is shown. $\square$

4   Avalanche Features in Terms of Differentials

Let $f \in \mathcal{GB}_n^{2^k}$ and $\mathbf{a} \in \mathbb{V}_n$, $c \in \mathbb{Z}_{2^k}$. We let

$$\delta(\mathbf{a},c) := |\{\mathbf{x} \in \mathbb{V}_n \mid D_{\mathbf{a}}f(\mathbf{x}) = c\}|,$$

and call the quantity $\Delta_f := \max_{(\mathbf{a},c)\in\mathbb{V}_n^*\times\mathbb{Z}_{2^k}} \delta_f(\mathbf{a},c)$ the *differential uniformity* of f (and f is a differentially Δ_f-uniform function). The multiset $\{\delta_f(\mathbf{a},c)\mid(\mathbf{a},c)\in \mathbb{V}_n\times\mathbb{Z}_{2^k}\}$ is called the *differential spectrum* of f. It is known that when f has values in $\mathbb{V}_n$, then $\Delta_f \geq 2$ (in odd characteristic, Δ_f can take the value 1), and functions achieving this bound are called *almost perfect nonlinear* (APN). Recall that bent functions f from a group A to a group B can be defined as functions for which $f(x+a) - f(x)$ is balanced for every nonzero $a \in A$, i.e. every $b \in B$ is taken on the same number $|A|/|B|$ times. A function f from $\mathbb{V}_n$ to $\mathbb{Z}_{2^k}$ is hence bent if and only if $\Delta_f = 2^{n-k}$. In terms of character sum values a bent function from $\mathbb{V}_n$ to $\mathbb{Z}_{2^k}$ is then a function for which

$$\mathcal{H}_f(\alpha, \mathbf{u}) = \sum_{\mathbf{x} \in \mathbb{V}_n} \zeta^{\alpha f(\mathbf{x})}(-1)^{\mathbf{u} \cdot \mathbf{x}} \tag{6}$$

has absolute value $2^{n/2}$ for all nonzero $\alpha \in \mathbb{Z}_{2^k}$ and $\mathbf{u} \in \mathbb{V}_n$. We next investigate differential properties of generalized bent functions from $\mathbb{V}_n$ to $\mathbb{Z}_{2^k}$, which satisfy the weaker property that $|\mathcal{H}_f(\mathbf{u})| = 2^{n/2}$ for all $\mathbf{u} \in \mathbb{V}_n$, see Corollary 3 below.

If $\mathbb{V}_n = \mathbb{F}_2^n$, then we say that $f \in \mathcal{GB}_n^{2^k}$ satisfies the *(generalized) propagation criterion of order* ℓ $(1 \leq \ell \leq n)$, denoted by $gPC(\ell)$, if and only if the auto-correlation $\mathcal{C}_f(\mathbf{v}) = \sum_{\mathbf{x} \in \mathbb{V}_n} \zeta^{f(\mathbf{x})-f(\mathbf{x} \oplus \mathbf{v})} = 0$, for all vectors $\mathbf{v} \in \mathbb{F}_2^n$ of weight $0 < wt(\mathbf{v}) \leq \ell$. If $\ell = 1$, we say that f satisfies the *(generalized) strict avalanche criterion (gSAC)*. With the standard calculations we see that f is gbent if and only if $\mathcal{C}_f(\mathbf{v}) = 0$ for all $\mathbf{v}$ (in this case we do not require that $\mathbb{V}_n = \mathbb{F}_2^n$).

Theorem 4. *Let* $f \in \mathcal{GB}_n^{2^k}$, *and* $A_j^{(\mathbf{w})} = \{\mathbf{x} | f(\mathbf{x} \oplus \mathbf{w}) - f(\mathbf{x}) = j\}$. *Then* f *is* $gPC(\ell)$ *if and only if*

$$|A_0^{(\mathbf{0})}| = 2^n, |A_j^{(\mathbf{0})}| = 0, |A_j^{(\mathbf{w})}| = |A_{j+2^{k-1}}^{(\mathbf{w})}|, \text{ for } 0 \leq j \leq 2^{k-1} - 1, 1 \leq wt(\mathbf{w}) \leq \ell.$$

Proof. First note that unconditionally we always have $|A_0^{(\mathbf{0})}| = 2^n, |A_j^{(\mathbf{0})}| = 0$. For $\mathbf{v} \in \mathbb{V}_n, \mathbf{v} \neq \mathbf{0}$, with the notations in the statement of the theorem and $\bar{\zeta} = \zeta^{-1}$ we have

$$\mathcal{C}_f(\mathbf{v}) = \sum_{\mathbf{x} \in \mathbb{V}_n} \zeta^{f(\mathbf{x})-f(\mathbf{x} \oplus \mathbf{v})} = \sum_{j=0}^{2^k-1} |A_j^{(\mathbf{v})}| \bar{\zeta}^j = \sum_{j=0}^{2^{k-1}-1} (|A_j^{(\mathbf{v})}| - |A_{j+2^{k-1}}^{(\mathbf{v})}|) \bar{\zeta}^j.$$

Since the set $\{\bar{\zeta}^j : 0 \leq j \leq 2^{k-1} - 1\}$ is a basis of $\mathbb{Q}(\zeta)$, hence is linearly independent, we have $\mathcal{C}_f(\mathbf{v}) = 0$ if and only if $|A_j^{(\mathbf{v})}| = |A_{j+2^{k-1}}^{(\mathbf{v})}|$ for $0 \leq j \leq 2^{k-1} - 1$. $\square$

Corollary 3. *Let* $f \in \mathcal{GB}_n^{2^k}$. *Then* f *is gbent if and only if*

$$|A_0^{(\mathbf{0})}| = 2^n, |A_j^{(\mathbf{0})}| = 0, |A_j^{(\mathbf{w})}| = |A_{j+2^{k-1}}^{(\mathbf{w})}|, \text{ for all } 0 \leq j \leq 2^{k-1} - 1, \mathbf{w} \neq 0.$$

Recall that a Boolean function $g : \mathbb{V}_n \to \mathbb{F}_2$ is called *partially bent* if $g(\mathbf{x} \oplus \mathbf{a}) \oplus g(\mathbf{x})$ is either balanced or constant for all $\mathbf{a} \in \mathbb{V}_n$. Partially bent functions from $\mathbb{V}_n$ to $\mathbb{F}_2$ are always s-plateaued, where s is the dimension of the linear space of g. In an analog way we can define (generalized) *partially bent* functions from $\mathbb{V}_n$ to $\mathbb{Z}_{2^k}$ as functions f for which $f(\mathbf{x} \oplus \mathbf{a}) - f(\mathbf{x})$ is either balanced or constant for all $\mathbf{a} \in \mathbb{V}_n$. With the standard proof for partially bent functions one can show that generalized partially bent functions $f : \mathbb{V}_n \to \mathbb{Z}_{2^k}$ as defined above, are plateaued with respect to their transform $\mathcal{H}_f(\alpha, u)$ of (6).

In Theorem 4 we characterized gbent functions via their differential properties. With this characterization the following definition of a partially gbent function is natural. As in Theorem 4, let $A_j^{(\mathbf{w})} = \{\mathbf{x} | f(\mathbf{x} \oplus \mathbf{w}) - f(\mathbf{x}) = j\}$. A function $f \in \mathcal{GB}_n^{2^k}$ is called *partially gbent*, if for all $\mathbf{w} \in \mathbb{V}_n$ we either have $|A_j^{(\mathbf{w})}| = |A_{j+2^{k-1}}^{(\mathbf{w})}|$ for all $0 \leq j \leq 2^{k-1} - 1$, or the derivative $f(\mathbf{x} \oplus \mathbf{w}) - f(\mathbf{x})$ is constant.

Proposition 4. *A partially gbent function $f \in \mathcal{GB}_n^{2^k}$ is s-plateaued, where s is the dimension of the linear space of f.*

Proof. Since $\mathcal{H}_{f+c}(\mathbf{u}) = \zeta^c \mathcal{H}_f(\mathbf{u})$, without loss of generality we can suppose that $f(\mathbf{0}) = 0$. For $\mathbf{u} \in \mathbb{V}_n$ we have

$$\mathcal{H}_f(\mathbf{u})\overline{\mathcal{H}_f(\mathbf{u})} = \sum_{\mathbf{z} \in \mathbb{V}_n} \zeta^{f(\mathbf{z})}(-1)^{\mathbf{u}\cdot\mathbf{z}} \sum_{\mathbf{x} \in \mathbb{V}_n} \zeta^{-f(\mathbf{x})}(-1)^{\mathbf{u}\cdot\mathbf{x}}$$
$$= \sum_{\mathbf{w} \in \mathbb{V}_n} \zeta^{-f(\mathbf{w})}(-1)^{\mathbf{u}\cdot\mathbf{w}} \sum_{\mathbf{x} \in \mathbb{V}_n} \zeta^{f(\mathbf{x}\oplus\mathbf{w})-f(\mathbf{x})+f(\mathbf{w})}.$$

Observe that $f(\mathbf{x} \oplus \mathbf{w}) - f(\mathbf{x}) + f(\mathbf{w}) = 0$ if $\mathbf{w}$ is a linear structure of f (we use that $f(\mathbf{0}) = 0$). If $\mathbf{w}$ is not a linear structure, then by our assumption $\sum_{\mathbf{x} \in \mathbb{V}_n} \zeta^{f(\mathbf{x}\oplus\mathbf{w})-f(\mathbf{x})+f(\mathbf{w})} = \zeta^{f(\mathbf{w})} \sum_{\mathbf{x} \in \mathbb{V}_n} \zeta^{f(\mathbf{x}\oplus\mathbf{w})-f(\mathbf{x})} = 0$. Hence putting $\Lambda = LS_{2^k}(f)$, $\mathcal{H}_f(\mathbf{u})\overline{\mathcal{H}_f(\mathbf{u})} = 2^n \sum_{\mathbf{w} \in \Lambda} \zeta^{-f(\mathbf{w})}(-1)^{\mathbf{u}\cdot\mathbf{w}}$. Let $\mathbf{z}$ be any element of $S_f = \{\mathbf{x} \in \mathbb{V}_n | \mathcal{H}_f(\mathbf{x}) \neq 0\}$. Then by Theorem 1 we have $\zeta^{f(\mathbf{w})} = (-1)^{\mathbf{z}\cdot\mathbf{w}}$ for every $\mathbf{w} \in \Lambda$. Therefore $\mathcal{H}_f(\mathbf{u})\overline{\mathcal{H}_f(\mathbf{u})} = 2^n \sum_{\mathbf{w} \in \Lambda}(-1)^{(\mathbf{z}\oplus\mathbf{u})\cdot\mathbf{w}}$, (independently from the choice of $\mathbf{z} \in S_f$). Consequently, if $\mathbf{z} \oplus \mathbf{u} \in \Lambda^{\perp}$, then $|\mathcal{H}_f(\mathbf{u})|^2 = 2^{n+s}$, where $s = \dim(\Lambda)$, otherwise $\mathcal{H}_f(\mathbf{u}) = 0$.   □

Remark 3. Observing that $\mathcal{H}_f(\mathbf{u})$ only depends on whether $\mathbf{z} \oplus \mathbf{u} \in \Lambda^{\perp}$, independent from the choice of $\mathbf{z} \in S_f$, we infer that S_f is a coset of $\Lambda^{\perp}$ (we also use that by Parseval's identity, $|S_f| = 2^{n-s}$). This coincides with the situation for conventional partially bent functions.

Similar as for conventional partially bent functions, cf. [4,5], we have the following corollary for partially gbent functions.

Corollary 4. *Let $\Lambda = LS_{2^k}(f)$ be the linear space of the function $f \in \mathcal{GB}_n^{2^k}$. Then f is partially gbent (partially bent) if and only if for any complement Λ^{comp} of Λ in $\mathbb{V}_n$, the function $f|\Lambda^{comp}$ is gbent (bent).*

Proof. With $D_{\mathbf{a}}f(\mathbf{x}) = f(\mathbf{x} \oplus \mathbf{a}) - f(\mathbf{x})$,

$$\mathcal{H}_{D_{\mathbf{a}}f}(\mathbf{0}) = \sum_{\mathbf{x} \in \mathbb{V}_n} \zeta^{f(\mathbf{x} \oplus \mathbf{a}) - f(\mathbf{x})} = \sum_{\mathbf{y} \in \Lambda} \sum_{\mathbf{z} \in \Lambda^{comp}} \zeta^{f(\mathbf{y} \oplus \mathbf{z} \oplus \mathbf{a}) - f(\mathbf{y} \oplus \mathbf{z})}.$$

Using that $f(\mathbf{x} \oplus \mathbf{y}) - f(\mathbf{x}) + f(\mathbf{y}) = 0$ if $\mathbf{y} \in \Lambda$, we see that $f(\mathbf{y} \oplus \mathbf{z} \oplus \mathbf{a}) - f(\mathbf{y} \oplus \mathbf{z}) = f(\mathbf{z} \oplus \mathbf{a}) - f(\mathbf{z})$, hence

$$\mathcal{H}_{D_{\mathbf{a}}f}(\mathbf{0}) = \sum_{\mathbf{y} \in \Lambda} \sum_{\mathbf{z} \in \Lambda^{comp}} \zeta^{f(\mathbf{z} \oplus \mathbf{a}) - f(\mathbf{z})} = |\Lambda| \sum_{\mathbf{z} \in \Lambda^{comp}} \zeta^{f(\mathbf{z} \oplus \mathbf{a}) - f(\mathbf{z})}. \tag{7}$$

First suppose that f is partially gbent. Then for $\mathbf{a} \notin \Lambda$ we have $\mathcal{H}_{D_{\mathbf{a}}f}(\mathbf{0}) = \sum_{\mathbf{x} \in \mathbb{V}_n} \zeta^{f(\mathbf{x} \oplus \mathbf{a}) - f(\mathbf{x})} = 0$, and hence with (7), $\sum_{\mathbf{z} \in \Lambda^{comp}} \zeta^{f(\mathbf{z} \oplus \mathbf{a}) - f(\mathbf{z})} = 0$. Consequently, for $\tilde{A}_j^{(\mathbf{a})} = \{\mathbf{z} \in \Lambda^{comp} | f(\mathbf{z} \oplus \mathbf{a}) - f(\mathbf{z}) = j\}$ we have $|\tilde{A}_j^{(\mathbf{a})}| = |\tilde{A}_{j+2^{k-1}}^{(\mathbf{a})}|$ for all $0 \leq j \leq 2^{k-1} - 1$ and all nonzero $\mathbf{a} \in \Lambda^{comp}$. By Theorem 4, f restricted to Λ^{comp} is gbent.

Conversely let $f|\Lambda^{comp}$ be gbent for any complement Λ^{comp} of Λ. Let $\mathbf{a} \notin \Lambda$ and let Λ^{comp} be a complement of Λ containing $\mathbf{a}$. By assumption, with Theorem 4 we have $\sum_{\mathbf{z} \in \Lambda^{comp}} \zeta^{f(\mathbf{z} \oplus \mathbf{a}) - f(\mathbf{z})} = 0$, hence by Eq. (7), $\mathcal{H}_{D_{\mathbf{a}}f}(\mathbf{0}) = 0$. Therefore $|A_j^{(\mathbf{a})}| = |A_{j+2^{k-1}}^{(\mathbf{a})}|$ for all $0 \leq j \leq 2^{k-1} - 1$, and f is partially gbent by definition. $\qquad \square$

Acknowledgement. This paper was started while the second and fourth named authors visited the third named author at the Institute of Algebra and Geometry, of Otto von Guericke University Magdeburg. They thank the host and the institute for hospitality and excellent working conditions.

References

1. Budaghyan, L.: Construction and Analysis of Cryptographic Functions. Springer, Cham (2014). https://doi.org/10.1007/978-3-319-12991-4
2. Carlet, C.: Boolean functions for cryptography and error correcting codes. In: Crama, Y., Hammer, P. (eds.) Boolean Methods and Models, pp. 257–397. Cambridge University Press, Cambridge (2010)
3. Carlet, C.: Vectorial Boolean functions for cryptography. In: Crama, Y., Hammer, P. (eds.) Boolean Methods and Models, pp. 398–472. Cambridge University Press, Cambridge (2010)
4. Carlet, C.: Partially bent functions. Des. Codes Cryptogr. **3**, 135–145 (1993)
5. Çeşmelioğlu, A., Meidl, W., Topuzoğlu, A.: Partially bent functions and their properties. Applied Algebra and Number Theory, pp. 22–38. Cambridge University Press, Cambridge (2014)
6. Cusick, T.W., Li, Y.: k-th order symmetric SAC Boolean functions and bisecting binomial coefficients. Discrete Appl. Math. **149**, 73–86 (2005)
7. Cusick, T.W., Stănică, P.: Cryptographic Boolean Functions and Applications, 2nd edn. Academic Press, San Diego (2017)
8. Dubuc, S.: Characterization of linear structures. Des. Codes Cryptogr. **22**, 33–45 (2001)

9. Evertse, J.-H.: Linear structures in blockciphers. In: Chaum, D., Price, W.L. (eds.) EUROCRYPT 1987. LNCS, vol. 304, pp. 249–266. Springer, Heidelberg (1988). https://doi.org/10.1007/3-540-39118-5_23

10. Freiman, G.A.: On solvability of a system of two Boolean linear equations. In: Chudnovsky, D.V., Chudnovsky, G.V., Nathanson, M.B. (eds.) Number Theory: New York Seminar 1991–1995, pp. 135–150. Springer, New York (1996)

11. von zur Gathen, J., Roche, J.: Polynomials with two values. Combinatorica **17**, 345–362 (1997)

12. Gopalakrishnan, K., Hoffman, D.G., Stinson, D.R.: A note on a conjecture concerning symmetric resilient functions. Inf. Process. Lett. **47**, 139–143 (1993)

13. Gouget, A.: On the propagation criterion of Boolean functions. In: Feng, K., Niederreiter, H., Xing, C. (eds.) Coding, Cryptography and Combinatorics, vol. 23, pp. 153–168. Birkhäuser, Basel (2004). https://doi.org/10.1007/978-3-0348-7865-4_9

14. Hodžić, S., Meidl, W., Pasalic, E.: Full characterization of generalized bent functions as (semi)-bent spaces, their dual and the Gray image. IEEE Trans. Inf. Theory **64**(7), 5432–5440 (2018)

15. Ionascu, E.J., Martinsen, T., Stănică, P.: Bisecting binomial coefficients. Discrete Appl. Math. **227**, 70–83 (2017)

16. Jefferies, N.: Sporadic partitions of binomial coefficients. Electron. Lett. **27**(15), 134–136 (1991)

17. Kumar, P.V., Scholtz, R.A., Welch, L.R.: Generalized bent functions and their properties. J. Comb. Theory Ser. A **40**, 90–107 (1985)

18. Lai, X.: Additive and linear structures of cryptographic functions. In: Preneel, B. (ed.) FSE 1994. LNCS, vol. 1008, pp. 75–85. Springer, Heidelberg (1995). https://doi.org/10.1007/3-540-60590-8_6

19. Lechner, R.L.: Harmonic analysis of switching functions. In: Mukhopadhyay, A. (ed.) Recent Developments in Switching Theory. Academic Press, New York (1971)

20. MacWilliams, F.J., Sloane, N.J.A.: The Theory of Error Correcting Codes. North-Holland, Amsterdam (1977)

21. Martinsen, T.: Correlation immunity, avalanche features, and other cryptographic properties of generalized Boolean functions. Ph.D. dissertation, Naval Postgraduate School, Monterey, CA (2017)

22. Martinsen, T., Meidl, W., Mesnager, S., Stanica, P.: Decomposing generalized bent and hyperbent functions. IEEE Trans. Inf. Theory **63**(12), 7804–7812 (2017)

23. Martinsen, T., Meidl, W., Stănică, P.: Generalized bent functions and their Gray images. In: Duquesne, S., Petkova-Nikova, S. (eds.) WAIFI 2016. LNCS, vol. 10064, pp. 160–173. Springer, Cham (2016). https://doi.org/10.1007/978-3-319-55227-9_12

24. Martinsen, T., Meidl, W., Stănică, P.: Partial spread and vectorial generalized bent functions. Des. Codes Cryptogr. **85**(1), 1–13 (2017)

25. Mesnager, S.: Bent Functions: Fundamentals and Results. Springer, Cham (2016). https://doi.org/10.1007/978-3-319-32595-8

26. Savicky, P.: On the bent Boolean functions that are symmetric. Eur. J. Comb. **15**, 407–410 (1994)

27. Schmidt, K.U.: Quaternary constant-amplitude codes for multicode CDMA. IEEE Trans. Inf. Theory **55**(4), 1824–1832 (2009)

28. Solé, P., Tokareva, N.: Connections between quaternary and binary bent functions. Prikl. Diskr. Mat. **1**, 16–18 (2009). http://eprint.iacr.org/2009/544.pdf

29. Stănică, P., Martinsen, T., Gangopadhyay, S., Singh, B.K.: Bent and generalized bent Boolean functions. Des. Codes Cryptogr. **69**, 77–94 (2013)

30. Tang, C., Xiang, C., Qi, Y., Feng, K.: Complete characterization of generalized bent and 2^k-bent Boolean functions. IEEE Trans. Inf. Theory **63**(7), 4668–4674 (2017)
31. Tokareva, N.: Bent Functions, Results and Applications to Cryptography. Academic Press, San Diego (2015)
32. Zheng, Y.L., Zhang, X.M.: On plateaued functions. IEEE Trans. Inf. Theory **47**(9), 1215–1223 (2001)

Characterizations of Partially Bent and Plateaued Functions over Finite Fields

Sihem Mesnager[1,2,3,4], Ferruh Özbudak[5,6], and Ahmet Sınak[2,3,6,7(⊠)]

[1] Department of Mathematics, University of Paris VIII, Saint-Denis, France
`smesnager@univ-paris8.fr`
[2] LAGA, UMR 7539, CNRS, University of Paris VIII, Saint-Denis, France
[3] LAGA, UMR 7539, CNRS, University of Paris XIII, Villetaneuse, France
[4] Telecom ParisTech, Paris, France
[5] Department of Mathematics, Middle East Technical University, Ankara, Turkey
`ozbudak@metu.edu.tr`
[6] Institute of Applied Mathematics, Middle East Technical University, Ankara, Turkey
[7] Department of Mathematics and Computer Sciences, Necmettin Erbakan University, Konya, Turkey
`asinak@konya.edu.tr`

Abstract. Partially bent and plateaued functions over finite fields have significant applications in cryptography, sequence theory, coding theory, design theory and combinatorics. They have been extensively studied due to their various desirable cryptographic properties. In this paper, we study on characterizations of partially bent and plateaued functions over finite fields, with the aim of clarifying their structure. We first redefine the notion of partially bent functions over any finite field $\mathbb{F}_q$, with q a prime power, and then provide a few characterizations of these functions in terms of their derivatives, Walsh power moments and autocorrelation functions. We next characterize partially bent (vectorial) functions over $\mathbb{F}_p$, with p a prime, by means of their derivatives and Walsh power moments. We finally characterize plateaued functions over $\mathbb{F}_p$ in terms of their Walsh power moments, derivatives and autocorrelation functions.

Keywords: p-ary functions · q-ary functions · Partially bent · Plateaued · Additive character

1 Introduction

The notion of bent Boolean functions, whose absolute Walsh transform takes only one nonzero value, had been initialized (1966) and published (1976) in [28] by Rothaus (also indicated (1974) by Dillon in his thesis [14]), since then they have been widely studied by several researchers (see *e.g.* [5,8,23]). In 1985, Kumar et al. [18] extended this notion to any residue class ring $\mathbb{Z}_k$ and the so-called *generalized bent functions* have been extensively studied in [1,7,15,19,21].

© Springer Nature Switzerland AG 2018
L. Budaghyan and F. Rodríguez-Henríquez (Eds.): WAIFI 2018, LNCS 11321, pp. 224–241, 2018.
https://doi.org/10.1007/978-3-030-05153-2_12

In 1991, Nyberg [27] introduced the notion of perfect nonlinear functions over $\mathbb{Z}_k$. It is worth mentioning that generalized bent and perfect nonlinear functions over $\mathbb{Z}_k$ are not equivalent for a positive integer k, in general. Nyberg, over $\mathbb{Z}_k$, showed that any perfect nonlinear function is a generalized bent function for any positive integer k, but the converse is true only if k is a prime number. In 1997, Coulter and Matthews [13] redefined bent functions over any finite field $\mathbb{F}_q$, with q a prime power, and discussed some of their properties and permutation behaviors. They showed that bent and perfect nonlinear functions are equivalent over $\mathbb{F}_q$. Additionally, Hou presented in [16] further new results on bent functions over $\mathbb{F}_q$.

In 1993, because of unbalancedness of bent functions, as an extension of the class of bent functions, Carlet introduced in [4] the class of *partially bent functions* and studied within these functions the propagation criterion. Partially bent functions are interesting in their own right and workable with respect to their significant cryptographic properties such as the balancedness, high nonlinearity, high correlation immunity and good propagation criterion. The interest of these functions is further from cryptographic point of view since they involve bent functions. In [12], the notion of partially bent functions was extended to $\mathbb{F}_p$, with p any prime, and they have been studied in [2,5,12,23,30], especially [12] gives a deeper understanding of p-ary partially bent functions in many contexts. In this paper we redefine the notion of partially bent functions over $\mathbb{F}_q$, with q a prime power.

In 1999, as an extension of partially bent functions, Zheng and Zhang [31] introduced the notion of *plateaued Boolean functions* which are functions whose squared Walsh transform takes two distinct values (one being zero). They have been deeply studied by several researchers (see *e.g.* [5,6,10,23]). This notion was extended to $\mathbb{F}_p$, with p any prime, and the *p-ary plateaued functions* have been studied in [9,11,17,22,24]. Recently, this notion has been studied over $\mathbb{F}_q$, with q any prime power [26].

This paper is organized as follows. Section 2 fixes main notations and recalls the necessary background. Section 3 redefines the notion of partially bent functions over $\mathbb{F}_q$, and then presents a few characterizations of these functions in terms of their derivatives, Walsh power moments and autocorrelation functions. We also highlight that q-ary bent and q-ary partially bent functions are q-ary plateaued functions. Section 4 characterizes p-ary partially bent (vectorial) functions by their derivatives and Walsh power moments. In Sect. 5, we study on characterizations of p-ary plateaued functions by means of their Walsh power moments, derivatives and autocorrelation functions.

2   Preliminaries

For any set E, $\#E$ denotes the size of E and $E^* = E \setminus \{0\}$. Given the complex number $z \in \mathbb{C}$, where $\mathbb{C}$ is the field of complex numbers, $|z|$ and $\bar{z}$ denote the absolute value and the conjugate of z, respectively. For a prime number p and a positive integer m, the finite field with p^m elements is denoted by $\mathbb{F}_q$, where $q = p^m$. For a positive integer n, the extension field $\mathbb{F}_{q^n}$ over $\mathbb{F}_q$ can be regarded as an n-dimensional vector space over $\mathbb{F}_q$, and denoted by $\mathbb{F}_q^n$. Hence, an element $\alpha \in \mathbb{F}_{q^n}$ can be viewed as a vector $\alpha = (\alpha_1, \alpha_2, \ldots, \alpha_n) \in \mathbb{F}_q^n$ where $\alpha_i \in \mathbb{F}_q$ for $1 \leq i \leq n$.

The *relative trace* of $\alpha \in \mathbb{F}_{q^n}$ over $\mathbb{F}_q$ is defined as $\mathrm{Tr}_q^{q^n}(\alpha) = \alpha + \alpha^q + \cdots + \alpha^{q^{n-1}}$ and the *absolute trace* of $\beta \in \mathbb{F}_{p^m}$ over $\mathbb{F}_p$ is defined as $\mathrm{Tr}_p^{p^m}(\beta) = \beta + \beta^p + \cdots + \beta^{p^{m-1}}$. The trace function is linear and satisfies the transitivity property in a chain of extension fields, that is,

$$\mathrm{Tr}_p^{q^n}(\alpha) = \mathrm{Tr}_p^q\left(\mathrm{Tr}_q^{q^n}(\alpha)\right)$$

for $\alpha \in \mathbb{F}_{q^n}$ where $q = p^m$. A *primitive p-th root of unity* in $\mathbb{C}$ is denoted by $\xi_p = e^{2\pi i/p}$ where $i = \sqrt{-1}$, and its complex conjugation is its inverse, i.e., $\overline{\xi_p} = \xi_p^{-1}$. The function χ from $\mathbb{F}_q$ to $\mathbb{C}$ defined as

$$\chi(x) = \xi_p^{\mathrm{Tr}_p^q(x)}, \quad \text{for } x \in \mathbb{F}_q \tag{1}$$

is called *the canonical additive character* of $\mathbb{F}_q$. Notice that for each $y \in \mathbb{F}_q$, the function χ_y defined as $\chi_y(x) = \chi(yx)$ for $x \in \mathbb{F}_q$ is an additive character of $\mathbb{F}_q$ and every additive character of $\mathbb{F}_q$ is obtained in this way. For each character χ of $\mathbb{F}_q$, there is associated the *conjugate* character $\overline{\chi}$ defined as $\overline{\chi}(x) := \overline{\chi(x)}$ for $x \in \mathbb{F}_q$. The canonical additive characters χ of $\mathbb{F}_q$ and ψ of $\mathbb{F}_q^n$ are connected by the identity $\chi(\mathrm{Tr}_q^{q^n}(\alpha)) = \psi(\alpha)$ for all $\alpha \in \mathbb{F}_q^n$. The following lemma gives the properties of additive characters of $\mathbb{F}_q$.

Lemma 1 [20]. *Let $\chi : \mathbb{F}_q \to \mathbb{C}$ be an additive character as in (1). Then for all $x_1, x_2 \in \mathbb{F}_q$, we have $\chi(x_1 + x_2) = \chi(x_1)\chi(x_2)$ and $\overline{\chi}(x_1) = \chi(-x_1)$.*

Let $f : \mathbb{F}_q^n \to \mathbb{F}_q$. We can define a corresponding function χ_f from $\mathbb{F}_q^n$ to $\mathbb{C}$ by

$$\chi_f(x) = \chi(f(x)) = \xi_p^{\mathrm{Tr}_p^q(f(x))}.$$

The *Walsh-(Hadamard) transform* of f at $\omega \in \mathbb{F}_q^n$ is the Fourier transform $\widehat{\chi_f} : \mathbb{F}_q^n \to \mathbb{C}$ of the function χ_f defined by

$$\widehat{\chi_f}(\omega) = \sum_{x \in \mathbb{F}_q^n} \chi_f(x)\overline{\chi}(\omega \cdot x), \tag{2}$$

where $\chi : \mathbb{F}_q \to \mathbb{C}$ is any non-trivial additive character of $\mathbb{F}_q$ in (1) and "$\cdot$" denotes an inner product in $\mathbb{F}_q^n$ over $\mathbb{F}_q$. It is worth noting that (2) can be also given without the conjugate of χ. If $\mathbb{F}_q^n$ is identified with $\mathbb{F}_{q^n}$, we can take $\omega \cdot x = \mathrm{Tr}_q^{q^n}(\omega x)$ for $\omega, x \in \mathbb{F}_q^n$. The Walsh support of f is the set $\{\omega \in \mathbb{F}_{q^n} : \widehat{\chi_f}(\omega) \neq 0\}$, denoted by $\mathrm{Supp}(\widehat{\chi_f})$. We denote $\mathcal{N}_{\widehat{\chi_f}} = \#\mathrm{Supp}(\widehat{\chi_f})$ and as is readily seen $\mathcal{N}_{\widehat{\chi_f}} \leq q^n$. For any nonnegative integer i, even Walsh power moment of f is defined by

$$S_i(f) = \sum_{\omega \in \mathbb{F}_q^n} |\widehat{\chi_f}(\omega)|^{2i}$$

with the convention $S_0(f) = q^n$. It is a well known fact that $S_1(f) = q^{2n}$, which is called *the Parseval identity*. A function f is k-th order correlation immune $(1 \leq k \leq n)$ if

$$\widehat{\chi_f}(\omega) = 0; \qquad 1 \leq \mathrm{wt}(\omega) \leq k,$$

where $\mathrm{wt}(\omega)$ denotes the Hamming weight of $\omega \in \mathbb{F}_q^n$. A function f is said to be *balanced (or, permutation polynomial)* over $\mathbb{F}_q$ if

$$\widehat{\chi_f}(0) = 0,$$

i.e., $\#\{x \in \mathbb{F}_q^n : f(x) = l\} = q^{n-1}$ for each $l \in \mathbb{F}_q$; otherwise, f is called unbalanced. The derivative of f at point $a \in \mathbb{F}_q^n$ is the map $\mathcal{D}_a f$ from $\mathbb{F}_q^n$ to $\mathbb{F}_q$ defined by

$$\mathcal{D}_a f(x) = f(x + a) - f(x)$$

for all $x \in \mathbb{F}_q^n$. The second-order derivative of f at point $(a, b) \in (\mathbb{F}_q^n)^2$ is given as $\mathcal{D}_b \mathcal{D}_a f(x) = f(x + a + b) - f(x + a) - f(x + b) + f(x)$ for all $x \in \mathbb{F}_q^n$. For $(a, b) \in (\mathbb{F}_q^n)^2$, readily $\mathcal{D}_b \mathcal{D}_a f(x) = \mathcal{D}_a \mathcal{D}_b f(x)$ for all $x \in \mathbb{F}_q^n$. The autocorrelation function of f is the map from $\mathbb{F}_q^n$ to $\mathbb{C}$ defined by

$$\Delta_f(a) = \sum_{x \in \mathbb{F}_q^n} \chi(\mathcal{D}_a f(x))$$

for all $a \in \mathbb{F}_q^n$, where χ is any non-trivial additive character of $\mathbb{F}_q$ in (1). A function f satisfies the propagation criterion $PC(k)$ of degree k $(1 \leq k \leq n)$ if

$$\Delta_f(a) = 0; \qquad 1 \leq \mathrm{wt}(a) \leq k.$$

We denote by $\mathrm{Supp}(\Delta_f)$ the set of elements $a \in \mathbb{F}_q^n$ such that $\mathcal{D}_a f$ is not balanced, i.e.,

$$\mathrm{Supp}(\Delta_f) = \{a \in \mathbb{F}_q^n : \Delta_f(a) \neq 0\}. \tag{3}$$

Denote by $\mathcal{N}_{\Delta_f}$ the size of $\mathrm{Supp}(\Delta_f)$. The following lemma can be easily proven (see [4], in characteristic 2).

Lemma 2. *Let $f : \mathbb{F}_q^n \to \mathbb{F}_q$. Then*

(i.) $\widehat{\widehat{\chi_f}}(\omega) = \overline{\widehat{\chi_f}(-\omega)}$ *for all $\omega \in \mathbb{F}_q^n$.*
(ii.) $|\widehat{\chi_f}(\omega)|^2 = \widehat{\Delta_f}(\omega)$ *for all $\omega \in \mathbb{F}_q^n$.*
(iii.) $|\widehat{\chi_f}(0)|^2 = \sum_{a \in \mathbb{F}_q^n} \Delta_f(a)$.

We end this section with the following definitions for a p-ary function $f :$ $\mathbb{F}_{p^n} \to \mathbb{F}_p$. A function f is said to be a *p-ary bent* if $|\widehat{\chi_f}(\omega)|^2 = p^n$ for every $\omega \in \mathbb{F}_{p^n}$, and f is a *p-ary partially bent* if the derivative $\mathcal{D}_a f$ is either balanced or constant for all $a \in \mathbb{F}_{p^n}$. Moreover, f is said to be a *p-ary s-plateaued* if $|\widehat{\chi_f}(\omega)|^2 \in \{0, p^{n+s}\}$ for every $\omega \in \mathbb{F}_{p^n}$, where s is an integer with $0 \leq s \leq n$. Notice that the symbol "$*$" denotes usual product.

3 On the q-ary Partially Bent Functions and Their Characterizations

This section first redefines the notion of partially bent functions over $\mathbb{F}_q$, where $q = p^m$ for a prime p and a positive integer m, and next gives few characterizations of these functions.

3.1 On the Notion of q-ary Partially Bent Functions

The generalized bent and perfect nonlinear functions over $\mathbb{Z}_k$ for a positive integer k were introduced by Kumar et al. [18] and Nyberg [27], respectively. Then, these notions were redefined over $\mathbb{F}_q$ as follows in [13], where $q = p^m$ for a prime p and an integer $m > 1$.

Definition 1. Let $f : \mathbb{F}_q^n \rightarrow \mathbb{F}_q$. Then, f is called a q-ary bent if $|\widehat{\chi_f}(\omega)|^2 = q^n$ for all $\omega \in \mathbb{F}_q^n$, and f is said to be a perfect nonlinear if the derivative $\mathcal{D}_a f$ is balanced for all nonzero $a \in \mathbb{F}_q^n$.

Proposition 1 ([13, Theorem 2.3]). Let $f : \mathbb{F}_q^n \rightarrow \mathbb{F}_q$. Then, f is q-ary bent if and only if f is perfect nonlinear.

The following corollary can be easily derived from Proposition 1.

Corollary 1. Let $f : \mathbb{F}_q^n \rightarrow \mathbb{F}_q$. Then, we have $\sum_{a \in \mathbb{F}_q^n} |\Delta_f(a)|^2 \geq q^{2n}$, with equality if and only if f is q-ary bent.

In [26], the notion of plateaued functions was redefined over $\mathbb{F}_q$ as follows.

Definition 2. Let $f : \mathbb{F}_q^n \rightarrow \mathbb{F}_q$ and s be an integer with $0 \leq s \leq n$. Then, f is called a q-ary s-plateaued if $|\widehat{\chi_f}(\omega)|^2 \in \{0, q^{n+s}\}$ for all $\omega \in \mathbb{F}_q^n$.

The Walsh distribution of a q-ary plateaued function is given as follows.

Lemma 3. Let $f : \mathbb{F}_q^n \rightarrow \mathbb{F}_q$ be an s-plateaued function. Then for $\omega \in \mathbb{F}_q^n$, $|\widehat{\chi_f}(\omega)|^2$ takes q^{n-s} times the value q^{n+s} and $q^n - q^{n-s}$ times the value 0.

The notion of linear translators for q-ary functions is given as follows.

Definition 3 [23]. Let $f : \mathbb{F}_q^n \rightarrow \mathbb{F}_q$. A nonzero element $\alpha \in \mathbb{F}_q^n$ is called a b-linear translator for a function f if the equation $f(x + u\alpha) - f(x) = ub$ holds for all $x \in \mathbb{F}_q^n$, $u \in \mathbb{F}_q$ and a fixed $b \in \mathbb{F}_q$.

The linear translators of q-ary functions have the following properties.

Lemma 4. Let $f : \mathbb{F}_q^n \rightarrow \mathbb{F}_q$ and let $\mathcal{L}_f$ be the set of linear translators of f. Let $\alpha \in \mathcal{L}_f$. Then we have the following.

(i.) $f(x + u\alpha) = f(x) + f(u\alpha) - f(0)$ for all $x \in \mathbb{F}_q^n$ and $u \in \mathbb{F}_q$.
(ii.) $\mathcal{L}_f$ is a linear subspace of $\mathbb{F}_q^n$ and it is called a linear space of f.
(iii.) $l(x) := f(x) - f(0)$ is a linear function on $\mathcal{L}_f$.

We are now going to redefine the notion of partially bent functions over $\mathbb{F}_q$. To do this, we first give a bound stating the trade-off between the number of nonzero values of the autocorrelation function and of the Walsh transform of q-ary functions. In the original paper of the notion of partially bent functions [4], Carlet, in characteristic 2, proved that this bound holds for every Boolean function, and partially bent functions are defined to be these functions which satisfy the equality (for the p-ary case, see [12]).

Proposition 2. *Let* $f : \mathbb{F}_q^n \to \mathbb{F}_q$. *Then*

$$q^n \leq \mathcal{N}_{\Delta_f} * \mathcal{N}_{\widehat{\chi_f}}, \tag{4}$$

with equality if and only if the derivative $D_a f$ *is either balanced or constant for all* $a \in \mathbb{F}_q^n$.

Proof. We have $|\widehat{\chi_f}(0)|^2 = \sum_{a \in \mathbb{F}_q^n} \Delta_f(a)$ by Lemma 2 (iii). Then by (3) we have $|\widehat{\chi_f}(0)|^2 \leq q^n \mathcal{N}_{\Delta_f}$. Notice that $\mathcal{N}_{\Delta_f}$ is invariant if $f(x)$ is replaced with $f(x) - \omega \cdot x$ for all $\omega \in \mathbb{F}_q^n$, and so we have $|\widehat{\chi_f}(\omega)|^2 \leq q^n \mathcal{N}_{\Delta_f}$ for all $\omega \in \mathbb{F}_q^n$. Hence, from the Parseval identity we have

$$q^{2n} \leq \max_{b \in \mathbb{F}_q^n}(|\widehat{\chi_f}(b)|^2) * \mathcal{N}_{\widehat{\chi_f}} \leq q^n \mathcal{N}_{\Delta_f} * \mathcal{N}_{\widehat{\chi_f}}. \tag{5}$$

For the equality case, assume that the equality in (4) holds. In view of (5), we have that $\max_{b \in \mathbb{F}_q^n}(|\widehat{\chi_f}(b)|^2) * \mathcal{N}_{\widehat{\chi_f}} = q^{2n}$. By the Parseval identity, for all $\omega \in \text{Supp}(\widehat{\chi_f})$ we have $|\widehat{\chi_f}(\omega)|^2 = \max_{b \in \mathbb{F}_q^n}(|\widehat{\chi_f}(b)|^2)$, that is, there exits an integer s such that $|\widehat{\chi_f}(\omega)|^2 = q^{n+s}$, i.e., f is s-plateaued. For all $\omega \in \text{Supp}(\widehat{\chi_f})$, by Lemma 2 (ii),

$$|\widehat{\chi_f}(\omega)|^2 = \sum_{a \in \text{Supp}(\Delta_f)} \sum_{x \in \mathbb{F}_q^n} \chi(D_a f(x) - \omega \cdot a),$$

where we used that $D_a f$ is balanced for all $a \in \mathbb{F}_q^n \setminus \text{Supp}(\Delta_f)$. By Lemma 3, we have $\mathcal{N}_{\widehat{\chi_f}} = q^{n-s}$ and hence by (4), we get $\mathcal{N}_{\Delta_f} = q^s$. Then for all $a \in \text{Supp}(\Delta_f)$, we have

$$\sum_{x \in \mathbb{F}_q^n} \chi(D_a f(x) - \omega \cdot a) = q^n$$

for all $\omega \in \text{Supp}(\widehat{\chi_f})$, that is, $D_a f$ is constant (since $D_a f(x) = \omega \cdot a$ for all $x \in \mathbb{F}_q^n$). Conversely, assume that $D_a f$ is either balanced or constant for all $a \in \mathbb{F}_q^n$. Then $\text{Supp}(\Delta_f) = \mathcal{L}_f$ and there exists an integer s such that $\mathcal{N}_{\Delta_f} = q^s$. Our next aim is to find $\mathcal{N}_{\widehat{\chi_f}}$. For all $\omega \in \mathbb{F}_q^n$, by Lemma 2 (ii) we have

$$|\widehat{\chi_f}(\omega)|^2 = \sum_{a \in \mathcal{L}_f} \sum_{x \in \mathbb{F}_q^n} \chi(D_a f(x) - \omega \cdot a).$$

By Lemma 4, if $a \in \mathcal{L}_f$, then $f(x + a) - f(x) = f(a) - f(0)$ for all $x \in \mathbb{F}_q^n$. Then for all $\omega \in \mathbb{F}_q^n$ we have

$$|\widehat{\chi_f}(\omega)|^2 = q^n \sum_{a \in \mathcal{L}_f} \chi(f(a) - f(0) - \omega \cdot a) = \begin{cases} q^{n+s}, & \text{if } f(a) - \omega \cdot a = f(0) \text{ on } \mathcal{L}_f, \\ 0, & \text{otherwise,} \end{cases}$$

where in the last equality we used that $f(a) - f(0) - \omega \cdot a$ is linear on $\mathcal{L}_f$. Then by the Parseval identity, we have $\mathcal{N}_{\widehat{\chi_f}} = q^{n-s}$. Hence, the equality in (4) holds. $\qquad\square$

Definition 4. *Let $f : \mathbb{F}_q^n \to \mathbb{F}_q$. Then, f is called a q-ary partially bent if the derivative $\mathcal{D}_a f$ is either balanced or constant for all $a \in \mathbb{F}_q^n$.*

The definition of q-ary partially bent functions can be slightly revisited as follows.

Remark 1. Let $f : \mathbb{F}_q^n \to \mathbb{F}_q$ be a function with linear space $\mathcal{L}_f$ such that $\dim(\mathcal{L}_f) = s$ where s is an integer with $0 \leq s \leq n$. Then, f is said to be a q-ary s-partially bent if the derivative $\mathcal{D}_a f$ is balanced for all $a \in \mathbb{F}_q^n \setminus \mathcal{L}_f$. Clearly, $\mathcal{D}_a f$ is constant for all $a \in \mathcal{L}_f$.

The absolute Walsh transform of q-ary partially bent functions takes only one nonzero value (also possibly the value 0).

Proposition 3. *Let $f : \mathbb{F}_q^n \to \mathbb{F}_q$ with linear space $\mathcal{L}_f$ and $\dim(\mathcal{L}_f) = s$. If f is q-ary s-partially bent, then $|\widehat{\chi_f}(\omega)|^2 \in \{0, q^{n+s}\}$ for all $\omega \in \mathbb{F}_q^n$.*

In view of Proposition 3, the Walsh distribution of q-ary partially bent functions follows from the Parseval identity.

Lemma 5. *Let $f : \mathbb{F}_q^n \to \mathbb{F}_q$ be an s-partially bent function. Then for $\omega \in \mathbb{F}_q^n$, $|\widehat{\chi_f}(\omega)|^2$ takes q^{n-s} times the value q^{n+s} and $q^n - q^{n-s}$ times the value 0.*

Remark 2. Any q-ary bent f is a q-ary partially bent function with $\mathcal{L}_f = \{0\}$ by Proposition 1. Over $\mathbb{F}_q$, any perfect nonlinear function is a partially bent function.

In the light of the above results, we can give the following natural consequence for q-ary functions.

Proposition 4. *Let $f : \mathbb{F}_q^n \to \mathbb{F}_q$ with linear space $\mathcal{L}_f$ and $\dim(\mathcal{L}_f) = s$. Then, f is q-ary s-partially bent if and only if f is q-ary s-plateaued. In particular, f is q-ary bent if and only if f is q-ary 0-plateaued.*

Remark 3. The set of q-ary bent functions is a proper subset of the set of q-ary partially bent functions. Similarly, the set of q-ary partially bent functions is a proper subset of the set of q-ary plateaued functions. Namely, a q-ary s-plateaued with $\dim(\mathcal{L}_f) < s$ is not a q-ary partially bent.

Remark 4. Let $f : \mathbb{F}_q^n \to \mathbb{F}_q$. Then, f is affine if and only if $\mathcal{N}_{\widehat{\chi_f}} = 1$ and f is q-ary bent if and only if $\mathcal{N}_{\Delta_f} = 1$, i.e., $\max_{a \in \mathbb{F}_{q^n}^*}(|\Delta_f(a)|) = 0$. Moreover, f is q-ary partially bent if and only if $|\Delta_f(a)| \in \{0, q^n\}$ for all $a \in \mathbb{F}_q^n$.

3.2   Characterizations of q-ary Partially Bent Functions

In this subsection, we obtain several characterizations of q-ary partially bent functions by means of their derivatives, Walsh power moments and autocorrelation functions.

A link between the autocorrelation function and the second-order derivative of a q-ary function is given as follows (see [5], in characteristic 2).

Proposition 5. *Let $f : \mathbb{F}_q^n \to \mathbb{F}_q$. Then*

$$\sum_{a \in \mathbb{F}_q^n} |\Delta_f(a)|^2 = \sum_{a,b,x \in \mathbb{F}_q^n} \chi(\mathcal{D}_a \mathcal{D}_b f(x)). \tag{6}$$

Proof. Since $|z|^2 = z\bar{z}$ for $z \in \mathbb{C}$, the left hand side of (6) is given as

$$\sum_{a \in \mathbb{F}_q^n} \sum_{b \in \mathbb{F}_q^n} \chi(\mathcal{D}_a f(b)) \sum_{x \in \mathbb{F}_q^n} \overline{\chi}(\mathcal{D}_a f(x)) = \sum_{a,b,x \in \mathbb{F}_q^n} \chi(\mathcal{D}_a f(b) - \mathcal{D}_a f(x)) = \sum_{a,b,x \in \mathbb{F}_q^n} \chi(\mathcal{D}_a \mathcal{D}_b f(x)),$$

where in the last equality we used the bijective change of variable: $b \mapsto b + x$. $\square$

The identity involving the fourth Walsh power moment and the second-order derivative of a q-ary function is constituted as follows (see [5] for binary case and [22] for p-ary case).

Proposition 6. *Let $f : \mathbb{F}_q^n \to \mathbb{F}_q$. Then*

$$S_2(f) = q^n \sum_{a,b,x \in \mathbb{F}_q^n} \chi(\mathcal{D}_a \mathcal{D}_b f(x)). \tag{7}$$

Proof. Since $|z|^4 = z^2 \bar{z}^2$ for $z \in \mathbb{C}$, we have

$$\sum_{\omega \in \mathbb{F}_q^n} |\widehat{\chi_f}(\omega)|^4 = \sum_{x,a,b,c \in \mathbb{F}_q^n} \chi(f(x) - f(a) + f(b) - f(c)) \sum_{\omega \in \mathbb{F}_q^n} \overline{\chi}(\omega \cdot (x - a + b - c))$$

$$= q^n \sum_{a,b,x \in \mathbb{F}_q^n} \chi(f(x) - f(a) + f(b) - f(x - a + b))$$

since $\sum_{\omega \in \mathbb{F}_q^n} \xi_p^{\mathrm{Tr}_p^{q^n}(-\omega(x-a+b-c))} = \begin{cases} q^n, & \text{if } c = x - a + b, \\ 0, & \text{otherwise.} \end{cases}$

Hence, since $(a, b, x) \mapsto (x + a, x + a + b, x)$ is the permutation of $(\mathbb{F}_q^n)^3$, then (7) holds. $\square$

The following link follows readily from Propositions 5 and 6.

Proposition 7. *Let $f : \mathbb{F}_q^n \to \mathbb{F}_q$. Then we have $S_2(f) = q^n \sum_{a \in \mathbb{F}_q^n} |\Delta_f(a)|^2$.*

We first give the following characterization of partially bent functions by means of their autocorrelation functions (see [5], in characteristic 2).

Proposition 8. *Let $f : \mathbb{F}_q^n \to \mathbb{F}_q$ with linear space $\mathcal{L}_f$ and $\dim(\mathcal{L}_f) = s$. Then*

$$\sum_{a \in \mathbb{F}_q^n} |\Delta_f(a)|^2 \geq q^{2n+s},$$

with equality if and only if f is q-ary s-partially bent.

Proof. Due to the fact that $\mathcal{D}_a f$ is constant for all $a \in \mathcal{L}_f$, we have $\sum_{a \in \mathcal{L}_f} |\Delta_f(a)|^2 = q^{2n+s}$. Moreover, we have

$$\sum_{a \in \mathbb{F}_q^n \setminus \mathcal{L}_f} |\Delta_f(a)|^2 \geq 0,$$

with equality if and only if $\Delta_f(a)$ is zero (i.e., $\mathcal{D}_a f$ is balanced) for all $a \in \mathbb{F}_q^n \setminus \mathcal{L}_f$. Hence, the proof is complete. $\qquad\square$

The following can be directly derived from Propositions 5 and 8.

Corollary 2. *Let $f : \mathbb{F}_q^n \to \mathbb{F}_q$ with linear space $\mathcal{L}_f$ and $\dim(\mathcal{L}_f) = s$. Then we have*

$$\sum_{a,b,x \in \mathbb{F}_q^n} \chi(\mathcal{D}_a \mathcal{D}_b f(x)) \geq q^{2n+s},$$

with equality if and only if f is q-ary s-partially bent.

We derive directly from Propositions 7 and 8 the following characterization of partially bent functions in terms of their fourth Walsh power moment.

Theorem 1. *Let $f : \mathbb{F}_q^n \to \mathbb{F}_q$ with linear space $\mathcal{L}_f$ and $\dim(\mathcal{L}_f) = s$. Then*

$$S_2(f) \geq q^{3n+s},$$

with equality if and only if f is q-ary s-partially bent.

The sequence of even Walsh power moments of a q-ary partially bent function is a simple geometric sequence.

Corollary 3. *Let $f : \mathbb{F}_q^n \to \mathbb{F}_q$ be a q-ary s-partially bent. Then for every positive integer i, we have $S_i(f) = q^{n(i+1)+s(i-1)}$ and for every integer $j \geq 2$,*

$$S_i(f)S_j(f) = S_{i+1}(f)S_{j-1}(f).$$

Proof. By Lemma 5, we have $S_i(f) = q^{n-s}(q^{n+s})^i = q^{n(i+1)+s(i-1)}$ for every positive integer i. The second assertion follows readily from the first assertion. $\qquad\square$

We recall the strong properties of the Fourier transform of complex valued functions. For $G : \mathbb{F}_q^n \to \mathbb{C}$, let $\widehat{G} : \mathbb{F}_q^n \to \mathbb{C}$ be its Fourier transform. Then we have $\widehat{\widehat{G}}(u) = q^n G(-u)$ for all $u \in \mathbb{F}_q^n$. As is readily seen, $G(u) = 0$ for all $u \in \mathbb{F}_q^n$ if and only if $\widehat{G}(v) = 0$ for all $v \in \mathbb{F}_q^n$. Hence for two functions $G_1, G_2 : \mathbb{F}_q^n \to \mathbb{C}$,

$$G_1(u) = G_2(u), \ \forall u \in \mathbb{F}_q^n \Longleftrightarrow \widehat{G_1}(v) = \widehat{G_2}(v), \ \forall v \in \mathbb{F}_q^n. \tag{8}$$

We are now going to give a powerful characterization of q-ary partially bent functions by means of their second-order derivatives (see [10] and [24] for bent Boolean functions and p-ary plateaued functions, respectively). In order to keep the paper self-contained we give short proof although the argument of the proof is similar to that of [24, Theorem 3].

Theorem 2. *Let $f : \mathbb{F}_q^n \to \mathbb{F}_q$ with linear space $\mathcal{L}_f$ and $\dim(\mathcal{L}_f) = s$. Set*

$$\theta_f(x) = \sum_{a,b \in \mathbb{F}_q^n} \chi(\mathcal{D}_a \mathcal{D}_b f(x))$$

for $x \in \mathbb{F}_{p^n}$. Then, f is q-ary s-partially bent if and only if $\theta_f(x) = q^{n+s}$ for all $x \in \mathbb{F}_q^n$. In particular, f is q-ary bent if and only if $\theta_f(x) = q^n$ for all $x \in \mathbb{F}_q^n$.

Proof. Put $\theta = q^{n+s}$. For a function f, $\theta_f(x) = \theta$ for all $x \in \mathbb{F}_q^n$ if and only if for all $x \in \mathbb{F}_q^n$

$$\sum_{a,b \in \mathbb{F}_q^n} \chi(f(a+b-x) - f(a) - f(b)) = \theta * \chi(-f(x)) \tag{9}$$

(by the bijective change of variables: $a \mapsto a - x$ and $b \mapsto b - x$). We can easily see that the Fourier transforms of the left-hand side of (9) at $\omega \in \mathbb{F}_q^n$ is given by $\widehat{\chi_f}(\omega) \ \overline{\widehat{\chi_f}(\omega)} \widehat{\chi_f}(-\omega)$ and of its right-hand side by $\theta * \overline{\widehat{\chi_f}}(\omega)$. By (8), for all $x \in \mathbb{F}_q^n$, (9) holds if and only if for all $\omega \in \mathbb{F}_q^n$,

$$\widehat{\chi_f}(\omega) \ \overline{\widehat{\chi_f}(\omega)} \widehat{\chi_f}(\omega) = \theta * \overline{\widehat{\chi_f}(\omega)};$$

equivalently, $|\widehat{\chi_f}(\omega)|^2 \in \{0, \theta\}$ for all $\omega \in \mathbb{F}_q^n$ where $\theta = q^{n+s}$, that is, f is q-ary s-partially bent. In particular, for $s = 0$, $\theta_f(x) = q^n$ for all $x \in \mathbb{F}_q^n$ if and only if $|\widehat{\chi_f}(\omega)|^2 = q^n$ for all $\omega \in \mathbb{F}_q^n$ by the Parseval identity, i.e., f is q-ary bent. $\square$

Notice that Theorem 2 says that any q-ary quadratic function is a q-ary partially bent function since the second-order derivative of a quadratic function is constant. We also deduce the following proposition by using the linear translators of a q-ary function.

Proposition 9. *Let $f : \mathbb{F}_q^n \to \mathbb{F}_q$ with linear space $\mathcal{L}_f$ and $\dim(\mathcal{L}_f) = s$. Then for all $x \in \mathbb{F}_q^n$, $\sum_{a,b \in \mathbb{F}_q^n} \chi(\mathcal{D}_a \mathcal{D}_b f(x)) \geq q^{n+s}$, with equality if and only if f is q-ary s-partially bent.*

Proposition 10 [26]. *Let $f : \mathbb{F}_q^n \to \mathbb{F}_q$. Then for all $x \in \mathbb{F}_q^n$,*

$$\sum_{\omega \in \mathbb{F}_q^n} \chi(f(x) - \omega \cdot x) \overline{\widehat{\chi_f}(\omega)} \, |\widehat{\chi_f}(\omega)|^2 = q^n \sum_{a,b \in \mathbb{F}_q^n} \chi(\mathcal{D}_a \mathcal{D}_b f(x)).$$

The following corollary follows directly from Propositions 9 and 10.

Corollary 4. *Let $f : \mathbb{F}_q^n \to \mathbb{F}_q$ with linear space $\mathcal{L}_f$ and $\dim(\mathcal{L}_f) = s$. Then we have for all $x \in \mathbb{F}_q^n$,*

$$\sum_{\omega \in \mathbb{F}_q^n} \chi(f(x) - \omega \cdot x) \overline{\widehat{\chi_f}(\omega)} \, |\widehat{\chi_f}(\omega)|^2 \geq q^{2n+s},$$

with equality if and only if f is q-ary s-partially bent.

In the following sections, we assume that $m = 1$ (i.e., $q = p$), namely, f is a p-ary function from $\mathbb{F}_{p^n}$ to $\mathbb{F}_p$ where p is a prime.

4   Characterizations of p-ary Partially Bent (vectorial) Functions

This section characterizes p-ary partially bent (vectorial) functions in terms of their second-order derivatives and Walsh power moments.

4.1   Characterizations of p-ary Partially Bent Functions

We first give a link between the fourth Walsh power moments and the zeros of the second-order derivatives. For a function $f : \mathbb{F}_{p^n} \to \mathbb{F}_p$, a corresponding function $f_\lambda := \lambda f : \mathbb{F}_{p^n} \to \mathbb{F}_p$ is defined as $x \mapsto \lambda f(x)$ for every $\lambda \in \mathbb{F}_p^*$. Then for any $\lambda \in \mathbb{F}_p^*$, we have $\mathcal{D}_b \mathcal{D}_a \lambda f(x) = \lambda(\mathcal{D}_b \mathcal{D}_a f(x))$ at $(a,b) \in \mathbb{F}_{p^n}^2$ for every $x \in \mathbb{F}_{p^n}$. We denote by $\mathfrak{N}(f)$ the size of the set $K = \{(a,b,x) \in \mathbb{F}_{p^n}^3 : \mathcal{D}_b \mathcal{D}_a f(x) = 0\}$.

Proposition 11. *Let $f : \mathbb{F}_{p^n} \to \mathbb{F}_p$. Then we have $\sum_{\lambda \in \mathbb{F}_p^*} S_2(\lambda f) = p^{n+1} \mathfrak{N}(f) - p^{4n}$.*

Proof. By Proposition 6, we have

$$\sum_{\lambda \in \mathbb{F}_p^*} S_2(\lambda f) = \sum_{\lambda \in \mathbb{F}_p^*} \left(p^n \sum_{a,b,x \in \mathbb{F}_{p^n}} \xi_p^{\mathcal{D}_b \mathcal{D}_a \lambda f(x)} \right)$$

$$= p^n \left(\sum_{\lambda \in \mathbb{F}_p^*} \sum_{(a,b,x) \in K} \xi_p^{\lambda \mathcal{D}_b \mathcal{D}_a f(x)} + \sum_{(a,b,x) \notin K} \sum_{\lambda \in \mathbb{F}_p^*} \xi_p^{\lambda \mathcal{D}_b \mathcal{D}_a f(x)} \right)$$

$$= p^n \Big((p-1)\mathfrak{N}(f) - (p^{3n} - \mathfrak{N}(f)) \Big) = p^{n+1} \mathfrak{N}(f) - p^{4n},$$

where in the third equality we used that $1 + \xi_p + \xi_p^2 + \cdots + \xi_p^{p-1} = 0$. $\square$

From Theorem 1 and Proposition 11, we derive a characterization of partially bent functions in terms of the zeros of their second-order derivatives.

Corollary 5. *Let* $f : \mathbb{F}_{p^n} \to \mathbb{F}_p$ *with linear space* $\mathcal{L}_f$ *and* $\dim(\mathcal{L}_f) = s$. *Then,* f *is p-ary s-partially bent if and only if* $\mathfrak{N}(f) = p^{3n-1} + p^{2n+s} - p^{2n+s-1}$.

Proof. Clearly, f is p-ary s-partially bent if and only if f_λ is p-ary s-partially bent for every $\lambda \in \mathbb{F}_p^*$. Then by Theorem 1, f is p-ary s-partially bent if and only if

$$\sum_{\lambda \in \mathbb{F}_p^*} S_2(f_\lambda) = (p-1)p^{3n+s};$$

equivalently, by Proposition 11, we have $\mathfrak{N}(f) = p^{3n-1} + p^{2n+s} - p^{2n+s-1}$.   □

A function $f : \mathbb{F}_{p^n} \to \mathbb{F}_p$ with linear space $\mathcal{L}_f$ is p-ary partially bent if and only if the derivative $\mathcal{D}_a f$ is balanced for all $a \in \mathbb{F}_{p^n} \setminus \mathcal{L}_f$. It would be interesting to prove directly the following theorem without using partially bent-ness of f. To do this, we need the following well-known lemma (see [24] for vectorial case).

Lemma 6. *Let* $h : \mathbb{F}_{p^n} \to \mathbb{F}_p$. *Then* $p^{2n-1} \leq \#\{(x_1, x_2) \in \mathbb{F}_{p^n}^2 : h(x_1) = h(x_2)\}$, *with equality if and only if* h *is balanced.*

Theorem 3. *Let* $f : \mathbb{F}_{p^n} \to \mathbb{F}_p$ *with linear space* $\mathcal{L}_f$ *and* $\dim(\mathcal{L}_f) = s$. *Then,* $\mathcal{D}_a f$ *is balanced for all* $a \in \mathbb{F}_{p^n} \setminus \mathcal{L}_f$ *if and only if* $\mathfrak{N}(f) = p^{3n-1} + p^{2n+s} - p^{2n+s-1}$.

Proof. Clearly, for all $(a, b, x) \in \mathbb{F}_{p^n}^3$, $\mathcal{D}_b \mathcal{D}_a f(x) = 0$ if and only if

$$\mathcal{D}_a f(x) = \mathcal{D}_a f(x + b). \tag{10}$$

For all $a \in \mathcal{L}_f$, since the derivative $\mathcal{D}_a f$ is constant, we have

$$\#\{(a, b, x) \in \mathbb{F}_{p^n}^3 : a \in \mathcal{L}_f \text{ and } \mathcal{D}_b \mathcal{D}_a f(x) = 0\} = p^{2n+s}. \tag{11}$$

For all $a \in \mathbb{F}_{p^n} \setminus \mathcal{L}_f$, by Lemma 6, $\mathcal{D}_a f$ is balanced if and only if the number of pairs $(b, x) \in \mathbb{F}_{p^n}^2$ satisfying (10) is equal to p^{2n-1}; equivalently,

$$\#\{(a, b, x) \in \mathbb{F}_{p^n}^3 : a \notin \mathcal{L}_f \text{ and } \mathcal{D}_b \mathcal{D}_a f(x) = 0\} = (p^n - p^s)p^{2n-1}. \tag{12}$$

Hence, combining (11) and (12) concludes the result.   □

4.2   Characterizations of p-ary Vectorial s-Partially Bent Functions

In this subsection, we first give the notion of vectorial p-ary s-partially bent functions and next present their characterizations.

Definition 5. *Let* $F : \mathbb{F}_{p^n} \to \mathbb{F}_{p^m}$ *be a vectorial function. For every* $\lambda \in \mathbb{F}_{p^m}^*$, *let* $F_\lambda : \mathbb{F}_{p^n} \to \mathbb{F}_p$, *defined by* $F_\lambda(x) = \mathrm{Tr}_p^{p^m}(\lambda F(x))$, *be its component function with linear space* $\mathcal{L}_{F_\lambda}$.

- *Then* F *is called a vectorial p-ary partially bent if* F_λ, $\lambda \in \mathbb{F}_{p^m}^*$, *is p-ary partially bent.*

– *Assume that there exits an integer s with $0 \leq s \leq n$ such that $\dim(\mathcal{L}_{F_\lambda}) = s$ for every $\lambda \in \mathbb{F}_{p^m}^\star$. Then, F is called a vectorial p-ary s-partially bent if F_λ, $\lambda \in \mathbb{F}_{p^m}^\star$, is p-ary s-partially bent.*

Remark 5. The notion of vectorial partially bent functions coincides the notion of strongly-plateaued functions introduced in [6]. More precisely, all derivatives of a function $f : \mathbb{F}_{p^n} \to \mathbb{F}_p$ are either constant or balanced if and only if for all $a \in \mathbb{F}_{p^n}$ and $v \in \mathbb{F}_p$, the size of the set $\{b \in \mathbb{F}_{p^n} : \mathcal{D}_a f(b) = \mathcal{D}_a f(x) + v\}$ is independent of $x \in \mathbb{F}_{p^n}$.

We can derive directly from Theorem 1 the following characterization of vectorial partially bent functions.

Theorem 4. *Let $F : \mathbb{F}_{p^n} \to \mathbb{F}_{p^m}$ and for every $\lambda \in \mathbb{F}_{p^m}^\star$ $F_\lambda : \mathbb{F}_{p^n} \to \mathbb{F}_p$ be its component function with linear space $\mathcal{L}_{F_\lambda}$ such that $\dim(\mathcal{L}_{F_\lambda}) = s$. Then F is vectorial p-ary s-partially bent if and only if*

$$\sum_{\lambda \in \mathbb{F}_{p^m}^\star} S_2(F_\lambda) = (p^m - 1)p^{3n+s} \tag{13}$$

Proof. Assume that F is p-ary s-partially bent. Then by Theorem 1, the assertion holds. Conversely, assume that (13) holds. By Theorem 1, for every $\lambda \in \mathbb{F}_{p^m}^\star$, we have

$$S_2(F_\lambda) \geq p^{3n+s},$$

with equality because of (13), which implies that F_λ is p-ary s-partially bent. This completes the proof. □

In [22], Mesnager showed that the left-hand side of (13) can be computed by counting the zeros of the second-order derivatives. We denote by $\mathfrak{N}(F)$ the size of the set $\{(a, b, x) \in \mathbb{F}_{p^n}^3 : \mathcal{D}_b \mathcal{D}_a F(x) = 0\}$.

Proposition 12 [22]. *Let $F : \mathbb{F}_{p^n} \to \mathbb{F}_{p^m}$ and for every $\lambda \in \mathbb{F}_{p^m}^\star$ $F_\lambda : \mathbb{F}_{p^n} \to \mathbb{F}_p$ be its component function. Then*

$$\sum_{\lambda \in \mathbb{F}_{p^m}^\star} S_2(F_\lambda) = p^{n+m}\mathfrak{N}(F) - p^{4n}.$$

We then deduce a characterization of vectorial partially bent functions by means of the zeros of their second-order derivatives.

Theorem 5. *Let $F : \mathbb{F}_{p^n} \to \mathbb{F}_{p^m}$ and for every $\lambda \in \mathbb{F}_{p^m}^\star$ $F_\lambda : \mathbb{F}_{p^n} \to \mathbb{F}_p$ be its component function with linear space $\mathcal{L}_{F_\lambda}$ such that $\dim(\mathcal{L}_{F_\lambda}) = s$. Then F is vectorial p-ary s-partially bent if and only if $\mathfrak{N}(F) = p^{2n+s} + p^{3n-m} - p^{2n+s-m}$.*

Proof. By Theorem 4 and Proposition 12, F is s-partially bent if and only if $p^{3n+s}(p^m - 1) = p^{n+m}\mathfrak{N}(F) - p^{4n}$. Hence, the proof is complete. □

The following characterization of plateaued functions was given in [25].

Proposition 13 [25, Theorem 7]. *Let* $F : \mathbb{F}_{p^n} \to \mathbb{F}_{p^m}$. *For* $v \in \mathbb{F}_{p^m}$, *let* $\mathcal{N}_F(v;x) = \#\{(a,b) \in \mathbb{F}_{p^n}^2 : \mathcal{D}_b\mathcal{D}_aF(x) = v\}$ *for* $x \in \mathbb{F}_{p^n}$. *Then there exists an integer* s *with* $0 \le s \le n$ *such that* F *is vectorial p-ary s-plateaued if and only if* $\mathcal{N}_F(v;x)$ *does not depend on* $x \in \mathbb{F}_{p^n}$, *nor on* $v \in \mathbb{F}_{p^m}^{\star}$.

We then deduce directly from Theorem 5 and Proposition 13 the following.

Corollary 6. *Let* $F : \mathbb{F}_{p^n} \to \mathbb{F}_{p^m}$. *Then* F *is vectorial p-ary s-partially bent if and only if* $\#\{(a,b) \in \mathbb{F}_{p^n}^2 : \mathcal{D}_b\mathcal{D}_aF(x) = 0\} = p^{n+s} + p^{2n-m} - p^{n+s-m}$ *for every* $x \in \mathbb{F}_{p^n}$.

5   Characterizations of *p*-ary Plateaued Functions

In this section, we characterize p-ary plateaued functions in terms of their Walsh power moments, the value distribution of their derivatives and autocorrelation functions. We first recall the following well-known inequality.

Theorem 6 (Hölder's Inequality) [29]. *Let* $p_1, p_2 \in (1, \infty)$ *with* $\frac{1}{p_1} + \frac{1}{p_2} = 1$. *Then, for all vectors* $(x_1, x_2, \ldots, x_m), (y_1, y_2, \ldots, y_m) \in \mathbb{R}^m$ *or* $\mathbb{C}^m$, *Hölder's Inequality states that*

$$\sum_{k=1}^{m} |x_k y_k| \le \left(\sum_{k=1}^{m} |x_k|^{p_1}\right)^{\frac{1}{p_1}} \left(\sum_{k=1}^{m} |y_k|^{p_2}\right)^{\frac{1}{p_2}}.$$

The above inequality becomes equality if and only if for every $k \in \{1, \ldots, m\}$, $|x_k|^{p_1} = d|y_k|^{p_2}$ *for some* $d \in \mathbb{R}^+$. *In particular, if* $p_1 = p_2 = 2$, *then this is called the Cauchy-Schwarz Inequality.*

We are now going to deduce from Hölder's Inequality the following characterizations of plateaued functions in terms of even power moments of their Walsh transform.

Applying the Cauchy-Schwarz Inequality, for $x_k = |\widehat{\chi_f}(\omega)|^2$ and $y_k = |\widehat{\chi_f}(\omega)|^{2i}$ for all $\omega \in \mathbb{F}_{p^n}$, $1 \le k \le p^n$, we have

$$\left(\sum_{\omega \in \mathbb{F}_{p^n}} |\widehat{\chi_f}(\omega)|^{2i+2}\right)^2 \le \sum_{\omega \in \mathbb{F}_{p^n}} |\widehat{\chi_f}(\omega)|^4 \sum_{\omega \in \mathbb{F}_{p^n}} |\widehat{\chi_f}(\omega)|^{4i},$$

that is, $S_{i+1}(f)^2 \le S_2(f)S_{2i}(f)$, where the equality holds for one (and hence for all) $i \ge 1$ if and only if for all $\omega \in \mathbb{F}_{p^n}$, we have $|\widehat{\chi_f}(\omega)|^2 = d\,|\widehat{\chi_f}(\omega)|^{2i}$ for some $d \in \mathbb{R}^+$; equivalently, $|\widehat{\chi_f}(\omega)|^2$ is either the same positive integer or 0, that is, f is p-ary plateaued. This proves the following.

Theorem 7. *Let* $f : \mathbb{F}_{p^n} \to \mathbb{F}_p$. *Then for every positive integer* i, *we have*

$$S_{i+1}(f)^2 \le S_2(f)S_{2i}(f),$$

with equality for one (and hence for all) $i \ge 1$ *if and only if* f *is p-ary plateaued.*

Theorem 8. *Let $f : \mathbb{F}_{p^n} \to \mathbb{F}_p$. Then for every integer $i \geq 2$, we have*

$$p^{2ni} \leq S_i(f) * \mathcal{N}_{\widehat{\chi_f}}^{(i-1)},$$

where the equality holds for one (and hence for all) $i \geq 2$ if and only if f is p-ary plateaued.

Proof. By Theorem 6, putting $x_k = |\widehat{\chi_f}(\omega)|^2$ for all $\omega \in \mathrm{Supp}(\widehat{\chi_f})$ and $y_k = 1$, $1 \leq k \leq \mathcal{N}_{\widehat{\chi_f}}$, with $p_1 = i$ and $p_2 = \frac{i}{i-1}$, we have

$$\sum_{\omega \in \mathrm{Supp}(\widehat{\chi_f})} |\widehat{\chi_f}(\omega)|^2 \leq \left(\sum_{\omega \in \mathrm{Supp}(\widehat{\chi_f})} |\widehat{\chi_f}(\omega)|^{2i} \right)^{\frac{1}{i}} \left(\sum_{\omega \in \mathrm{Supp}(\widehat{\chi_f})} 1 \right)^{\frac{i-1}{i}},$$

namely by the Parseval identity, $p^{2ni} \leq S_i(f) * \mathcal{N}_{\widehat{\chi_f}}^{(i-1)}$, where the equality holds for one (and hence for all) $i \geq 2$ if and only if for all $\omega \in \mathrm{Supp}(\widehat{\chi_f})$, $|\widehat{\chi_f}(\omega)|^2 = d$ for some $d \in \mathbb{R}^+$; equivalently, f is p-ary plateaued. The proof is complete. $\square$

Proposition 14. *Let $f : \mathbb{F}_{p^n} \to \mathbb{F}_p$. Then for every positive integer i, we have*

$$S_i(f)^2 \leq S_{2i}(f) * \mathcal{N}_{\widehat{\chi_f}},$$

with equality for one (and hence for all) $i \geq 1$ if and only if f is p-ary plateaued.

Proof. Applying the Cauchy-Schwarz Inequality, for $x_k = |\widehat{\chi_f}(\omega)|^{2i}$ for all $\omega \in \mathrm{Supp}(\widehat{\chi_f})$ and $y_k = 1$, $1 \leq k \leq \mathcal{N}_{\widehat{\chi_f}}$, as in the case of Theorem 8, the proof is complete. $\square$

In particular, for $i = 1$, Proposition 14 (also for $i = 2$, Theorem 8) introduces the following bound stating the trade-off between the number of nonzero Walsh transform values and the value of fourth Walsh power moment of a p-ary function, and this bound is satisfied only by plateaued functions.

Corollary 7. *Let $f : \mathbb{F}_{p^n} \to \mathbb{F}_p$. Then we have $p^{4n} \leq S_2(f) * \mathcal{N}_{\widehat{\chi_f}}$, with equality if and only if f is p-ary plateaued.*

The following bound can be clearly derived from Proposition 6 and Corollary 7.

Corollary 8. *Let $f : \mathbb{F}_{p^n} \to \mathbb{F}_p$. Set $\theta_f = \sum_{a,b,x \in \mathbb{F}_{p^n}} \xi_p^{D_a D_b f(x)}$. Then we have $p^{3n} \leq \theta_f * \mathcal{N}_{\widehat{\chi_f}}$, with equality if and only if f is p-ary plateaued.*

We derive from Proposition 5 and Corollary 8 the following bound, which was first observed in [31], in characteristic 2.

Corollary 9. *Let $f : \mathbb{F}_{p^n} \to \mathbb{F}_p$. Set $\mathcal{A}_{\Delta_f} = \sum_{a \in \mathbb{F}_{p^n}} |\Delta_f(a)|^2$. Then we have $p^{3n} \leq \mathcal{A}_{\Delta_f} * \mathcal{N}_{\widehat{\chi_f}}$, with equality if and only if f is p-ary plateaued.*

The following result was first given in [31], in characteristic 2.

Proposition 15. *Let* $f : \mathbb{F}_{p^n} \to \mathbb{F}_p$. *Then*

$$p^{2n} \leq \mathcal{N}_{\widehat{\chi_f}} * \max_{b \in \mathbb{F}_{p^n}} (|\widehat{\chi_f}(b)|^2), \tag{14}$$

with equality if and only if f is p-ary plateaued.

Proof. Clearly, by the Parseval identity we have

$$p^{2n} = \sum_{\omega \in \mathbb{F}_p^n} |\widehat{\chi_f}(\omega)|^2 \leq \mathcal{N}_{\widehat{\chi_f}} * \max_{b \in \mathbb{F}_{p^n}} (|\widehat{\chi_f}(b)|^2).$$

For the equality case, assume that the bound (14) is satisfied. By the Parseval identity, for all $\omega \in \mathrm{Supp}(\widehat{\chi_f})$, we have $|\widehat{\chi_f}(\omega)|^2 = \max_{b \in \mathbb{F}_{p^n}}(|\widehat{\chi_f}(b)|^2)$, that is, there exits an integer s such that $|\widehat{\chi_f}(\omega)|^2 = p^{n+s}$, i.e., f is s-plateaued. Conversely, by Lemma 5, we have $\mathcal{N}_{\widehat{\chi_f}} = p^{n-s}$ and $|\widehat{\chi_f}(\omega)|^2 = p^{n+s}$ for all $\omega \in \mathrm{Supp}(\widehat{\chi_f})$. Hence, the proof is complete. □

We now deduce from the Parseval identity a characterization of plateaued functions in terms of their fourth Walsh power moment.

Proposition 16. *Let* $f : \mathbb{F}_{p^n} \to \mathbb{F}_p$. *Then*

$$S_2(f) \leq p^{2n} \max_{b \in \mathbb{F}_{p^n}} (|\widehat{\chi_f}(b)|^2),$$

with equality if and only if f is p-ary plateaued.

Proof. Clearly, we have

$$\sum_{\omega \in \mathbb{F}_{p^n}} |\widehat{\chi_f}(\omega)|^4 = \sum_{\omega \in \mathbb{F}_{p^n}} |\widehat{\chi_f}(\omega)|^2 |\widehat{\chi_f}(\omega)|^2 \leq \sum_{\omega \in \mathbb{F}_{p^n}} |\widehat{\chi_f}(\omega)|^2 \max_{b \in \mathbb{F}_{p^n}} (|\widehat{\chi_f}(b)|^2); \tag{15}$$

equivalently, $S_2(f) \leq S_1(f) \max_{b \in \mathbb{F}_{p^n}} (|\widehat{\chi_f}(b)|^2)$. For the equality case, assume that f is plateaued. By Lemma 3, we conclude that the bound is satisfied. Conversely, by (15) for all $\omega \in \mathrm{Supp}(\widehat{\chi_f})$, we have $|\widehat{\chi_f}(\omega)|^2 = \max_{b \in \mathbb{F}_{p^n}} (|\widehat{\chi_f}(b)|^2)$, that is, f is plateaued. □

We can derive directly from Propositions 6 and 16 the following bounds.

Proposition 17. *Let* $f : \mathbb{F}_{p^n} \to \mathbb{F}_p$. *Set* $\theta_f = \sum_{a,b,x \in \mathbb{F}_{p^n}} \xi_p^{\mathcal{D}_a \mathcal{D}_b f(x)}$. *Then we have*

$$\theta_f \leq p^n \max_{b \in \mathbb{F}_{p^n}} (|\widehat{\chi_f}(b)|^2), \tag{16}$$

with equality if and only if f is p-ary plateaued. Set $\mathcal{A}_{\Delta_f} = \sum_{a \in \mathbb{F}_{p^n}} |\Delta_f(a)|^2$. *Equivalently, by Proposition 5 we have*

$$\mathcal{A}_{\Delta_f} \leq p^n \max_{b \in \mathbb{F}_{p^n}} (|\widehat{\chi_f}(b)|^2),$$

with equality if and only if f is p-ary plateaued.

Notice that the equality case of (16), in characteristic 2, was observed in [3].

6 Conclusion

Some plateaued functions have attracted attention since their introduction in the literature due to their role in diverse domains of Boolean and vectorial functions for sequences and cryptography. In this paper, we provided several characterizations of p-ary plateaued functions via their Walsh power moments, second-order derivatives and autocorrelation functions. We also characterized p-ary partially bent (vectorial) functions by their second-order derivatives and fourth Walsh power moments. Furthermore, for any prime power q, we redefined the notion of partially bent functions over $\mathbb{F}_q$ and next presented some of their characterizations in terms of their second-order derivatives, Walsh power moments and autocorrelation functions.

Acknowledgment. The authors would like to thank the anonymous reviewers of WAIFI-2018 for their valuable comments and suggestions. The third author is supported by the Scientific and Technological Research Council of Turkey (TÜBİTAK), program no: BİDEB 2214/A.

References

1. Ambrosimov, A.: Properties of bent functions of q-valued logic over finite fields (1994)
2. Anbar, N., Meidl, W.: Quadratic functions and maximal Artin-Schreier curves. Finite Fields Appl. **30**, 49–71 (2014)
3. Canteaut, A., Carlet, C., Charpin, P., Fontaine, C.: On cryptographic properties of the cosets of r (1, m). IEEE Trans. Inf. Theory **47**(4), 1494–1513 (2001)
4. Carlet, C.: Partially-bent functions. Des. Codes Crypt. **3**(2), 135–145 (1993)
5. Carlet, C.: Boolean functions for cryptography and error correcting codes. Boolean Models Methods Math. Comput. Sci. Eng. **2**, 257–397 (2010)
6. Carlet, C.: Boolean and vectorial plateaued functions and APN functions. IEEE Trans. Inf. Theory **61**(11), 6272–6289 (2015)
7. Carlet, C., Dubuc, S.: On generalized bent and q-ary perfect nonlinear functions. In: Jungnickel, D., Niederreiter, H. (eds.) Finite Fields and Applications, pp. 81–94. Springer, Heidelberg (2001). https://doi.org/10.1007/978-3-642-56755-1_8
8. Carlet, C., Mesnager, S.: Four decades of research on bent functions. Des. Codes Crypt. **78**(1), 5–50 (2016)
9. Carlet, C., Mesnager, S., Özbudak, F., Sınak, A.: Explicit characterizations for plateaued-ness of p-ary (vectorial) functions. In: El Hajji, S., Nitaj, A., Souidi, E.M. (eds.) C2SI 2017. LNCS, vol. 10194, pp. 328–345. Springer, Cham (2017). https://doi.org/10.1007/978-3-319-55589-8_22
10. Carlet, C., Prouff, E.: On plateaued functions and their constructions. In: Johansson, T. (ed.) FSE 2003. LNCS, vol. 2887, pp. 54–73. Springer, Heidelberg (2003). https://doi.org/10.1007/978-3-540-39887-5_6
11. Çeşmelioğlu, A., Meidl, W.: A construction of bent functions from plateaued functions. Des. Codes Crypt. **66**, 231–242 (2013)
12. Çesmelioglu, A., Meidl, W., Topuzoglu, A.: Partially bent functions and their properties. In: Applied Algebra and Number Theory. Cambridge University Press, Cambridge (2014)

13. Coulter, R.S., Matthews, R.W.: Bent polynomials over finite fields. Bull. Aust. Math. Soc. **56**(3), 429–437 (1997)
14. Dillon, J.F.: Elementary hadamard difference sets. Ph.D. thesis. University of Maryland (1974)
15. Hou, X.D.: q-ary bent functions constructed from chain rings. Finite Fields Appl. **4**(1), 55–61 (1998)
16. Hou, X.D.: p-ary and q-ary versions of certain results about bent functions and resilient functions. Finite Fields Appl. **10**(4), 566–582 (2004)
17. Hyun, J.Y., Lee, J., Lee, Y.: Explicit criteria for construction of plateaued functions. IEEE Trans. Inf. Theory **62**(12), 7555–7565 (2016)
18. Kumar, P.V., Scholtz, R.A., Welch, L.R.: Generalized bent functions and their properties. J. Comb. Theory Ser. A **40**(1), 90–107 (1985)
19. Langevin, P.: On generalized bent functions. In: Camion, P., Charpin, P., Harari, S. (eds.) Eurocode 92. ICMS, vol. 339, pp. 147–152. Springer, Vienna (1993). https://doi.org/10.1007/978-3-7091-2786-5_13
20. Lidl, R., Niederreiter, H.: Finite Fields, vol. 20. Cambridge University Press, Cambridge (1997)
21. Logachev, O.A., Salnikov, A., Yashchenko, V.V.: Bent functions on a finite Abelian group. Discrete Math. Appl. **7**(6), 547–564 (1997)
22. Mesnager, S.: Characterizations of plateaued and bent functions in characteristic p. In: Schmidt, K.-U., Winterhof, A. (eds.) SETA 2014. LNCS, vol. 8865, pp. 72–82. Springer, Cham (2014). https://doi.org/10.1007/978-3-319-12325-7_6
23. Mesnager, S.: Bent Functions: Fundamentals and Results. Springer, Heidelberg (2016). https://doi.org/10.1007/978-3-319-32595-8
24. Mesnager, S., Özbudak, F., Sınak, A.: Results on characterizations of plateaued functions in arbitrary characteristic. In: Pasalic, E., Knudsen, L.R. (eds.) Balkan-CryptSec 2015. LNCS, vol. 9540, pp. 17–30. Springer, Cham (2016). https://doi.org/10.1007/978-3-319-29172-7_2
25. Mesnager, S., Özbudak, F., Sınak, A.: On the p-ary (cubic) bent and plateaued (vectorial) functions. Des. Codes Crypt. **86**, 1865–1892 (2017)
26. Mesnager, S., Özbudak, F., Sınak, A., Cohen, G.: On q-ary plateaued functions over $\mathbb{F}_q$ and their explicit characterizations. Eur. J. Comb. (2018). Elsevier
27. Nyberg, K.: Constructions of bent functions and difference sets. In: Damgård, I.B. (ed.) EUROCRYPT 1990. LNCS, vol. 473, pp. 151–160. Springer, Heidelberg (1991). https://doi.org/10.1007/3-540-46877-3_13
28. Rothaus, O.S.: On "bent" functions. J. Comb. Theory Ser. A **20**(3), 300–305 (1976)
29. Rudin, W.: Principles of Mathematical Analysis, vol. 3. McGraw-Hill, New York (1964)
30. Wang, X., Zhou, J.: Generalized partially bent functions. In: Future Generation Communication and Networking (FGCN 2007), vol. 1, pp. 16–21. IEEE (2007)
31. Zheng, Y., Zhang, X.-M.: Plateaued functions. In: Varadharajan, V., Mu, Y. (eds.) ICICS 1999. LNCS, vol. 1726, pp. 284–300. Springer, Heidelberg (1999). https://doi.org/10.1007/978-3-540-47942-0_24

Codes of Length Two Correcting Single Errors of Limited Size II

Torleiv Kløve[(✉)]

Department of Informatics, University of Bergen, 5020 Bergen, Norway
Torleiv.Klove@uib.no

Abstract. Linear codes of length 2 over the integers modulo some integer q that can correct single errors of limited size are considered. A code can be determined by a check pair of integers. The errors e considered are in the range $-\mu \le e \le \lambda$, such a code can only exist for q sufficiently large. The main content of this note is to make this statement precise, that is, to determine "q sufficiently large" in terms of the integers $-\mu$ and λ.

Keywords: Error correcting code · Errors of limited size
Integers modulo n

1 Introduction

We consider linear codes that can correct unbalanced errors i.e. a symbol a over the alphabet $\mathbb{Z}_q = \{0, 1, \ldots, q-1\}$ may be modified during transmission into another symbol $b \in \mathbb{Z}_q$, where $-\mu \le b - a \le \lambda$, and $\mu \ge 0$ and $\lambda \ge 1$ are integers, see [10]. Without loss of generality, we may assume that $\mu \le \lambda$ (see [11]).

Codes for $\mu = 0$ have been considered e.g. in [3,4,7,8]. Codes for $\mu = \lambda$ have been considered e.g. in [3,5,9,10]. Codes for the general unbalanced case have been considered in [1,10,11]. A basic building block for many of these code constructions are sets which we have called $B[-\mu, \lambda](q)$ sets. They correspond to check vectors. In this note, we consider such sets of size two, corresponding to codes of length two.

We let $q_L(-\mu, \lambda)$ be the smallest integer q such that there exists a linear code in $\mathbb{Z}_q^2$ that can correct a single error from $[-\mu, \lambda]$.

In [6] we gave some observations and conjectures based on the values of $q_L(-\mu, \lambda)$ for small values of μ and λ.

In Sect. 2 we give some definitions and known results from [6]. In particular, we quote some upper bounds on $q_L(-\mu, \lambda)$ for $\mu < \lambda < 2\mu$.

In Sect. 3 we give some upper bounds on $q_L(-\mu, \lambda)$ for $\lambda > 2\mu$. This is the main result of this paper.

2 Definitions and Known Results

Let

$$[-\mu, \lambda] = \{-\mu, -\mu + 1, \ldots, \lambda - 1, \lambda\}$$

© Springer Nature Switzerland AG 2018
L. Budaghyan and F. Rodríguez-Henríquez (Eds.): WAIFI 2018, LNCS 11321, pp. 242–249, 2018.
https://doi.org/10.1007/978-3-030-05153-2_13

and
$$[-\mu, \lambda]^* = \{-\mu, -\mu+1, \ldots, -1\} \cup \{1, 2, \ldots, \lambda\}.$$

Let q be a positive integer. Let λ and μ be integers, where $0 \leq \mu \leq \lambda < q - \mu$. We consider the following channel: Our alphabet is $\mathbb{Z}_q$. An element $a \in \mathbb{Z}_q$ may be changed into $a + e$, where $e \in [-\mu, \lambda]$.

Consider a set $P = \{a, b\}$ where $a, b \in \mathbb{Z}_q$. The corresponding linear code of length 2 is
$$C = \{(x, y) \in \mathbb{Z}_q^2 \mid xa + yb = 0\}.$$

The size of the code is
$$|C| = dq \quad \text{where} \quad d = \gcd(a, b, q).$$

Let
$$S = \{up \mid u \in [-\mu, \lambda], p \in P\}.$$

This is the set of syndroms of C. Clearly, $|S| \leq 2(\mu + \lambda) + 1$. If $|S| = 2(\mu + \lambda) + 1$, then P is called a $B[-\mu, \lambda](q)$ set; the corresponding code C can correct a single error from $[-\mu, \lambda]$.

A number of constructions of $B[-\mu, \lambda](q)$ sets are known, in particular for $\mu = 0$ and for $\mu = \lambda$, see [1–10].

We can reformulate the definition of $B[-\mu, \lambda](q)$ sets of size 2 by specifying the conditions to check.

Definition 1. *A set $\{a, b\} \subseteq \mathbb{Z}_q$ is a $B[-\mu, \lambda](q)$ set if and only if*

$$xa \not\equiv 0 \,(mod\,q) \text{ for all } x \in [-\mu, \lambda]^*, \tag{1}$$
$$xa \not\equiv ya \,(mod\,q) \text{ for all } x, y \in [-\mu, \lambda]^*, x < y, \tag{2}$$
$$xb \not\equiv 0 \,(mod\,q) \text{ for all } x \in [-\mu, \lambda]^*, \tag{3}$$
$$xb \not\equiv yb \,(mod\,q) \text{ for all } x, y \in [-\mu, \lambda]^*, x < y, \tag{4}$$
$$and \ xa \not\equiv yb \,(mod\,q) \text{ for all } x, y \in [-\mu, \lambda]^*. \tag{5}$$

Definition 2. *Given μ and λ, $q_L(-\mu, \lambda)$ is the smallest q for which a $B[-\mu, \lambda](q)$ set of size two exists.*

In [8] we showed that $q_L(0, \lambda) = 2\lambda + 1$ and a corresponding $B[0, \lambda](q)$ set is $\{1, q - 1\}$. In [9], we showed that $q_L(-\lambda, \lambda) = (\lambda + 1)^2 + 1$ and a corresponding $B[-\lambda, \lambda](q)$ set is $\{1, \lambda + 1\}$.

Let
$$p_{-\mu, \lambda} = (\lambda + 1)^2 - (\lambda - \mu)^2 = (\mu + 1)(2\lambda + 1 - \mu).$$

We have shown the following results:

Theorem 1.(a) [6, Theorem 1]: $q_L(-\mu, \lambda) \geq p_{-\mu, \lambda}$ for all μ, λ.
(b) [6, Theorem 2]: $q_L(-\mu, \lambda) = p_{-\mu, \lambda}$ if $\gcd(\lambda + 1, \lambda - \mu) = 1$.

We have computed $q_L(-\mu, \lambda)$ by complete search for $0 \leq \mu < \lambda \leq 20$. For these values, we gave the following observations in [6]:

1. If $\gcd(\lambda + 1, \lambda - \mu) > 1$ and $\mu < \lambda < 2\mu$, then $q_L(-\mu, \lambda) = p_{-\mu,\lambda} + \lambda - \mu$.
2. If $\gcd(\lambda + 1, \lambda - \mu) > 1$ and $\lambda > 2\mu$, then $q_L(-\mu, \lambda) = p_{-\mu,\lambda} + \mu + 1$.

Possibly these expressions are true for all μ, λ.

Upper bounds are obtained by explicit constructions. For $\mu + 1 < \lambda < 2\mu$ we gave the following result.

Theorem 2. [6, Theorem 3]: *For all μ, λ such that $\mu + 1 < \lambda < 2\mu$, we have $q_L(-\mu, \lambda) \leq p_{-\mu,\lambda} + \lambda - \mu$. If $\gcd(\lambda + 1, \lambda - \mu) > 1$, then one $B[-\mu, \lambda](p_{-\mu,\lambda} + \lambda - \mu)$ set is $\{2\lambda - \mu, 2\lambda - \mu + 1\}$.*

The goal of the following paper is to give a similar result for $\lambda > 2\mu$.

Remark. We have a related channel for the integers: any $a \in [0, q - 1]$ can be changed to $b \in [0, q - 1$ where $-\mu \leq b - a \leq \lambda$. Error in flash memories can be modeled by this channel, see e.g. [2,10]. We see that codes correcting single errors over the channel defined over $\mathbb{Z}_q$ in particular corrects errors from $[0, q-1]$ in the corresponding channels over the integers.

3 Upper Bounds on $q_L(-\mu, \lambda)$ for $\lambda > 2\mu$

The main result in the present paper is the following upper bound:

Theorem 3. *If $\mu \geq 1$ and $\lambda > 2\mu$, then*

$$q_L(-\mu, \lambda) \leq p_{-\mu,\lambda} + \mu + 1 = (\mu + 1)(2\lambda + 2 - \mu).$$

In [6, Theorem 4] we proved this in a special case, namely when $\lambda + 1$ is multiple of $\mu + 1$. In that case, $\{1, 2\lambda + 1 - \mu\}$ is a $B[\mu, \lambda]((\mu + 1)(2\lambda + 2 - \mu))$ set.

To prove Theorem 3 in general, we treat μ even and μ odd separately. For both cases, we let

$$t = 2\lambda + 2 - \mu \text{ and } q = (\mu + 1)t.$$

Lemma 1. *If μ is even, $\lambda > 2\mu$, $a = 2\lambda + 1 - \mu$, and $b = a + 2$, then $\{a, b\}$ is a $B[\mu, \lambda](q)$ set for $q = (\mu + 1)(2\lambda + 2 - \mu)$.*

Proof: We check (1)–(5) in Definition 1. We have

$$a = t - 1 \text{ and } b = t + 1.$$

Hence, we clearly get the following relations:

If $x \in [-\mu, -1]$, then $xb \pmod{t} = t + x$ and $xa \pmod{t} = -x$.
If $x \in [1, \lambda]$, then $xb \pmod{t} = x$ and $xa \pmod{t} = t - x$.

We see that $xa \pmod{t} \neq 0$. In particular, $xa \pmod{q} \neq 0$. Hence (1) is satisfied.

Since $\lambda + \mu < t$, we see that if $x, y \in [-\mu, \lambda]$ and $x < y$, then $xa \not\equiv ya \pmod{t}$. In particular, $xa \not\equiv ya \pmod{q}$, that is, (2) is satisfied.

Similarly, (3) and (4) are satisfied.

Finally, suppose that $x, y \in [-\mu, \lambda]$ and $yb \equiv xa \pmod{q}$. Then $y \equiv -x \pmod{t}$ and so $y = -x$ and so

$$x, y \in [-\mu, \mu]^*. \tag{6}$$

Hence

$$2xt = x(a + b) = xa - yb \equiv 0 \pmod{(\mu + 1)t}$$

and so

$$2x \equiv 0 \pmod{(\mu + 1)}.$$

Since $\mu + 1$ is odd, this implies that $x \equiv 0 \pmod{(\mu + 1)}$, but this contradicts (6). Hence, (5) is satisfied. □

We give a closer look at the code determined by the $B[\mu, \lambda](q)$ set in Lemma 1, that is

$$C = \{(x, y) \mid (2\lambda + 1 - \mu)x + (2\lambda + 3 - \mu)y \equiv 0 \pmod{q}\}$$

where $q = (\mu + 1)(2\lambda + 2 - \mu)$. We remark that $\gcd(\lambda + 1, \lambda - \mu) = \gcd(\lambda + 1, \mu + 1)$.

Lemma 2. *If μ is even, then*

(a) $\gcd(2\lambda + 1 - \mu, q) = \gcd(\lambda + 1, \mu + 1)$,
(b) $\gcd(2\lambda + 3 - \mu, q) = \gcd(\lambda + 1, \mu + 2)$.

Proof: (a) Let $a = 2\lambda + 1 - \mu$. We observe that $q = (\mu + 1)(a + 1)$. Hence

$$\gcd(a, q) = \gcd(a, (\mu + 1)a + \mu + 1) = \gcd(a, \mu + 1)$$
$$= \gcd(2\lambda + 2 - (\mu + 1), \mu + 1) = \gcd(2(\lambda + 1), \mu + 1) = \gcd(\lambda + 1, \mu + 1)$$

since $\mu + 1$ is odd.

The proof of (b) is similar since $b = a + 2$:

$$\gcd(b, q) = \gcd(a + 2, (\mu + 1)(a + 2) - \mu - 1) = \gcd(a + 2, -\mu - 1)$$
$$= \gcd(a + 2, \mu + 1) = \gcd(2\lambda + 4 - (\mu + 1), \mu + 1)$$
$$= \gcd(2(\lambda + 2), \mu + 1) = \gcd(\lambda + 2, \mu + 1).$$

□

Since we $q_L(-\mu, \lambda) = p_{-\mu, \lambda}$ for $\gcd(\lambda + 1, \mu + 1) = 1$, the construction in Lemma 1 is mainly of interest when $\gcd(\lambda + 1, \mu + 1) > 1$. By Lemma 2, a does not have an inverse modulo q and b has an inverse exactly when $\gcd(\lambda + 1, \mu + 2) = 1$.

A simple count shows that there are 124500 pair (λ, μ) such that μ is even and $0 < 2\mu < \lambda \leq 1000$. For 100615 of these pair, we have $\gcd(\lambda + 1, \mu + 1) = 1$. For the remaining 23885 pairs, we have $\gcd(\lambda + 1, \mu + 2) = 1$ in 10392 cases.

Example 1. The smallest example for which μ is even, $\gcd(\lambda+1, \mu+1) > 1$, and $\gcd(b, q) = 1$, is $\mu = 2$ and $\lambda = 5$. Consider the construction in Lemma 1. We have $a = 9$, $b = 11$, $q = 30$. Since $-9 \cdot 11^{-1} \equiv 21 \pmod{30}$, $9x + 11y \equiv 0 \pmod{30}$ is equivalent to $y \equiv 21x \pmod{30}$. Hence, the code is

$$C = \{(x, 21x) \mid x \in \mathbb{Z}_{30}\}.$$

The simplest corresponding encoding is, of course, $z \mapsto (z, 21z)$.

For $(x, y) \in \mathbb{Z}_{30}^2$, the corresponding syndrom is $9x + 11y$. For $(x, y) \in C$ and $e \in [-2, 5]$, the syndrom corresponding to the error $(e, 0)$ is

$$9(x + e) + 11y = 9x + 11y + 9e = 9e$$

and the syndrom corresponding to the error $(0, e)$ is $11e$. We give the values of the syndroms in the following table.

e	-2	-1	1	2	3	4	5
$9e$	12	21	9	18	27	6	15
$11e$	8	19	11	22	3	14	25

They are all distinct, that is, the set $\{9, 11\}$ is indeed a $B[-2, 5](30)$ set.

Example 2. The smallest example for which μ is even, $\gcd(\lambda+1, \mu+1) > 1$ and $\gcd(b, q) > 1$, is $\mu = 14$ and $\lambda = 34$. Here $q = 840$, $a = 55$, $b = 57$. We see that $\gcd(a, q) = 5$ and $\gcd(b, q) = 3$. From $55x + 57y \equiv 0 \pmod{840}$ we get $y \equiv 0 \pmod 5$, that is $y = 5y_1$, and similarly we get $x = 3x_1$. Hence

$$5 \cdot 11 \cdot 3x_1 + 3 \cdot 19 \cdot 5y_1 \equiv 0 \pmod{3 \cdot 5 \cdot 56},$$

that is,

$$11x_1 + 19y_1 \equiv 0 \pmod{56}$$

and so

$$x_1 \equiv 39y_1 \pmod{56}.$$

Therefore,

$$C = \{(3(x_1 + 56\alpha), 5y_1) \mid y_1 \in [0, 167], x_1 \in [0, 55], \alpha \in [0, 4], x_1 \equiv 39y_1 \pmod{56}\}.$$

Let $z \in [0, 839]$. Then $z = 168\alpha + \beta$ where $\alpha \in [0, 4]$ and $\beta \in [0, 167]$. Let $\gamma \in [0, 55]$ be defined by $\gamma \equiv 39\beta \pmod{56}$. Then, by the expressions above, we see that a possible encoding is the following:

$$z \mapsto (3(\gamma + 56\alpha), 5\beta).$$

The set of syndroms corresponding to single non-zero errors are

$$\bigcup_{e \in [-14, 34]^*} \{55e, 57e\}.$$

The values of the syndroms are given in the following table.

e	-14	-13	-12	-11	-10	-9	-8	-7	-6	-5	-4	-3	-2	-1
$55e$	70	125	180	235	290	345	400	455	510	565	620	675	730	785
$57e$	42	99	156	213	270	327	384	441	498	555	612	669	726	783

e	1	2	3	4	5	6	7	8	9	10	11	12	13	14	15	16	17
$55e$	55	110	165	220	275	330	385	440	495	550	605	660	715	770	825	40	95
$57e$	57	114	171	228	285	342	399	456	513	570	627	684	741	798	15	72	129

e	18	19	20	21	22	23	24	25	26	27	28	29	30	31	32	33	34
$55e$	150	205	260	315	370	425	480	535	590	645	700	755	810	25	80	135	190
$57e$	186	243	300	357	414	471	528	585	642	699	756	813	30	87	144	201	258

For μ odd we find a similar, but more complicated, construction.

Lemma 3. *If $\mu = 2\nu + 1$ is odd, $\lambda > 2\mu$, $a = \mu\lambda - \theta$ where $\theta = 2\nu^2$, and $b = a + 1$, then $\{a, b\}$ is a $B[\mu, \lambda](q)$ set.*

Proof: First we note that

$$2a = 2\mu\lambda - (\mu - 1)^2 = \mu(2\lambda - \mu + 2) - 1 = \mu t - 1, \tag{7}$$
$$2b = 2a + 2 = \mu t + 1. \tag{8}$$

Further

$$a + b = 2a + 1 = \mu t \equiv 0 \,(\text{mod}\, t). \tag{9}$$

Hence,

$$2a \equiv t - 1 = 2\lambda + 2 - \mu - 1 = 2\lambda + 2 - 2\nu - 1 - 1 = 2(\lambda - \nu)\,(\text{mod}\, t)$$

and so, since t is odd, we get

$$a \equiv \lambda - \nu \,(\text{mod}\, t). \tag{10}$$

Let $\ell = \lfloor \lambda/2 \rfloor$. From (7) and (10) we get the following relations:

If $x \in [-\nu, -1]$,	then $2xa \,(\text{mod}\, t)$	$= -x$.
If $x \in [-\nu - 1, -1]$,	then $(2x + 1)a \,(\text{mod}\, t)$	$= \lambda - \nu - x$.
If $x \in [1, \ell]$,	then $2xa \,(\text{mod}\, t)$	$= t - x$.
If $x \in [0, \ell]$,	then $(2x + 1)a \,(\text{mod}\, t)$	$= \lambda - \nu - x$.

Hence, (1) is satisfied.
Further,

$\{2xa \,(\text{mod}\, t) \mid x \in [-\nu, -1]\}$	$= [1, \nu]$
$\{(2x + 1)a \,(\text{mod}\, t) \mid x \in [0, \ell]\}$	$= [\lambda - \nu - \ell, \lambda - \nu]$
$\{(2x + 1)a \,(\text{mod}\, t) \mid x \in [-\nu - 1, -1]\}$	$= [\lambda - \nu + 1, \lambda + 1]$
$\{2xa \,(\text{mod}\, t) \mid x \in [1, \ell]\}$	$= [t - \ell, t - 1]$

We have

$$(t - \ell) - (\lambda + 1) = (\lambda - \nu - \ell) - \nu = \lambda - \ell - 2\nu,$$

and

$$2(\lambda - \ell - 2\nu) = 2\lambda - 2\ell - 2(\mu - 1)$$
$$= (\lambda - 2\ell) + (\lambda - 2\mu - 1) + 3 \geq 1 + 0 + 3 > 0.$$

Hence, we see that if $x, y \in [-\mu, \lambda]$ and $x < y$, then $xa \not\equiv ya \pmod{t}$. In particular, $xa \not\equiv ya \pmod{q}$. Hence (2) is satisfied.

From (9) we get $b \equiv -a \pmod{t}$. Hence, (3) is satisfied. Further we see that if $x, y \in [-\mu, \lambda]$ and $x < y$, then $xb \not\equiv yb \pmod{t}$. In particular, $xb \not\equiv yb \pmod{q}$. Hence (4) is satisfied.

Finally, if $x, y \in [-\mu, \lambda]$ and $xa \equiv yb \pmod{q}$, then $xa \equiv yb \equiv -ya \pmod{t}$. Since t is odd, we have

$$\gcd(a, t) = \gcd(2a, t) = \gcd(2\mu t - 1, t) = 1.$$

Hence $x \equiv -y \pmod{t}$ and so $x = -y$. Therefore,

$$x, y \in [-\mu, \mu]^*. \tag{11}$$

Further, we get $xa \equiv -xb \pmod{q}$ and so $x(a + b) \equiv 0 \pmod{q}$. Hence

$$x\mu t \equiv 0 \pmod{(\mu + 1)t}.$$

Therefore,

$$x\mu \equiv 0 \pmod{\mu + 1}$$

and so

$$x \equiv 0 \pmod{\mu + 1}$$

which is impossible by (11). Hence (5) is satisfied. $\qquad\square$

We now take a closer look at the code determined by the $B[\mu, \lambda](q)$ set in Lemma 3, that is

$$C = \{(x, y) \mid ax + by \equiv 0 \pmod{q}\}$$

where $q = (\mu + 1)(2\lambda + 2 - \mu)$, $\theta = (\mu - 1)^2/2$, $a = \mu\lambda - \theta$, and $b = \mu\lambda - \theta + 1$.

Lemma 4: *If μ is odd, then we have*

(a) $\gcd(b, q) = \gcd(\lambda + 1, \mu + 1)$,
(b) $\gcd(a, q) = \gcd(\lambda + 2, \mu + 1)$.

Proof: (a) Let

$$\beta_1 = q - 2b = 2\lambda - \mu^2 + \mu + 2\theta,$$
$$\beta_2 = b - \frac{\mu - 1}{2}\beta_1 = \lambda + 1,$$
$$\beta_3 = 2\beta_2 - \beta_1 = \mu + 1.$$

Then $\gcd(b, q) = \gcd(\beta_2, \beta_3) = \gcd(\lambda + 1, \mu + 1)$. This proves (a).

(b) Let

$$\alpha_1 = q - 2a = 2\lambda - \mu + 3,$$
$$\alpha_2 = a - \frac{\mu - 1}{2}\alpha_1 = \lambda + 1 - \mu,$$
$$\alpha_3 = \alpha_1 - 2\alpha_2 = \mu + 1.$$

Then $\gcd(a, q) = \gcd(\alpha_2, \alpha_3) = \gcd(\alpha_2 + \alpha_3, \alpha_3) = \gcd(\lambda + 2, \mu + 1)$. This proves (b).

$\square$

The smallest example for which μ is odd, $\gcd(\lambda+1, \mu+1) > 1$, and $\gcd(a, q) = 1$ is $\mu = 3$ and $\lambda = 7$.

The smallest example for which μ is odd, $\gcd(\lambda+1, \mu+1) > 1$, and $\gcd(a, q) > 1$ is $\mu = 5$ and $\lambda = 13$.

References

1. Battaglioni, M., Chiaraluce, F., Kløve, T.: On non-linear codes correcting errors of limited size. In: Proceedings of the Globecom, Singapore, 4–8 December 2017, pp. 1–7. IEEE Xplore (2017)
2. Dolecek, L., Cassuto, Y.: Channel coding for nonvolatile memory technologies: theoretical advances and practical considerations. Proc. IEEE **105**(9), 1705–1724 (2017)
3. Elarief, N., Bose, B.: Optimal, systematic, q-ary codes correcting all asymmetric and symmetric errors of limited magnitude. IEEE Trans. Inf. Theory **56**, 979–983 (2010)
4. Jiang, A., Mateescu, R., Schwartz, M., Bruck, J.: Rank modulation for flash memories. IEEE Trans. Inf. Theory **55**, 2659–2673 (2009)
5. Kløve, T.: Codes of length 2 correcting single errors of limited size. In: Groth, J. (ed.) IMACC 2015. LNCS, vol. 9496, pp. 190–201. Springer, Cham (2015). https://doi.org/10.1007/978-3-319-27239-9_12
6. Kløve, T.: Codes of length two correcting single errors of limited size. In: Cryptography and Communications (to appear)
7. Kløve, T., Elarief, N., Bose, B.: Systematic, single limited magnitude error correcting codes for flash memories. IEEE Trans. Inf. Theory **57**, 4477–4487 (2011)
8. Kløve, T., Luo, J., Naydenova, I., Yari, S.: Some codes correcting asymmetric errors of limited magnitude. IEEE Trans. Inf. Theory **57**, 7459–7472 (2011)
9. Kløve, T., Luo, J., Yari, S.: Codes correcting single errors of limited magnitude. IEEE Trans. Inf. Theory **58**, 2206–2219 (2012)
10. Schwartz, M.: Quasi-cross lattice tilings with applications to flash memory. IEEE Trans. Inf. Theory **58**, 2397–2405 (2012)
11. Yari, S., Kløve, T., Bose, B.: Some linear codes correcting single errors of limited magnitude for flash memories. IEEE Trans. Inf. Theory **59**, 7278–7287 (2013)

Fractional Jumps: Complete Characterisation and an Explicit Infinite Family

Federico Amadio Guidi and Giacomo Micheli[✉]

Mathematical Institute, University of Oxford, Oxford, UK
{federico.amadio,giacomo.micheli}@maths.ox.ac.uk

Abstract. In this paper we provide a complete characterisation of transitive fractional jumps. In particular, we prove that they can only arise from transitive projective automorphisms apart from a couple of degenerate cases which we entirely classify. Furthermore, we prove that such construction is feasible for arbitrarily large dimension by exhibiting an infinite class of projectively primitive polynomials whose companion matrix can be used to define a full orbit sequence over an affine space.

1 Introduction

The study of dynamical systems over finite fields have a long history (see for example [2,4–6,9,12,13,18]) and is an interesting and still hot topic (see for example [7,8,10,14–17,19]), both for its number theoretical impact in finite fields theory, and for its practical applications, in particular for random number generation.

Let q be a prime power, let $\mathbb{F}_q$ denote the finite field with q elements, and let m be a positive integer. One of the most interesting questions for applications consists of constructing sequences over the m-dimensional affine space over $\mathbb{F}_q$ defined by iterations of rational maps $f : \mathbb{F}_q^m \to \mathbb{F}_q^m$ satisfying the following conditions:

1. The period of the recursive sequence $\{f^k(0)\}_{k\in\mathbb{N}}$ they define is "long".
2. Their iterations as rational maps have "low degree growth".

The motivation for (1) is rather clear: since we generally want to use these sequences for pseudorandom number generation, we do not want to revisit an element twice too soon, or otherwise the entire sequence will repeat. The motivation for (2) is a little more subtle and comes from the uniformity conditions we want the sequence to satisfy (for additional information on this see [16]).

In [1] we introduced the theory of fractional jumps to address this problem by showing a natural way to build full orbit sequences from projective automorphisms, recovering as a particular case the construction of the Inversive Congruential Generator.

L. Budaghyan and F. Rodríguez-Henríquez (Eds.): WAIFI 2018, LNCS 11321, pp. 250–263, 2018.
https://doi.org/10.1007/978-3-030-05153-2_14

In this paper we complete the theory of fractional jumps by both proving the uniqueness of the construction, i.e. transitive fractional jumps can only arise from transitive projective automorphisms (except from a couple of degenerate cases which we entirely classify), and by providing an explicit infinite class of projectively primitive polynomials, see definition [1, Definition 3.1], whose companion matrix can be used to define a full orbit sequence over $\mathbb{F}_p^{p-1}$, for p a prime. For this family of fractional jumps, which we call *Artin-Schreier fractional jumps*, we show that the computation of the $(k+1)$-th affine point of the full orbit sequence they define, given the k-th one, is as expensive as reading out a look-up table once for each entry.

This latter construction entirely addresses points (1) and (2) above, since the corresponding sequences have full orbit (they cover the entire affine space) and they have zero degree growth. The main technique we use is the fractional jump construction provided in [1].

1.1 Notation

We denote by $\mathbb{N}$ the set of natural numbers, and by $\mathbb{Z}$ the set of integers. Given $a \in \mathbb{Z}$, we let $\mathbb{Z}_{\geq a}$ denote the set of integers $k \in \mathbb{Z}$ such that $k \geq a$.

Given a commutative ring with unity R, we let R^* be the (multiplicative) group of invertible elements in R.

For a prime power q, we denote by $\mathbb{F}_q$ the finite field of cardinality q. For $m \in \mathbb{N}$, we denote the m-dimensional affine space $\mathbb{F}_q^m$ by $\mathbb{A}^m$, and the m-dimensional projective space over $\mathbb{F}_q$ by $\mathbb{P}^m$. More generally, for any vector space V over $\mathbb{F}_q$ we denote by $\mathbb{P}V$ the projectivisation of V. Also, we denote by $\mathbb{F}_q[x_1, \ldots, x_m]$ the ring of polynomials in m variables with coefficients in $\mathbb{F}_q$.

For $m \in \mathbb{N}$, let us denote by $\mathrm{GL}_m(\mathbb{F}_q)$ the general linear group over $\mathbb{F}_q$, that is the group of $m \times m$ invertible matrices with entries in $\mathbb{F}_q$. Also, we denote by $\mathrm{PGL}_m(\mathbb{F}_q)$ the group of automorphisms of $\mathbb{P}^{m-1}$. Recall that $\mathrm{PGL}_m(\mathbb{F}_q)$ can be identified with the quotient group $\mathrm{GL}_m(\mathbb{F}_q)/\mathbb{F}_q^*\mathrm{Id}_m$, where $\mathbb{F}_q^*\mathrm{Id}$ is the subgroup of $\mathbb{F}_q^*$-multiples of the identity matrix Id_m. For $M \in \mathrm{GL}_m(\mathbb{F}_q)$, we denote by $[M]$ its class in $\mathrm{PGL}_m(\mathbb{F}_q)$.

We say that a polynomial $\chi(T) \in \mathbb{F}_q[T]$ of degree $\deg \chi(T) = d$ is *projectively primitive* if it is irreducible and if given any root α in $\mathbb{F}_{q^d} \cong \mathbb{F}_q[T]/(\chi(T))$ the class $\overline{\alpha}$ of α in the quotient group $G = \mathbb{F}_{q^d}^*/\mathbb{F}_q^*$ generates G.

Let X be a set, and let G be a group acting on it. For any $x \in X$ we denote by $\mathcal{O}_G(x)$ the orbit of x with respect to the action of G on X. Given a bijective map $f : X \to X$, for any $x \in X$ we set $\mathcal{O}_f(x) = \mathcal{O}_{\langle f \rangle}(x)$, where $\langle f \rangle$ denotes the cyclic subgroup of the group of maps from X to itself generated by f, and we define $o_f(x) = |\mathcal{O}_f(x)|$. We say that a bijective map $f : X \to X$ *acts transitively* on X, or simply that it is *transitive*, if for any $x, y \in X$ there exists $k \in \mathbb{Z}$ such that $y = f^k(x)$. Equivalently, f acts transitively on X if and only if for any $x_0 \in X$, the f-orbit of x_0 has size $o_f(x_0) = |X|$. Finally, we say that a sequence $\{x_k\}_{k \in \mathbb{N}}$ in X has *full orbit* if $\{x_k : k \in \mathbb{N}\} = X$.

2   Transitive Fractional Jumps

For the sake of completeness, we recall the definition of fractional jump of a projective automorphism, as introduced in [1].

Fix the standard projective coordinates $X_0, \ldots, X_n$ on $\mathbb{P}^n$, and fix the canonical decomposition

$$\mathbb{P}^n = U \cup H,$$

where

$$U = \{[X_0 : \ldots : X_n] \in \mathbb{P}^n \ : \ X_n \neq 0\},$$
$$H = \{[X_0 : \ldots : X_n] \in \mathbb{P}^n \ : \ X_n = 0\}.$$

Fix also the isomorphism

$$\pi : \mathbb{A}^n \xrightarrow{\sim} U, \quad (x_1, \ldots, x_n) \mapsto [x_1 : \ldots : x_n : 1].$$

Let now Ψ be an automorphism of $\mathbb{P}^n$. For $P \in U$, we define the *fractional jump index* of Ψ at P as

$$\mathfrak{J}_P = \min \left\{ k \in \mathbb{Z}_{\geq 1} \ : \ \Psi^k(P) \in U \right\}.$$

The *fractional jump* of Ψ is then defined as the map

$$\psi : \mathbb{A}^n \to \mathbb{A}^n, \quad x \mapsto \pi^{-1} \Psi^{\mathfrak{J}_{\pi(x)}} \pi(x).$$

Essentially, the map ψ is defined on a point $x \in \mathbb{A}^n$ as follows: we firstly send x in $\mathbb{P}^n$ via the canonical map π, then we iterate Ψ on $\pi(x)$ until we end up with a point in U, and finally we take its image in $\mathbb{A}^n$ via π^{-1}.

When Ψ acts transitively on $\mathbb{P}^n$, its fractional jump ψ admits an explicit description in terms of multivariate linear fractional transformations. More precisely, we have the following:

Theorem 1 ([1, Sect. 5]). *Let Ψ be a transitive automorphism of $\mathbb{P}^n$, and let ψ be its fractional jump. Then, for $i \in \{1, \ldots, n+1\}$ there exist*

$$a_1^{(i)}, \ldots, a_n^{(i)}, b^{(i)} \in \mathbb{F}_q[x_1, \ldots, x_n]$$

of degree 1 such that, if

$$U_1 = \left\{ x \in \mathbb{A}^n \ : \ b^{(1)}(x) \neq 0 \right\},$$

$$U_i = \left\{ x \in \mathbb{A}^n \ : \ b^{(i)}(x) \neq 0, \ and \ b^{(j)}(x) = 0, \ \forall j \in \{1, \ldots, i-1\} \right\},$$

$$for \ i \in \{2, \ldots, n+1\},$$

and

$$f^{(i)} = \left(\frac{a_1^{(i)}}{b^{(i)}}, \ldots, \frac{a_n^{(i)}}{b^{(i)}} \right),$$

$$for \ i \in \{1, \ldots, n+1\},$$

then $\psi(x) = f^{(i)}(x)$ if $x \in U_i$. Moreover, the rational maps $f^{(i)}$ can be explicitly computed.

Proof (sketch). Let us denote by K the field $\mathbb{F}_q(x_1, \ldots, x_n)$ of rational functions on $\mathbb{A}^n$. We construct a map

$$\imath : \mathrm{PGL}_{n+1}(\mathbb{F}_q) \to K^n$$

in the following way. Let $\Phi \in \mathrm{PGL}_{n+1}(\mathbb{F}_q)$, and write

$$\Phi = [F_0 : \ldots : F_n],$$

for $F_0, \ldots, F_n \in \mathbb{F}_q[X_0, \ldots, X_n]$ homogeneous polynomials of degree 1. Define then $\imath(\Phi) \in K^n$ to be the n-tuple of elements of K whose j-th entry for $j \in \{1, \ldots, n\}$ is given by

$$\imath(\Phi)_j = \frac{F_{j-1}(x_1, \ldots, x_n, 1)}{F_n(x_1, \ldots, x_n, 1)}.$$

It is immediate to check that $\imath$ is well defined, that for any $f = (f_1, \ldots, f_n)$ in the image of $\imath$ all the f_j's are rational functions of degree 1, whose denominators are all equal up to a non-zero constants, and that $\imath(\Phi_1 \circ \Phi_2) = \imath(\Phi_1) \circ \imath(\Phi_2)$, where $\imath(\Phi_1) \circ \imath(\Phi_2)$ is simply defined by plugging in the components of $\imath(\Phi_2)$ into the variables of $\imath(\Phi_1)$.

Let now $\Psi \in \mathrm{PGL}_{n+1}(\mathbb{F}_q)$ be transitive. Define $f^{(i)} = \imath(\Psi^i)$ for $i \in \mathbb{Z}_{\geq 1}$. Then, by construction for any $i \in \mathbb{Z}_{\geq 1}$ there exist $a_1^{(i)}, \ldots, a_n^{(i)}, b^{(i)} \in \mathbb{F}_q[x_1, \ldots, x_n]$ of degree 1 such that

$$f^{(i)} = \left(\frac{a_1^{(i)}}{b^{(i)}}, \ldots, \frac{a_n^{(i)}}{b^{(i)}} \right).$$

It can be proved, see [1, Sect. 5] for the details, that the transitivity of Ψ implies that

$$\bigcap_{i=1}^{n+2} \left\{ x \in \mathbb{A}^n : b^{(i)}(x) = 0 \right\} = \emptyset. \tag{2.1}$$

Define then

$$U_1 = \left\{ x \in \mathbb{A}^n : b^{(1)}(x) \neq 0 \right\},$$

$$U_i = \left\{ x \in \mathbb{A}^n : b^{(i)}(x) \neq 0, \text{ and } b^{(j)}(x) = 0, \forall j \in \{1, \ldots, i-1\} \right\},$$

$$\text{for } i \in \{2, \ldots, n+1\}.$$

By (2.1) we have that $\{U_i\}_{i \in \{1, \ldots, n+1\}}$ is a disjoint covering of $\mathbb{A}^n$. Also, we clearly have that $\psi(x) = f^{(i)}(x)$ if $x \in U_i$.

Remark 1. The reader should notice that the $b^{(i)}$ are equal on each component, and therefore the evaluation of ψ only requires one inversion in the base field.

Remark 2. Another important fact to notice is that the definition of ψ depends uniquely on the rows of M^i's, where $M \in \mathrm{GL}_{n+1}(\mathbb{F}_q)$ is any matrix in the class of Ψ. In fact, notice that if the last row of M^i is $(m^{(i)}_{n+1,1}, \ldots, m^{(i)}_{n+1,n+1})$, then $b^{(i)} = m^{(i)}_{n+1,n+1} + \sum_{j=1}^{n} m^{(i)}_{n+1,j} x_j$. On the other hand, for any $j \in \{1, \ldots, n\}$, if $(m^{(i)}_{j,1}, \ldots, m^{(i)}_{j,n+1})$ is the j-th row of M^i, then $a^{(i)}_j = m^{(i)}_{j,n+1} + \sum_{j=1}^{n} m^{(i)}_{j,n+1} x_j$. What is done here is essentially dehomogenising the projective map induced by the class of M^i and then restricting that to the affine points.

We now provide a simple example to fix the ideas.

Example 1. Let $q = 5$ and $n = 2$. Consider the automorphism of $\mathbb{P}^2$ defined by

$$\Psi([X_0 : X_1 : X_2]) = [3X_0 + 2X_1 + X_2 : 3X_0 + 3X_1 + X_2 : 3X_1 + 4X_2].$$

A representative for Ψ in $\mathrm{GL}_3(\mathbb{F}_5)$ is given by

$$M = \begin{pmatrix} 3 & 2 & 1 \\ 3 & 3 & 1 \\ 0 & 3 & 4 \end{pmatrix},$$

whose characteristic polynomial

$$\chi_M(T) = T^3 + 4T + 3$$

is projectively primitive, since it is irreducible, and $(5^3 - 1)/(5 - 1) = 31$ is prime. By [1, Theorem 3.4], it follows that Ψ acts transitively on $\mathbb{P}^2$, and then Theorem 1 applies to the fractional jump ψ of Ψ. Direct computations show that for

$$U_1 = \left\{(x_1, x_2) \in \mathbb{A}^2 : 3x_2 + 4 \neq 0\right\},$$
$$U_2 = \left\{(x_1, x_2) \in \mathbb{A}^2 : 3x_2 + 4 = 0, \text{ and } 4x_1 + x_2 + 4 \neq 0\right\},$$
$$U_3 = \left\{(1, 2)\right\},$$

and

$$f^{(1)}(x_1, x_2) = \left(\frac{3x_1 + 2x_2 + 1}{3x_2 + 4}, \frac{3x_1 + 3x_2 + 1}{3x_2 + 4}\right),$$

$$f^{(2)}(x_1, x_2) = \left(\frac{4}{4x_1 + x_2 + 4}, \frac{3x_1 + 3x_2}{4x_1 + x_2 + 4}\right),$$

$$f^{(3)}(x_1, x_2) = \left(\frac{2x_2 + 1}{3x_2 + 1}, \frac{3x_1 + 1}{3x_2 + 1}\right),$$

we have that $\{U_i\}_{i \in \{1,2,3\}}$ is a disjoint covering of $\mathbb{A}^2$ such that $\psi(x) = f^{(i)}(x)$ if $x \in U_i$.

The purpose of this section is to show that transitive fractional jumps can only arise from transitive projective automorphisms, except from some very special cases, which can be entirely classified. Before proving the main theorem, let us recall a standard linear algebra fact, which follows from the results in [11, XIV, Sects. 2 and 3].

Lemma 1. *Let* $\Bbbk$ *be a field, let* V *be a finite dimensional vector space over* $\Bbbk$, *and let* M *be a* $\Bbbk$*-linear endomorphism of* V. *Assume that the minimal polynomial and the characteristic polynomial of* M *are equal. Then, there exists* $v_0 \in V$ *such that the set* $\left\{ M^k v_0 : k \in \mathbb{Z}_{\geq 0} \right\}$ *spans* V *over* $\Bbbk$.

We also need the following lemma:

Lemma 2. *Let* $p(T) \in \mathbb{F}_q[T]$ *be an irreducible polynomial, and let* $e \geq 1$ *be a positive integer. Let* $[T]$ *be the class of* T *in* $\Gamma = (\mathbb{F}_q[T]/(p(T)^e))^*$, *and let* $[[T]]$ *be the class of* T *in* $G = \Gamma/\mathbb{F}_q^*$. *Then, the order of* $[[T]]$ *in* G *equals the order of* $[T]^{q-1}$ *in* Γ.

Proof. Let k be the order of $[[T]]$ in G and let h be the order of $[T]^{q-1}$ in Γ. Then, $[[T]]^k = 1$ in G gives $[T]^k \in \mathbb{F}_q^*$. But then $1 = ([T]^k)^{q-1} = ([T]^{q-1})^k$, and so $h \mid k$.

On the other hand, let us firstly show that if $s \in \mathbb{F}_q[T]/(p(T)^e)$ satisfies $s^{q-1} - 1 = 0$, then $s \in \mathbb{F}_q^*$. In fact, by reducing s modulo $p(T)$ we get that

$$s = c + k(T)p(T) \mod p(T)^e, \quad \text{for } c \in \mathbb{F}_q^* \text{ and } k(T) \in \mathbb{F}_q[T].$$

Now, by multiplying the equation $s^{q-1} - 1 = 0$ by s, and plugging in the above special form for s, we get

$$(c + k(T)p(T))^q - (c + k(T)p(T)) \equiv (k(T)p(T))^q - k(T)p(T)$$

$$\equiv k(T)p(T)((k(T)p(T))^{q-1} - 1) \equiv 0 \mod p(T)^e.$$

But now $k(T)p(T))^{q-1} - 1$ is invertible modulo $p(T)^e$, and so $k(T)p(T)$ must be zero modulo $p(T)^e$, which forces s to be c modulo $p(T)^e$.

It then follows that $1 = ([T]^{q-1})^h = ([T]^h)^{q-1}$ in Γ gives $[T]^h \in \mathbb{F}_q^*$, from which we get $[[T]]^h = 1$ in G, and so $k \mid h$.

The main result of this section is the following:

Theorem 2. *Let* Ψ *be an automorphism of* $\mathbb{P}^n$ *and let* ψ *be its fractional jump. Then,* Ψ *acts transitively on* $\mathbb{P}^n$ *if and only if* ψ *acts transitively on* $\mathbb{A}^n$, *unless* q *is prime and* $n = 1$, *or* $q = 2$ *and* $n = 2$, *with explicit examples in both cases.*

Proof. For any q and n, it is immediate to show that if Ψ is transitive then ψ is transitive. In the case of q prime and $n = 1$ or $q = 2$ and $n = 2$ there exist explicit examples of transitive affine transformations, namely

$$\varphi_1(x) = x + 1, \qquad\qquad \text{if } q \text{ is prime and } n = 1,$$

$$\varphi_2(x_1, x_2) = \begin{pmatrix} 1 & 1 \\ 0 & 1 \end{pmatrix} \cdot \begin{pmatrix} x_1 \\ x_2 \end{pmatrix} + \begin{pmatrix} 1 \\ 1 \end{pmatrix}, \qquad \text{if } q = 2 \text{ and } n = 2.$$

Define then

$$\Phi_1([X_0 : X_1]) = [X_0 + X_1 : X_1], \qquad \text{if } q \text{ is prime and } n = 1,$$

$$\Phi_2([X_0 : X_1 : X_2]) = [X_0 + X_1 + X_2 : X_1 + X_2 : X_2], \qquad \text{if } q = 2 \text{ and } n = 2.$$

Clearly, φ_i is the fractional jump of Φ_i for $i \in \{1, 2\}$. However, it is immediate to see that Φ_i fixes the hyperplane at infinity, so cannot be transitive for $i \in \{1, 2\}$.

Let us now assume that we are not in the above pathological cases, and that ψ is transitive. Write $\Psi = [M] \in \mathrm{PGL}_{n+1}(\mathbb{F}_q)$ for some $M \in \mathrm{GL}_{n+1}(\mathbb{F}_q)$, and let $\chi_M(T), \mu_M(T) \in \mathbb{F}_q[T]$ be respectively the characteristic polynomial and the minimal polynomial of M. The vector space $V = \mathbb{F}_q^{n+1}$ over $\mathbb{F}_q$ has a natural structure of $\mathbb{F}_q[T]$-module given by

$$f(T)v = f(M)v, \quad \text{for } f(T) \in \mathbb{F}_q[T], \text{ and } v \in V.$$

Let $\mathbb{F}_q[M]$ be the subalgebra of the algebra of $\mathbb{F}_q$-linear endomorphisms of V generated by M, and let G_Ψ be the quotient (multiplicative) group $\mathbb{F}_q[M]^* / \mathbb{F}_q^*$.

We firstly prove that $\mu_M(T) = \chi_M(T)$. Assume by contradiction $\mu_M(T) \neq \chi_M(T)$, so that $\deg \mu_M(T) \leq n$. Then, given any $P \in U$, and any $x \in \mathbb{A}^n$ such that $P = \pi(x)$, we have

$$
\begin{aligned}
q^n = o_\psi(x) &\leq o_\Psi(P) \\
&\leq |G_\Psi| \\
&\leq \frac{q^n - 1}{q - 1} < q^n,
\end{aligned}
$$

a contradiction, which implies $\mu_M(T) = \chi_M(T)$.

Define now
$$N = \left\{ P \in H : \Psi^i(P) \in H, \forall i \in \mathbb{Z} \right\}.$$

We want to show that $N = \emptyset$. Note that this would immediately imply that Ψ is transitive. To see this, given any $P, Q \in \mathbb{P}^n$, if $N = \emptyset$ then there exist $i, j \in \mathbb{Z}$ such that $P' = \Psi^i(P), Q' = \Psi^j(Q) \in U$. Let $x', y' \in \mathbb{A}^n$ be such that $P' = \pi(x')$ and $Q' = \pi(y')$. As ψ acts transitively on $\mathbb{A}^n$ by hypothesis, there exists $\ell \in \mathbb{Z}$ such that $y' = \psi^\ell(x')$. Then, by the definition of ψ, there exists an integer $k \geq \ell$ such that $Q' = \Psi^k(P')$. In conclusion, we get $Q = \Psi^{i+k-j}(P)$, and so we have that if $N = \emptyset$ then Ψ is transitive.

Assume by contradiction that $N \neq \emptyset$. Define
$$W = \left\{ v \in V : (M^i v)_{n+1} = 0, \forall i \in \mathbb{Z} \right\},$$

where $(M^i v)_{n+1}$ denotes the $(n+1)$-th component of $M^i v$. It is immediate to check that W is a subspace of V, and that $N = \mathbb{P}W$. Also, W is clearly $\mathbb{F}_q[M]$-invariant, and so it is an $\mathbb{F}_q[T]$-submodule of V. Let $g(T) \in \mathbb{F}_q[T]$ is a monic generator of the annihilator $\mathrm{Ann}_{\mathbb{F}_q[T]}(W)$ of W as $\mathbb{F}_q[T]$-module. We have that $g(T) \mid \mu_M(T)$, since $\mu_M(M)w = 0$ for any $w \in W$. Also, $g(T) \neq 1$ as $N \neq \emptyset$ by assumption, and $g(T) \neq \mu_M(T)$, since $N \subseteq H$ gives $\deg g(T) \leq n$. This shows that if $N \neq \emptyset$ the $\mu_M(T)$ is reducible.

Let us now prove instead that $\mu_M(T)$ is irreducible, so that we get a contradiction. We firstly prove that $\mu_M(T) = p(T)^e$ for some irreducible polynomial $p(T) \in \mathbb{F}_q[T]$ and some integer $e \geq 1$.

Since $\mu_M(T) = \chi_M(T)$, then by Lemma 1 we know that there exists $v_0 \in V$ such that the set $\left\{ M^k v_0 : k \in \mathbb{Z}_{\geq 0} \right\}$ spans V over $\mathbb{F}_q$. Clearly, $v_0 \notin W$, since otherwise we would have $W = V$, as W is $\mathbb{F}_q[M]$-invariant, which is a contradiction as $N \subseteq H$. We show now that $d(M)v_0 \in W \setminus \{0\}$ for any $d(T) \in \mathbb{F}_q[T]$ such that $d(T) \mid \mu_M(T)$, and $d(T) \neq 1, \mu_M(T)$. Let $d(T)$ be any of such polynomials. Clearly $d(M)v_0 \neq 0$, as otherwise the span of $\{M^k v_0 : k \in \mathbb{Z}_{\geq 0}\}$ over $\mathbb{F}_q$ would have dimension less or equal than $\deg d(T)$, which is less or equal than n by assumption. Define then W_d to be the span of $\{M^k d(M)v_0 : k \in \mathbb{Z}_{\geq 0}\}$ over $\mathbb{F}_q$. It is immediate to see that W_d is an $\mathbb{F}_q[M]$-invariant subspace of V of dimension less or equal than $\deg(\mu_M(T)/d(T))$, which is less or equal than n by assumption. Assume by contradiction $d(M)v_0 \notin W$. Then, if we let P_d be the class of $d(M)v_0$ in $\mathbb{P}^n$, we have $P_d \notin N$, and so there exists $i \in \mathbb{Z}$ such that $Q_d = \Psi^i(P_d) \in U$. Let $y_d \in \mathbb{A}^n$ be such that $Q_d = \pi(y_d)$. Then,

$$
\begin{aligned}
q^n = o_\psi(y_d) &\leq o_\Psi(Q_d) \\
&= |\mathcal{O}_\Psi(P_d)| \\
&\leq |\mathbb{P} W_d| \\
&\leq \frac{q^n - 1}{q - 1} < q^n,
\end{aligned}
$$

a contradiction. This proves that $d(M)v_0 \in W \setminus \{0\}$ for any $d(T) \in \mathbb{F}_q[T]$ such that $d(T) \mid \mu_M(T)$, and $d(T) \neq 1, \mu_M(T)$.

Recall that we want to prove that $\mu_M(T) = p(T)^e$ for some irreducible polynomial $p(T) \in \mathbb{F}_q[T]$ and some integer $e \geq 1$. Assume then by contradiction that there exist $p_1(T), p_2(T) \in \mathbb{F}_q[T]$ distinct irreducible polynomials such that $p_1(T), p_2(T) \mid \mu_M(T)$. Then, by Bézout's identity, there exist $a(T), b(T) \in \mathbb{F}_q[T]$ such that $a(T)p_1(T) + b(T)p_2(T) = 1$, and so $a(M)p_1(M)v_0 + b(M)p_2(M)v_0 = v_0$. Now, $p_i(M)v_0 \in W$ for $i \in \{1, 2\}$ by the claim above, and so $v_0 \in W$, as W is an $\mathbb{F}_q[M]$-invariant subspace of W, which is a contradiction. Therefore, we conclude that $\mu_M(T) = p(T)^e$ for some irreducible $p(T) \in \mathbb{F}_q[T]$ and some $e \geq 1$.

We finally show that $\mu_M(T)$ is irreducible, that is $e = 1$. Let us set $f = \deg p(T)$, and let $[[T]]$ be the class of T in G_Ψ. We want to show that the order of $[[T]]$ in G_Ψ divides

$$
A(q, e, f) = q^{\lceil \log_q e \rceil} \frac{q^f - 1}{q - 1}.
$$

Let $[T]$ be the class of T in $\mathbb{F}_q[M]^*$. As $\mathbb{F}_q[M]^* \cong (\mathbb{F}_q[T]/(p(T)^e))^*$, by Lemma 2 it is enough to show that the order of $[T]^{q-1}$ in $\mathbb{F}_q[M]^*$ divides $A(q, e, f)$. Now, since $[T]^{q^f - 1} \equiv 1 \mod p(T)$, we have $[T]^{q^f - 1} = 1 + k(T)p(T)$ for some $k(T) \in \mathbb{F}_q[T]$, and so

$$
\begin{aligned}
([T]^{q-1})^{A(q,e,f)} &= ([T]^{q^f - 1})^{q^{\lceil \log_q e \rceil}} \\
&= [1 + k(T)p(T)]^{q^{\lceil \log_q e \rceil}} \\
&= [1 + k(T)^{q^{\lceil \log_q e \rceil}} p(T)^{q^{\lceil \log_q e \rceil}}] = 1 \quad \text{in } \mathbb{F}_q[M]^*,
\end{aligned}
$$

as $q^{\lceil \log_q e \rceil} \geq e$.

Let $P \in U$, and let $x \in \mathbb{A}^n$ be such that $P = \pi(x)$. Then

$$q^n = o_\psi(x) \le o_\Psi(P)$$
$$\le A(q, e, f),$$

since the size of $\mathcal{O}_\Psi(P)$ is less or equal than the order of $[[T]]$ in G_Ψ. Notice also that here $n = ef - 1$, since $\mu_M(T) = p(T)^e$ and $f = \deg p(T)$.

Assume by contradiction that $e \ge 2$. We firstly prove that this forces $f = 1$. Rewrite the inequality $q^{ef-1} \le A(q, e, f)$ as

$$q^{ef-1-\lceil \log_q e \rceil} \le \frac{q^f - 1}{q - 1}. \tag{2.2}$$

Since the quantity

$$q^{ef-1-\lceil \log_q e \rceil} - \frac{q^f - 1}{q - 1}$$

is increasing in e and f, it is enough to show that (2.2) is never verified for $e = 2$ and $f = 2$. Now, inequality (2.2) for $e = 2$ and $f = 2$ becomes

$$q^2 \le q + 1,$$

which is false for every q. Then $f = 1$.

We want now to show that for $f = 1$ the inequality (2.2) forces q to be prime and $n = 1$, or $q = 2$ and $n = 2$, which are exactly the pathological cases we excluded. For $f = 1$, inequality (2.2) becomes

$$q^{e-1-\lceil \log_q e \rceil} \le 1,$$

which is equivalent to

$$e - 1 - \lceil \log_q e \rceil \le 0.$$

The quantity $e - 1 - \lceil \log_q e \rceil$ is clearly increasing in e. Then, for $e \ge 4$ it is enough to show that it never holds for $e = 4$. In this case, in fact, we have $\lceil \log_q 4 \rceil \le 2$ for every q, and so $4 - 1 - \lceil \log_q 4 \rceil \ge 1$ for every q. For $e = 3$, in which case $n = 2$, we have $\lceil \log_2 3 \rceil = 2$, and $\lceil \log_q 3 \rceil = 1$ otherwise. Then, the inequality is satisfied for $q = 2$, and never satisfied for $q \ne 2$. Finally, for $e = 2$ we have $n = 1$. Since for $n = 1$ if Ψ sends a point of U to the point at infinity, then ψ transitive gives Ψ transitive by [1, Proposition 2.6], and so $e = 1$ by [1, Theorem 3.4], a contradiction. We have then that Ψ maps no point of U to the point at infinity, and so ψ is an affine map. But then, since ψ is transitive (and in particular the inequality holds) then q is prime by [1, Theorem 2.7]. In conclusion, we proved that if $e \ge 2$ then q is prime and $n = 1$, or $q = 2$ and $n = 2$, which are the pathological cases excluded at the beginning. Therefore $e = 1$, and so $\mu_M(T)$ is irreducible.

3   Artin-Schreier Fractional Jumps

Let $q = p$ be a prime number. In this section we consider fractional jumps of automorphisms of $\mathbb{P}^{p-1}$ defined by companion matrices of Artin-Schreier polynomials

$$\alpha_c(T) = T^p - T - c \in \mathbb{F}_p[T], \quad \text{for } c \in \mathbb{F}_p^*.$$

Proposition 1. *The polynomial $\alpha_c(T)$ is projectively primitive for every $c \in \mathbb{F}_p^*$.*

Proof. Notice that it is well known that the polynomials $\alpha_c(T)$ are irreducible for every $c \in \mathbb{F}_p^*$ by the theory of Artin-Schreier extensions. Let now $c \in \mathbb{F}_p^*$ be fixed. We want to show that $\alpha_c(T)$ is projectively primitive. Let $c' \in \mathbb{F}_p^*$ be such that c/c' generates $\mathbb{F}_p^*$. Then, the polynomial $T^p - T - c/c'$ is primitive by [3, Theorem 1.2], and so projectively primitive. Now, this implies that the polynomial $c'T^p - c'T - c = (c'T)^p - c'T - c$ is projectively primitive, and so $\alpha_c(T)$ is projectively primitive. $\square$

Fix $c \in \mathbb{F}_p^*$, let $M \in \mathrm{GL}_p(\mathbb{F}_q)$ be the companion matrix of $\alpha_c(T)$, let $\Psi = [M]$, and let ψ be the fractional jump of Ψ. Let $x_0 \in \mathbb{A}^{p-1}$, and let $\{x^{(k)}\}_{k \in \mathbb{N}}$ be the sequence recursively defined by $x^{(k+1)} = \psi(x^{(k)})$. By [1, Theorem 3.4] we know that the sequence $\{x^{(k)}\}_{k \in \mathbb{N}}$ has full orbit.

3.1   Explicit Description

In what follows we want to give the explicit description of the Artin-Schreier fractional jump ψ.

For $i \in \{1, \ldots, p-1\}$ we have that

$$M^i = \left(\begin{array}{c|c} 0_{i,p-i} & J_i(c)^t \\ \hline \mathrm{Id}_{p-i} & E_{p-i,i}^{(1,i)} \end{array}\right),$$

where $0_{i,p-i}$ is the $i \times (p-i)$ zero matrix, $J_i(c)^t$ is the transpose of a Jordan block of size $i \times i$ and eigenvalue c, that is

$$J_i(c)^t = \begin{pmatrix} c & 0 & \cdots\cdots & 0 \\ 1 & c & 0 & \vdots \\ 0 & 1 & c & \vdots \\ \vdots & & \ddots & \ddots & 0 \\ 0 & \cdots & 0 & 1 & c \end{pmatrix},$$

the matrix Id_{p-i} is the $(p-i) \times (p-i)$ identity, and $E_{p-i,i}^{(1,i)}$ is the $(p-i) \times i$ matrix with $(1,i)$-entry equal to 1, and all the other entries equal to zero. For $i = p$ we clearly have $M^p = M + c\mathrm{Id}_p$.

Following Remark 2, let us now compute explicitly the polynomials $b^{(i)}$'s and the sets U_i's, for $i \in \{1, \ldots, p\}$, by looking at the last row of M^i.

$$b^{(i)} = x_{p-i},$$
$$\text{for } i \in \{1, \ldots, p-2\},$$
$$b^{(p-1)} = x_1 + 1,$$
$$b^{(p)} = x_{p-1} + c,$$

which gives

$$U_1 = \{x \in \mathbb{A}^{p-1} : x_{p-1} \neq 0\},$$
$$U_i = \{x \in \mathbb{A}^{p-1} : x_{p-i} \neq 0, \text{ and } x_{p-j} = 0, \forall j \in \{1, \ldots, i-1\}\},$$
$$\text{for } i \in \{2, \ldots, p-2\},$$
$$U_{p-1} = \{x \in \mathbb{A}^{p-1} : x_1 + 1 \neq 0, \text{ and } x_{p-j} = 0, \forall j \in \{1, \ldots, p-2\}\},$$
$$U_p = \{x \in \mathbb{A}^{p-1} : x_{p-1} + c \neq 0, x_1 + 1 = 0, \text{ and } x_{p-j} = 0, \forall j \in \{1, \ldots, p-2\}\}$$
$$= \{(-1, 0, \ldots, 0)\}.$$

The polynomials $a_j^{(i)}$, for $i \in \{1, \ldots, p\}$ and $j \in \{1, \ldots, p-1\}$, are easily computed as well by looking at the j-th row of M^i.

- for $i = 1$ we have that
  - if $j = 1$ then $a_1^{(1)} = c$,
  - if $j = 2$ then $a_2^{(1)} = x_1 + 1$,
  - for any $j \in \{3, \ldots, p-1\}$ then $a_j^{(1)} = x_{j-1}$.
- for $i \in \{2, \ldots p-1\}$ we have that
  - if $j = 1$ then $a_1^{(i)} = c x_{p-i+1}$,
  - if $j \in \{2, \ldots, i-1\}$ then $a_j^{(i)} = x_{p-i+j-1} + c x_{p-i+j}$,
  - if $j = i$ then $a_i^{(i)} = x_{p-1} + c$,
  - if $j = i+1$ then $a_{i+1}^{(i)} = x_1 + 1$,
  - if $j \in \{i+2, \ldots, p-1\}$ then $a_j^{(i)} = x_{j-i}$.
- for $i = p$ we have that
  - if $j = 1$ then $a_1^{(p)} = c x_1 + c$,
  - if $j = 2$ then $a_2^{(p)} = x_1 + c x_2 + 1$,
  - if $j \in \{3, \ldots, p-1\}$ then $a_j^{(p)} = x_{j-1} + c x_j$.

By Theorem 1 this provides the explicit structure of ψ.

3.2   Computational Complexity

Now that we have the explicit description of the fractional jump, we are ready to establish the expected complexity of computing a random term in the sequence $\{x^{(k)}\}_{k \in \mathbb{N}}$ given by iterating the Artin-Schreier fractional jump ψ.

The expected complexity of computing $x^{(k+1)}$ given a term $x^{(k)}$ chosen uniformly at random in the sequence is

$$\mathbb{E} = \sum_{i=1}^{p} p_i c_i,$$

where p_i is the probability that $x^{(k)} \in U_i$, which is

$$p_i = \begin{cases} p^{-i}(p-1), & \text{if } i \in \{1, \ldots, p-1\}, \\ p^{1-p}, & \text{if } i = p, \end{cases}$$

and c_i is the complexity of evaluating ψ at $x^{(k)}$ when $x^{(k)} \in U_i$.

We want now to evaluate c_i for $i \in \{1, \ldots, p\}$. If $x^{(k)} \in U_i$, the number of sums needed to compute $x^{(k+1)} = \psi(x^{(k)}) = f^{(i)}(x^{(k)})$ is $s_i = \sum_{j=1}^{p}(r_j^{(i)} - 1)$, where $r_j^{(i)}$ is the number of non-zero entries in the j-th row of the matrix M^i.

Since the denominators of the components of $f^{(i)}$ are all equal, the number of inversions needed is always 1.

Also, the number of multiplications needed is given by the number m_i of entries different from 0 and 1 in the $p \times (p-1)$ submatrix of M^i given by dropping the last column (this can be seen as the last component of the projective point is set to 1 in the fractional jump) plus the number of multiplications of $b^{(i)}(x^{(k)})^{-1}$ by the $a_j^{(i)}$'s, which is simply $p - 1$.

Since the length of the orbit p^{p-1} is superexponential, the size of p can be chosen relatively small in such a way that one can build look-up tables for the operations in $\mathbb{F}_p$ (so they will all have the same cost) and still get a huge orbit. Therefore

$$c_i = \underbrace{s_i}_{\text{sums}} + \underbrace{1}_{\text{inversions}} + \underbrace{m_i + p - 1}_{\text{multiplications}} \, .$$

It remains to compute s_i and m_i. Given the explicit description previously provided, we have $s_i = i$ for $i \in \{1, \ldots, p-1\}$ and $s_p = p+1$, and $m_i = i - 1$ for $i \in \{1, \ldots, p\}$. Therefore, we have $c_i = p + 2i - 1$ for $i \in \{1, \ldots, p-1\}$ and $c_p = 3p$.

The expected complexity is then

$$\mathbb{E} = 3p^{2-p} + \sum_{i=1}^{p-1} p^{-i}(p-1)(p+2i-1)$$

$$= 3p^{2-p} - \frac{3p^3 - (p^2+1)p^p - 4p^2 + 3p}{p^p(p-1)} = p + O\left(\frac{1}{p}\right).$$

This means that the expected complexity of computing the $(k+1)$-th vector of the sequence roughly consists of p checks of the look-up tables, one for each component: morally, we are filling out each component of the term of the sequence by directly reading the look-up table, which is why the process is very efficient.

Remark 3. Clearly, the expected complexity can be further optimised by using the equations defining the U_i's, but this will not affect the asymptotic behaviour of $\mathbb{E}$.

4   Conclusions and Further Research

In this paper we proved that the transitivity of the fractional jumps and the transitivity of the projective automorphisms inducing them are equivalent conditions, except from some degenerate cases which are entirely classified. This puts the last stone for the foundational theory of this new construction: for fixed base field and fixed dimension, the problem of finding all transitive fractional jump is now reduced to finding transitive projective automorphisms. In addition, using the theory of Artin-Schreier polynomials, we showed that the construction is systematically feasible when the dimension of the projective space is prime and equal to the characteristic of the field. The question now arising is:

Question 1. Can one give new explicit classes of projectively primitive polynomials?

Such new classes will allow to use companion matrices of such polynomials (or their conjugates) to build full orbit fractional jump sequences. In particular, it would be of interest to do this for any fixed dimension and in characteristic 2, and with sparse polynomials.

Acknowledgment. The authors are grateful to Andrea Ferraguti for preliminary reading of this manuscript, and for useful discussions and suggestions. The second author is thankful to the Swiss National Science Foundation grant number 171248.

References

1. Amadio Guidi, F., Lindqvist, S., Micheli, G.: Full orbit sequences in affine spaces via fractional jumps and pseudorandom number generation. arXiv preprint arXiv:1712.05258v2 (2017). (To appear in Mathematics of Computation)
2. Brandstätter, N., Winterhof, A.: Some notes on the two-prime generator of order 2. IEEE Trans. Inf. Theory **51**(10), 3654–3657 (2005)
3. Cao, X.: On the order of the polynomial $x^p - x - a$. Cryptology ePrint Archive, Report 2010/034 (2010). https://eprint.iacr.org/2010/034.pdf
4. Chou, W.-S.: On inversive maximal period polynomials over finite fields. Appl. Algebra Eng. Commun. Comput. **6**(4), 245–250 (1995)
5. Eichenauer-Herrmann, J.: Inversive congruential pseudorandom numbers avoid the planes. Math. Comp. **56**(193), 297–301 (1991)
6. El-Mahassni, E.D., Gomez, D.: On the distribution of nonlinear congruential pseudorandom numbers of higher orders in residue rings. In: Bras-Amorós, M., Høholdt, T. (eds.) AAECC 2009. LNCS, vol. 5527, pp. 195–203. Springer, Heidelberg (2009). https://doi.org/10.1007/978-3-642-02181-7_21
7. Ferraguti, A., Micheli, G., Schnyder, R.: On sets of irreducible polynomials closed by composition. In: Duquesne, S., Petkova-Nikova, S. (eds.) WAIFI 2016. LNCS, vol. 10064, pp. 77–83. Springer, Cham (2016). https://doi.org/10.1007/978-3-319-55227-9_6

8. Gómez-Pérez, D., Ostafe, A., Shparlinski, I.E.: Algebraic entropy, automorphisms and sparsity of algebraic dynamical systems and pseudorandom number generators. Math. Comput. **83**(287), 1535–1550 (2014)
9. Gutierrez, J., Shparlinski, I.E., Winterhof, A.: On the linear and nonlinear complexity profile of nonlinear pseudorandom number generators. IEEE Trans. Inf. Theory **49**(1), 60–64 (2003)
10. Heath-Brown, D.R., Micheli, G.: Irreducible polynomials over finite fields produced by composition of quadratics. arXiv preprint arXiv:1701.05031 (2017)
11. Lang, S.: Algebra - Revised Third Edition. Graduate Texts in Mathematics, vol. 211. Springer, New York (2002). https://doi.org/10.1007/978-1-4613-0041-0
12. Niederreiter, H., Shparlinski, I.E.: Recent advances in the theory of nonlinear pseudorandom number generators. In: Fang, K.T., Niederreiter, H., Hickernell, F.J. (eds.) Monte Carlo and Quasi-Monte Carlo Methods 2000, pp. 86–102. Springer, Heidelberg (2002). https://doi.org/10.1007/978-3-642-56046-0_6
13. Niederreiter, H., Shparlinski, I.E.: Dynamical systems generated by rational functions. In: Fossorier, M., Høholdt, T., Poli, A. (eds.) AAECC 2003. LNCS, vol. 2643, pp. 6–17. Springer, Heidelberg (2003). https://doi.org/10.1007/3-540-44828-4_2
14. Ostafe, A.: Pseudorandom vector sequences derived from triangular polynomial systems with constant multipliers. In: Hasan, M.A., Helleseth, T. (eds.) WAIFI 2010. LNCS, vol. 6087, pp. 62–72. Springer, Heidelberg (2010). https://doi.org/10.1007/978-3-642-13797-6_5
15. Ostafe, A., Pelican, E., Shparlinski, I.E.: On pseudorandom numbers from multivariate polynomial systems. Finite Fields Appl. **16**(5), 320–328 (2010)
16. Ostafe, A., Shparlinski, I.E.: On the degree growth in some polynomial dynamical systems and nonlinear pseudorandom number generators. Math. Comput. **79**(269), 501–511 (2010)
17. Ostafe, A., Shparlinski, I.E.: On the length of critical orbits of stable quadratic polynomials. Proc. Am. Math. Soc. **138**(8), 2653–2656 (2010)
18. Topuzoğlu, A., Winterhof, A.: Pseudorandom sequences. In: Garcia, A., Stichtenoth, H. (eds.) Topics in Geometry, Coding Theory and Cryptography. AA, vol. 6, pp. 135–166. Springer, Dordrecht (2006). https://doi.org/10.1007/1-4020-5334-4_4
19. Winterhof, A.: Recent results on recursive nonlinear pseudorandom number generators. In: Carlet, C., Pott, A. (eds.) SETA 2010. LNCS, vol. 6338, pp. 113–124. Springer, Heidelberg (2010). https://doi.org/10.1007/978-3-642-15874-2_9

Some Sextics of Genera Five and Seven Attaining the Serre Bound

Motoko Qiu Kawakita$^{(\boxtimes)}$

Department of Mathematics, Shiga University of Medical Science,
Seta Tsukinowa-cho, Otsu, Shiga 520-2192, Japan
kawakita@belle.shiga-med.ac.jp

Abstract. We define two families of sextics. By computer search on one family, we find new curves of genus 5 attaining the Hasse–Weil–Serre bound over $\mathbb{F}_{71}$, $\mathbb{F}_{191}$ and $\mathbb{F}_{11^5}$, and we update 3 entries of genus 5 in manYPoints.org. Among another family, we find new curves of genus 7 attaining the Hasse–Weil–Serre bound over $\mathbb{F}_{p^3}$ for some primes p. We determine the precise condition on the finite field over which the sextics attain the Hasse–Weil–Serre bound.

Keywords: Algebro-geometric codes · Rational points · Serre bound

1 Introduction

Goppa discovered algebro-geometric codes in 1970s, where we can construct efficient codes from explicit curves with many rational points; see [11]. For a curve C of genus $g(C)$ over a finite field $\mathbb{F}_q$, we have the Hasse–Weil bound $\#C(\mathbb{F}_q) \leq q + 1 + 2g(C)\sqrt{q}$. A curve attaining this bound is said to be maximal. Here p is a prime number and q is a power of p, $\#C(\mathbb{F}_q)$ is the number of rational points of C over $\mathbb{F}_q$. By a curve we mean a projective geometrically irreducible nonsingular curve. In 1983, Serre improved this bound as $\#C(\mathbb{F}_q) \leq q + 1 + g(C)\lfloor 2\sqrt{q} \rfloor$, which we call the Serre bound. Here $\lfloor \cdot \rfloor$ means round down.

Many properties of maximal curves have been widely investigated; see [2], [4] and references therein. However, this is not the case of non-maximal curves attaining the Serre bound with its genera ≥ 4. There are known only examples of genera 4 and 10 in [6], genus 6 in [7–9], genus 11 in [10].

The purpose of this research is to find more explicit examples. In the process of studying the sextics in [7,8], we get an idea to define two families of sextics in Sects. 2 and 4. Among them by computer search, we find new non-maximal curves of genera 5 and 7 attaining the Serre bound in Sects. 3 and 5 respectively.

2 A Family of Sextics of Genus ≤ 5

Let k be a field of characteristic $p \neq 2, 3, 5$ in this section, and $\bar{k}$ be its algebraic closure.

Partially supported by JSPS Grant-in-Aid for Scientific Research (C) 17K05344.

© Springer Nature Switzerland AG 2018
L. Budaghyan and F. Rodríguez-Henríquez (Eds.): WAIFI 2018, LNCS 11321, pp. 264–271, 2018.
https://doi.org/10.1007/978-3-030-05153-2_15

Definition 1. *We set a sextic C over a field k with the following equation:*

$$x^3y^3 + x^5 + y^5 + ax^2y^2 + bxy + c = 0,$$

where $a, b, c \in k$ and $c \neq 0$.

Let J_C be the Jacobian variety of a curve C. Theorem B of [5] plays an important role when we decompose a Jacobian variety of a curve in this article.

Theorem 1 (Theorem B, [5]). *Given a curve X, let $G \leq \mathrm{Aut}(X)$ be a finite group such that $G = H_1 \cup \cdots \cup H_m$ where the subgroups H_i satisfy $H_i \cap H_j = 1_G$ if $i \neq j$. Then we have the following isogeny relation*

$$J_X^{m-1} \times J_{X/G}^g \sim J_{X/H_1}^{h_1} \times \cdots \times J_{X/H_m}^{h_m}$$

where $g = |G|$ and $h_i = |H_i|$ and J_r means the product of J with itself r times.

Proposition 1. *Assume that there exists $\zeta \in k$, such that $\zeta^5 = 1$. The Jacobian variety of C decomposes over k have the following isogeny relation*

$$J_C \sim J_{C_\sigma}^2 \times J_{C_\tau},$$

where $C_\sigma : f(x, y) = 0$ and $C_\tau : y^2 = h(x)$ with

$$f(x, y) = x^5 - 5x^3y + 5xy^2 + y^3 + ay^2 + by + c,$$
$$h(x) = (x^3 + ax^2 + bx + c)^2 - 4x^5.$$

Proof. For $\sigma : (x, y) \mapsto (y, x)$, we have the quotient as

$$C/\langle \sigma \rangle : x^5 - 5x^3y + 5xy^2 + y^3 + ay^2 + by + c = 0.$$

For $\tau : (x, y) \mapsto (\zeta x, \zeta^{-1} y)$, we have

$$C/\langle \tau \rangle : x^2 + (y^3 + ay^2 + by + c)x + y^5 = 0,$$

which is birational equivalent to $y^2 = (x^3 + ax^2 + bx + c)^2 - 4x^5$.
   Set $G = \langle \sigma, \tau \rangle$. We have $G = \langle \sigma \rangle \cup \langle \tau \rangle \cup \langle \sigma\tau \rangle \cup \langle \sigma\tau^2 \rangle \cup \langle \sigma\tau^3 \rangle \cup \langle \sigma\tau^4 \rangle$. From Theorem 1,

$$J_C^5 \times J_{C/G}^{10} \sim J_{C/\langle \sigma \rangle}^2 \times J_{C/\langle \tau \rangle}^5 \times J_{C/\langle \sigma\tau \rangle}^2 \times J_{C/\langle \sigma\tau^2 \rangle}^2 \times J_{C/\langle \sigma\tau^3 \rangle}^2 \times J_{C/\langle \sigma\tau^4 \rangle}^2.$$

The genus of C/G is 0. Further $C/\langle \sigma\tau^i \rangle$ for $i = 1, 2, 3, 4$ are birational equivalent to $C/\langle \sigma \rangle$, therefore $J_C \sim J_{C/\langle \sigma \rangle}^2 \times J_{C/\langle \tau \rangle}$. Setting $C/\langle \sigma \rangle$ and $C/\langle \tau \rangle$ as C_σ and C_τ respectively, which completes the proof.

Corollary 1. *Let $q = 1(\mathrm{mod}\, 5)$. We have that*

$$\#C(\mathbb{F}_q) = 2\#C_\sigma(\mathbb{F}_q) + \#C_\tau(\mathbb{F}_q) - 2q - 2.$$

Proof. It is well known that $\#C(\mathbb{F}_q) = q+1-t$, where t is the trace of Frobenius acting on a Tate module of J_C. Proposition 1 implies that this Tate module is isomorphic to a direct sum of two copies of the Tate module of J_{C_σ} and C_τ. Hence $t = 2t_1 + t_2$, where t_1 and t_2 are the trace of Frobenius on the Tate module of J_{C_σ} and C_τ respectively. Since $t_1 = q+1 - \#C_\sigma(\mathbb{F}_q)$ and $t_2 = q+1 - \#C_\tau(\mathbb{F}_q)$, the result follows.

For polynomials $u(x)$ and $v(x)$, we set the resultant $\mathrm{Res}(u,v)$ as the determinant of the Sylvester matrix.

Lemma 1. *Let α, β be roots of $1 - 3x + x^2 = 0$ in $\bar{k}$, $f_y(x,y)$ be the partial derivative of f with respect to y. Set $u_\alpha(x) = f(x, \alpha x^2)$, $v_\alpha(x) = f_y(x, \alpha x^2)$. If $\mathrm{Res}(u_\alpha, v_\alpha) = \mathrm{Res}(u_\beta, v_\beta) = 0$, then the genus $g(C_\sigma) \leq 2$.*

Proof. The infinity of C_σ is a singular point, hence the genus $g(C_\sigma) \leq 4$. If $\mathrm{Res}(u_\alpha, v_\alpha) = 0$, then there exists $s \in \bar{k}$, such that $u_\alpha(s) = v_\alpha(s) = 0$. It means that $f(s, \alpha s^2) = f_y(s, \alpha s^2) = 0$. The partial derivative of f with respect to x is $f_x(x,y) = 5(x^4 - 3x^2 y + y^2)$. Thus $f_x(s, \alpha s^2) = 0$, which means that $(s, \alpha s^2)$ is a singular point on the affine piece. Similarly, if $\mathrm{Res}(u_\beta, v_\beta) = 0$ then there exists another singular point $(t, \beta t^2)$ on the affine piece. Therefore the genus $g(C_\sigma) \leq 2$.

Lemma 2. *Set $h'(x)$ as the differentiation of $h(x)$. If $\mathrm{Res}(h, h') = 0$, then the genus $g(C_\tau) \leq 1$.*

Proof. If $\mathrm{Res}(h, h') = 0$, then there exists $s \in \bar{k}$ such that $h(x) = (x-s)^2 h_1(x)$ where $\deg h_1 = 4$. Hence C_τ is birational to $y^2 = h_1(x)$, which means $g(C_\tau) \leq 1$.

Proposition 2. *If $\mathrm{Res}(u_\alpha, v_\alpha) = \mathrm{Res}(u_\beta, v_\beta) = \mathrm{Res}(h, h') = 0$, then the genus $g(C) \leq 5$.*

Proof. From Proposition 1, we have that $g(C) = 2g(C_\sigma) + g(C_\tau)$. Lemmas 1 and 2 imply the result immediately.

We remark that the condition of Proposition 2 is simple to implement in computer search.

3   Curves of Genus 5 Attaining the Serre Bound

We search by MAGMA [1] among C over $\mathbb{F}_q$ for $q \equiv 1 \pmod 5$, under the condition of Proposition 2, using Corollary 1. New curves of genus 5 are found, which update three entries in [3], whom we list in Table 1. In [3] the tables record for a pair (q, g) an entry $\alpha - \beta$ where β is the best upper bound for the maximum number of points of a curve of genus g over $\mathbb{F}_q$ and α gives a lower bound obtained from an explicit example of a curve defined over $\mathbb{F}_q$ with α (or at least α) rational points.

Example 1. $x^3 y^3 + x^5 + y^5 + 2x^2 y^2 + 4xy + 25 = 0$ has 82 rational points over $\mathbb{F}_{31}$.

Table 1. Curves of genus 5 with many points

$\mathbb{F}_q$	$\#C(\mathbb{F}_q)$	old entry
31	82	-82
71	152	-152
11^5	165062	-165062

Example 2. The sextic C attains the Serre bound over $\mathbb{F}_q$, when $(q, a, b, c) = (71, 4, 46, 36)$, $(191, 134, 126, 2)$, $(11^5, 10, 9, 10)$.

Simultaneously, we find maximal curves of genus 5.

Example 3. The sextic C is maximal over $\mathbb{F}_{p^2}$, when $(p, a, b, c) = (29, 17, 28, 28)$, $(31, 1, 3, 7)$, $(41, 28, 29, 31)$, $(59, 9, 16, 28)$, $(61, 11, 9, 10)$, $(71, 0, 62, 64)$, $(79, 5, 10, 12)$, $(89, 8, 20, 8)$, $(101, 46, 89, 38)$, $(109, 4, 87, 7)$, $(131, 0, 107, 97)$, $(139, 2, 43, 122)$, $(149, 5, 43, 59)$, $(151, 5, 41, 115)$, $(179, 7, 152, 90)$, $(181, 67, 41, 18)$, $(191, 2, 9, 17)$, $(199, 17, 196, 24)$, etc.

We list them in Table 2. We note that we practice for $p \le 269$ in this case.

Table 2. Maximal curves of genus 5 over $\mathbb{F}_{p^2}$

7	11	13	17	19	23	29	31	37
						C	C	
41	43	47	53	59	61	67	71	73
C				C	C		C	
79	83	89	97	101	103	107	109	113
C		C		C			C	
127	131	137	139	149	151	157	163	167
	C		C	C	C			
173	179	181	191	193	197	199		
	C	C	C			C		

From Table 2, we have a conjecture.

Conjecture 1. Let $p > 23$. If $p \equiv \pm 1 \pmod 5$, then there exists a sextic C of genus 5, which is maximal over $\mathbb{F}_{p^2}$.

4   A Family of Sextics of Genus 7

Let k be a field of characteristic $p \ne 2, 3$ in this section.

Definition 2. *We set a sextic W over k with the following equation:*

$$x^4y^2 + y^4 + x^2 + x^2y^4 + y^2 + x^4 + bx^2y^2 = 0,$$

where $b \in k$.

We decompose the Jacobian variety, where the idea comes from Proposition 10 in [7].

Proposition 3. *The sextic W over a field k have the following isogeny relation:*

$$J_W \times H_2^2 \sim J_H^3,$$

where the curves are defined by

$$H_2 : x^2y + y^2 + x + xy^2 + y + x^2 + bxy = 0,$$
$$H : x^2y^2 + y^4 + x + xy^4 + y^2 + x^2 + bxy^2 = 0.$$

Proof. Since $\sigma : (x,y) \mapsto (-x,y)$, $\tau : (x,y) \mapsto (x,-y)$ are automorphisms of W, from Theorem 1, we have that

$$J_W \times J_{W/\langle \sigma, \tau \rangle}^2 \sim J_{W/\langle \sigma\tau \rangle} \times J_{W/\langle \sigma \rangle} \times J_{W/\langle \tau \rangle}.$$

$W/\langle \sigma, \tau \rangle$ is birational equivalent to H_2. Further, $W/\langle \sigma\tau \rangle$, $W/\langle \sigma \rangle$ and $W/\langle \tau \rangle$ are birational equivalent to H, which show the isogeny relation.

Afterward, set $b \neq 2, 3, -6$.

Proposition 4. *The jacobian variety of the curve H over a field k have the following isogeny relation:*

$$J_H \sim E_1 \times E_2 \times E_3,$$

where the elliptic curves $E_i : y^2 = xf_i(x)$ for $i = 1, 2, 3$ are given by

$$f_1(x) = x^2 - bx - (b - 3),$$
$$f_2(x) = (x - 1)(x - (b - 2)),$$
$$f_3(x) = x^2 + (b^2 - 12)x - 16(b - 3).$$

Proof. Since $\sigma : (x,y) \mapsto (x/y^2, 1/y)$, $\tau : (x,y) \mapsto (x,-y)$ are automorphisms of H, from Theorem 1, we have

$$J_H \times J_{H/\langle \sigma, \tau \rangle}^2 \sim J_{H/\langle \sigma\tau \rangle} \times J_{H/\langle \sigma \rangle} \times J_{H/\langle \tau \rangle}.$$

Now, an explicit quotient map $H \to H/\langle \sigma\tau \rangle$ is given by

$$(x,y) \mapsto (x + x/y^2, y - 1/y),$$

where one gets

$$H/\langle \sigma\tau \rangle : x^2 + xy^2 + bx + 2x + y^2 + 4 = 0,$$

which is birational equivalent to E_1.

Next, an explicit quotient map $H \to H/\langle\sigma\rangle$ is given by

$$(x, y) \mapsto (x/y, y + 1/y),$$

where we have

$$H/\langle\sigma\rangle : -(x^3 + y^3 - 3y) + (x + y)(x^2 + y^2 - 2) + bx = 0,$$

which is birational equivalent to E_2.

$H/\langle\tau\rangle$ is birational equivalent to E_3, and the genus of $H/\langle\sigma, \tau\rangle$ is 0, which give the desired result.

Theorem 2. *The sextic W over a field k have the following isogeny relation*

$$J_W \sim E_1^3 \times E_2^3 \times E_3.$$

And the genus $g(W) = 7$.

Proof. H_2 is birational equivalent to E_3, hence Propositions 3 and 4 show the result. Moreover, E_1, E_2 and E_3 are nonsingular when $b \neq 2, 3, -6$.

Corollary 2. *We have that*

$$\#W(\mathbb{F}_q) = 3\#E_1(\mathbb{F}_q) + 3\#E_2(\mathbb{F}_q) + \#E_3(\mathbb{F}_q) - 6q - 6.$$

Proof. It is well known that $\#W(\mathbb{F}_q) = q + 1 - t$, where t is the trace of Frobenius acting on a Tate module of J_W. Theorem 2 implies that this Tate module is isomorphic to a direct sum of three copies of the Tate module of E_1, E_2 and E_3. Hence $t = 3t_1 + 3t_2 + t_3$, where t_1, t_2 and t_3 are the trace of Frobenius on the Tate module of E_1, E_2 and E_3 respectively. Since $t_i = q + 1 - \#E_i(\mathbb{F}_q)$ for $i = 1, 2, 3$, the result follows.

Note that the j-invariants of E_1, E_2, E_3 are respectively

$$\frac{2^8(b^2 + 3b - 9)^3}{(b-2)(b-3)^2(b+6)}, \quad \frac{2^8(b^2 - 5b + 7)}{(b-2)^2(b-3)^2}, \quad \frac{b^3(b^3 - 24b + 48)^3}{(b-2)^3(b-3)^2(b+6)}.$$

5   Curves of Genus 7 Attaining the Serre Bound

We search by MAGMA [1] among W over $\mathbb{F}_q$, using Corollary 2. For an elliptic curve E, we implement the next algorithm to compute n_i with $i \geq 2$ from n_1, where $n_i = \#E(\mathbb{F}_{p^i})$. It is based on the theory of Zeta function.

Algorithm.
  INPUT: n_1, i.
  OUTPUT: $n_2, n_3, \cdots, n_i$.
  1. $a_1 \leftarrow p + 1 - n_1$.
  2. $a_2 \leftarrow a_1^2 - 2p$.
  3. $n_2 \leftarrow p^2 + 1 - a_2$.
  4. for $j = 3$ to i do:
        $a_j \leftarrow a_1 a_{j-1} - p a_{j-2}$
        $n_j \leftarrow p^j + 1 - a_j$.
  5. Return $n_2, n_3, \cdots, n_i$.

We find curves of genus 7 attaining the Serre bound.

Example 4. The sextic W is maximal over $\mathbb{F}_{p^2}$, when $(p, b) = (23, 13)$, $(47, 26)$, $(71, 1)$, $(167, 137)$, $(191, 45)$, $(239, 27)$, $(263, 87)$, $(383, 358)$, $(431, 267)$, $(479, 309)$, etc.

We note that we practice for $p \leq 99991$ in this case.

Afterward we consider the finite field $\mathbb{F}_p$ as $\mathbb{Z}/(p)$, which is the residue classes of the integers modulo the ideal generated by a prime p. Set $m = (p - 1)/2$. Denote the coefficients of x^m in $f_i(x)^m$ by $\overline{A}_i$ for $i = 1, 2, 3$, which means that

$$\overline{A}_1 = \sum_{i=0}^{\lfloor \frac{m}{2} \rfloor} \frac{m!}{(i!)^2 (m - 2i)!} (-1)^{m-i} b^{m-2i} (b - 3)^i,$$

$$\overline{A}_2 = H_p(b - 2) = \sum_{i=0}^{m} \binom{m}{i}^2 (b - 2)^i,$$

$$\overline{A}_3 = \sum_{i=0}^{\lfloor \frac{m}{2} \rfloor} \frac{m!}{(i!)^2 (m - 2i)!} (-16)^i (b^2 - 12)^{m-2i} (b - 3)^i.$$

Theorem 3. *Let $b \in \mathbb{F}_p$. W is maximal over $\mathbb{F}_{p^2}$ if and only if*

$$\overline{A}_1 \equiv \overline{A}_2 \equiv \overline{A}_3 \equiv 0 (\bmod p).$$

Proof. It follows from Sect. 5.4 of [12] and Theorem 2.

Example 5. The sextic W attaining the Serre bound over $\mathbb{F}_{p^3}$, when $(p, b) = (21313, 3663)$, $(30269, 10886)$, $(61519, 56766)$, $(76163, 6230)$, etc.

We note that we practice for $p \leq 131363$ in this case.

For $\overline{A} \in \mathbb{F}_p$, set A as the integer such that $\overline{A} \equiv A (\bmod p)$ and $0 \leq A < p$.

Theorem 4. *Let $p \geq 11$ and $b \in \mathbb{F}_p$. W over $\mathbb{F}_{p^3}$ attains the Serre bound if and only if*

$$A_1^3 - 3pA_1 = A_2^3 - 3pA_2 = A_3^3 - 3pA_3 = -\lfloor 2p\sqrt{p} \rfloor.$$

Proof. It follows from Theorem 4 in [7] and Theorem 2.

Acknowledgements. The author wishes to express her thanks to Massimo Giulietti, Gary McGuire, Maria Montanucci and Carlos Moreno for their valuable comments on this research.

References

1. Bosma, W., Cannon, J., Playoust, C.: The Magma algebra system I: The user language. J. Symbolic Comput. **24**, 235–265 (1997)
2. Garcia, A., Güneri, G., Stichtenoth, H.: A generalization of the Giulietti–Korchmáros maximal curve. Adv. Geom. **10**(3), 427–434 (2010)
3. van der Geer, G., Howe, E., Lauter, K., Ritzenthaler, C.: Table of curves with many points. http://www.manypoints.org
4. Giulietti, M., Montanucci, M., Zini, G.: On maximal curves that are not quotients of the Hermitian curve. Finite Fields Appl. **41**, 72–88 (2016)
5. Kani, E., Rosen, M.: Idempotent relations and factors of Jacobians. Math. Ann. **284**(2), 307–327 (1989)
6. Kawakita, M.Q.: On quotient curves of the Fermat curve of degree twelve attaining the Serre bound. Int. J. Math. **20**(5), 529–539 (2005)
7. Kawakita, M.Q.: Wiman's and Edge's sextic attaining Serre's bound II. In: Ballet, S., Perret, M., Zaytsev, A. (eds.) Algorithmic Arithmetic, Geometry, and Coding Theory. Contemporary Mathematics, vol. 637, pp. 191–203 (2015)
8. Kawakita, M.Q.: Certain sextics with many rational points. Adv. Math. Commun. **11**(2), 289–292 (2017)
9. Kawakita, M.Q.: Wiman's and Edge's sextic attaining Serre's bound. Euro. J. Math. **4**(1), 330–334 (2018)
10. Miura, S.: Algebraic geometric codes on certain plane curves. IEICE Trans. Fundam. **J75–A**(11), 1735–1745 (1992). (in Japanese)
11. Moreno, C.: Algebraic Curves over Finite Fields. Cambridge Tracts in Mathematics, vol. 97. Cambridge University Press, Cambridege (1991)
12. Silverman, J.H.: The Arithmetic of Elliptic Curves. Graduate Texts in Mathematics, vol. 106, 2nd edn. Springer, Heidelberg (2009). https://doi.org/10.1007/978-0-387-09494-6

Cryptography

Direct Constructions of (Involutory) MDS Matrices from Block Vandermonde and Cauchy-Like Matrices

Qiuping Li[1], Baofeng Wu[2(✉)], and Zhuojun Liu[3]

[1] University of Chinese Academy of Sciences, Beijing, China
[2] State Key Laboratory of Information Security, Institute of Information Engineering, Chinese Academy of Sciences, Beijing, China
wubaofeng@iie.ac.cn
[3] Key Laboratory of Mathematics Mechanization Academy of Mathematics and Systems Science, Chinese Academy of Sciences, Beijing, China

Abstract. MDS matrices are important components in the design of linear diffusion layers of many block ciphers and hash functions. Recently, there have been a lot of work on searching and construction of lightweight MDS matrices, most of which are based on matrices of special types over finite fields. Among all those work, Cauchy matrices and Vandermonde matrices play an important role since they can provide direct constructions of MDS matrices. In this paper, we consider constructing MDS matrices based on block Vandermonde matrices. We find that previous constructions based on Vandermonde matrices over finite fields can be directly generalized if the building blocks are pairwise commutative. Different from previous proof method, the MDS property of a matrix constructed by two block Vandermonde matrices is confirmed adopting a Lagrange interpolation technique, which also sheds light on a relationship between it and an MDS block Cauchy matrix. Those constructions generalize previous ones over finite fields as well, but our proofs are much simpler. Furthermore, we present a new type of block matrices called block Cauchy-like matrices, from which MDS matrices can also be constructed. More interestingly, those matrices turn out to have relations with MDS matrices constructed from block Vandermonde matrices and the so-called reversed block Vandermonde matrices. For all these constructions, we can also obtain involutory MDS matrices under certain conditions. Computational experiments show that lightweight involutory MDS matrices can be obtained from our constructions.

Keywords: MDS matrix · Involutory matrix
Block Vandermonde matrix · Block Cauchy-like matrix

1 Introduction

In the design of modern cryptographic primitives, confusion and diffusion are two basic requirements and they are defined by Shannon [27]. Confusion means

© Springer Nature Switzerland AG 2018
L. Budaghyan and F. Rodríguez-Henríquez (Eds.): WAIFI 2018, LNCS 11321, pp. 275–290, 2018.
https://doi.org/10.1007/978-3-030-05153-2_16

that every character of the ciphertext should depend on several parts of the plaintext and the key, obscuring the connections between them; diffusion means that changing a single character of the input will influence many characters of the output. In general, the confusion layers of a cipher are non-linear substitutions, while the diffusion layers are linear permutations. Focusing on the diffusion layer, the branch number of it can be used to estimate the number of actives S-boxes in differential and linear analysis of, for example, a two-round SPN cipher, thus reflecting the ability of a cipher to resist these two main cryptanalysis methods and to some extent, the diffusion power. Therefore, a security design target of a diffusion layer is to make its branch number as large as possible, and this is usually known as the "wide trail" design strategy [8]. Linear diffusion layers achieving maximal branch numbers are called MDS (maximal distance separable), and the matrices representing them are called MDS matrices. Such matrices are used in the design of many block ciphers and Hash functions such as AES [25], SHARK [26], Anubis [24], Twofish [31], Maelstrom [12] and Grøstl [14]. In addition, they also have deep relationships with MDS codes in coding theory.

In recent years, there has been a lot of work on finding MDS matrices. Some designers directly search them from special matrices such as circulant matrices [21,22], Hadamard matrices [28], Toeplitz matrices [30], etc. We call this a *structured searching* approach. An obvious advantage of these types of matrices is that all of their rows are similar, thus can reduce the search space. However, it is difficult to check the MDS properties of them in the searching process. Therefore, instead of structured searching, some other designers devote to directly constructing MDS matrices. With this approach, Cauchy matrices [7,11,34], Vandermonde matrices [20,29], rotational-XOR matrices [15] and some matrices obtained from certain famous MDS codes such as BCH codes [2,13], Gabidulin codes [4], etc., play an important role. An advantage of this way is that they can obtain MDS matrices of arbitrary dimension.

With the rapid development of lightweight cryptography, good hardware efficiency has become an important design goal. In order to save implementation costs, the MDS matrices used in the design of a cipher should also be as light as possible. A commonly and most frequently used metric to evaluate the weight of a matrix is its XOR count [19], which roughly speaking is the number of XOR operations needed to perform multiplication of the matrix with any vector (this is called the d-XOR count in [17]). After this, [5] proposed the idea of reusing intermediate results to decrease the XOR count, resulting in a new metric called s-XOR count [17]. Very recently, Kranz et al. [18] presented a new technique to further optimize the implementation costs based on shorter linear straight-line programs for MDS matrices. In this paper, we focus on theoretical constructions, so we only compute the s-XOR count to assure the validity of our theoretical results. In addition to these quantitative metrics, some other strategies are also adopted to save implementation costs, and this is often achieved by imposing new structures on the MDS matrices. For example, in the design of the PHOTON family of Hash function [9] and the LED block cipher [10], Guo et al. proposed to use recursive MDS matrices, which can be implemented by linear feedback shift registers (LFSRs). Another important idea is to use involutory MDS matrices. A matrix is called involutory if its inverse is itself. Obviously, this property saves hardware gates in implementation because the same structure can be used in both encryption and decryption. There were some work on constructing or searching of involutory MDS matrices. For example, as early as in 1997, Youssef et al. [34] provided a method to construct involutory MDS matrices with Cauchy matrices. After that, some other special matrices including Vandermonde matrices [29], (generalized) circulant matrices [21,22] and Hadamard matrices [22,28] were adopted to find involutory MDS matrices. Notably, [29] and [11] presented novel ideas on constructing MDS involutions based on Vandermonde matrices and Cauchy matrices over finite fields, respectively, and their constructions contain involutory MDS Hadamard matrices. As a matter of fact, in the sense of deriving involutory MDS Hadamard matrices, the work of [11] can be seen as special cases of that of [29]. However, the approach to prove the main results in [11] is much simpler than the one used in [29].

It should be pointed out that the MDS property is actually defined for block matrices (see Definition 2 in Sect. 2) over the binary field $\mathbb{F}_2$, but most work on constructing or searching of (involutory) MDS matrices is based on matrices over a finite extension of $\mathbb{F}_2$ (see e.g., [7,11,13,21,28–30]). As we all know, every element of a finite field $\mathbb{F}_{2^m}$ has a matrix representation over $\mathbb{F}_2$ of size $m \times m$, so those MDS matrices over $\mathbb{F}_{2^m}$ actually correspond to MDS block matrices over $\mathbb{F}_2$ with block size $m \times m$. However, we also know that not every matrix over $\mathbb{F}_2$ can represent an element of $\mathbb{F}_{2^m}$ (this depends on whether its minimal polynomial is irreducible or not), so it seems that we will lose some MDS matrices (and maybe some with good implementation features) when searching or constructing MDS matrices only considering matrices over $\mathbb{F}_{2^m}$. In fact, focusing on block matrices, some designers can find MDS matrices with good properties [1,22,35] that may not be obtained from matrices over finite fields. However, there are few papers on this topic and little previous work on direct constructions of MDS block matrices.

Our Contribution. In this paper, we devote to constructing MDS matrices and involutory MDS matrices from block matrices of special types. To simplify the analysis and make use of the special structures of the matrices considered, we only focus on those block matrices whose building blocks are pairwise commutative. The contributions include the following:

- We define block Vandermonde matrices and prove that for two such matrices V_1 and V_2, $V_1 V_2^{-1}$ turns out to be MDS or involutory MDS under certain conditions, which generalizes the results of [20,29]. However, for the involutory construction the proof technique used in [20,29] dost not work any more. We proceed our proof based on Lagrange interpolation, and this technique can shed light on the deeper structure of matrices of the form $V_1 V_2^{-1}$. More precisely, we find that $V_1 V_2^{-1} = D_1 C D_2$, where C is a block Cauchy matrix and D_1, D_2 are two block diagonal matrices. As a result, it can be seen in a more simple and clear way that under certain conditions involutory MDS matrices of the form $V_1 V_2^{-1}$ coincide with those constructed from block Cauchy matrices;

- We present a new type of block matrices called block Cauchy-like matrices, from which MDS matrices can also be constructed. Most interestingly, those matrices turn out to have relations with MDS matrices constructed as $V_1 V_2^{-1}$ where V_1 and V_2 are a block Vandermonde matrix and a reversed block Vandermonde matrix, respectively. By a reversed block Vandermonde matrix, we mean a matrix modified from a block Vandermonde matrix by reversing the order of its block columns. By modifying the matrix $V_1 V_2^{-1}$, involutory MDS matrices can be obtained as well;
- For all our constructions of involutory MDS matrices, by choosing the blocks to be polynomials of a given matrix, we can obtain pairwise commutative blocks, and computational experiments show that lightweight involutory MDS matrices exist. More precisely, we can find 4×4 block matrices with block size 8×8 that have XOR count 160 from the construction based on two block Vandermonde matrices, and have XOR count 151 from the construction based on a block Vandermonde matrix and a reversed block Vandermonde matrix.

The rest of the paper is organized as follows. In Sect. 2, we give some basic definitions and properties related to MDS matrices and their XOR counts. After that we briefly introduce some properties of Cauchy matrices and Vandermonde matrices over an arbitrary field. In Sect. 3, we construct MDS matrices and involutory MDS matrices with block Vandermonde matrices and block Cauchy-like matrices. Concluding remarks are given in Sect. 4.

2   Preliminaries

In this section, we first give some notations that will be used throughout the paper. Secondly, we give the definition of MDS matrices and some properties of them. We also recall the definition of XOR count of a matrix. At last, we simply state some properties of Cauchy matrices and Vandermonde matrices.

In this paper, the matrices considered are all square matrices and a block matrix means that the entries of the matrix are also matrices of a smaller dimension. The matrices $A_i \in \mathcal{M}_m(\mathbb{F}_2)(0 \leq i \leq n-1)$ are pairwise commutative that imply $A_i A_j = A_j A_i$ for all $0 \leq i, j \leq n-1$.

2.1   Notations

2.2   MDS Matrices and Their Properties

Given a vector $v = (v_0, v_1, \cdots, v_{n-1})^T \in (\mathbb{F}_2^m)^n$, where each component $v_i^T \in \mathbb{F}_2^m$ $(0 \leq i \leq n-1)$ is also a vector, its bundle weight $wt_b(v)$ is defined as the number of non-zero components. The branch number of an $n \times n$ diffusion matrix M is defined as follows.

Definition 1 (See [23]). *Let M be an $n \times n$ matrix over $\mathcal{M}_m(\mathbb{F}_2)$ (i.e., M is an $mn \times mn$ block matrix with block size $m \times m$). The differential branch number of M is defined as*

$$B_d(M) = \min_{v \neq 0} \{wt_b(v) + wt_b(Mv)\},$$

m: dimension of the blocks of a block matrix
n: dimension of the square matrix considered
$M_{i,j}$ or $M[i,j]$: (i,j)-entry of an $n \times n$ matrix M, where $0 \leq i,\ j \leq n-1$
$(M_{i,j})$ or $(M[i,j])$: $n \times n$ matrix whose (i,j)-entry is $M_{i,j}$
$\det(M)$: determinant of a matrix M
$\mathbb{F}_2$: the binary finite field
$\mathbb{F}_{2^m}$: the finite field with 2^m elements
$\mathcal{M}_m(\mathbb{F}_2)$: matrix ring formed by all $m \times m$ matrices over $\mathbb{F}_2$
$\frac{A}{B}$: matrix multiplication AB^{-1}, where B is invertible

and the linear branch number of M is defined as

$$B_\ell(M) = \min_{v \neq 0} \{wt_b(v) + wt_b(M^T v)\}.$$

For an $n \times n$ matrix M' over $\mathbb{F}_{2^m}$, the definition of its branch number is similar to Definition 1. It just needs to replace bundle weight by Hamming weight over finite fields. It can be easily seen that an upper bound of B_d and B_ℓ of any matrix is $n + 1$. Thus we have:

Definition 2. *An $n \times n$ matrix M over $\mathcal{M}_m(\mathbb{F}_2)$ is called an MDS matrix if $B_d(M) = B_\ell(M) = n + 1$.*

From the definition, we can see an MDS matrix has the maximal branch number, so the diffusion layer designed from it is also called an optimal diffusion layer. The following theorem is an important way to characterize MDS matrices from a pure linear algebra point of view.

Theorem 1 (See [6]). *An $n \times n$ matrix M over $\mathcal{M}_m(\mathbb{F}_2)$ is MDS if and only if all square block sub-matrices of M are non-singular.*

We can immediately have the following lemma from Theorem 1.

Lemma 1. *Let $M = (M_{i,j})$ be an $n \times n$ MDS matrix over $\mathcal{M}_m(\mathbb{F}_2)$, and $D = diag(D_0, D_1, \ldots, D_{n-1})$ be a block diagonal matrix over $\mathcal{M}_m(\mathbb{F}_2)$. If $\det(D) \neq 0$, then $D \cdot M$ and $M \cdot D$ are MDS matrices.*

Proof. Since $\det(D) \neq 0$, we have $\det(D_i) \neq 0$, $0 \leq i \leq n-1$. Since D is a block diagonal matrix and M is an MDS matrix, for any $k \times k (1 \leq k \leq n)$ sub-matrix of $D \cdot M$, we have

$$\det \begin{pmatrix} D_{i_0}M_{i_0,j_0} & \cdots & D_{i_0}M_{i_0,j_{k-1}} \\ D_{i_1}M_{i_1,j_0} & \cdots & D_{i_1}M_{i_1,j_{k-1}} \\ \vdots & & \vdots \\ D_{i_{k-1}}M_{i_{k-1},j_0} & \cdots & D_{i_{k-1}}M_{i_{k-1},j_{k-1}} \end{pmatrix}$$
$$= \det(D_{i_0})\det(D_{i_1})\cdots\det(D_{i_{k-1}})\det(M_{k\times k}) \neq 0,$$

where $M_{k\times k}$ is a $k \times k$ sub-matrix of M. This shows any $k \times k$ sub-matrix of $D \cdot M$ is non-singular, thus $D \cdot M$ is an MDS matrix by Theorem 1. Similarly, we also have $M \cdot D$ is an MDS matrix.

Theorem 1 provides a general way to check whether a matrix is MDS or not, but it may not be so efficient. Especially for a block matrix over $\mathcal{M}_m(\mathbb{F}_2)$, a $k \times k$ sub-matrix is actually a $km \times km$ matrix over $\mathbb{F}_2$, and we should compute determinant of this matrix. However, when the blocks of a matrix are pairwise commutative, we can compute the determinants of sub-matrices in a simpler manner thanks to the following theorem.

Lemma 2 (See [32]). *Let $\mathbb{F}$ be a field and $A = (a_{i,j})$ be an $n \times n$ matrix, where $a_{i,j} \in \mathcal{M}_m(\mathbb{F})$ are pairwise commutative, $0 \leq i, j \leq n - 1$. Then*

$$\det(A) = \det\left(\sum_{j_0\cdots j_{n-1}} (-1)^{\tau(j_0\cdots j_{n-1})} a_{0,j_0}\cdots a_{n-1,j_{n-1}}\right),$$

where $\tau(j_0\cdots j_{n-1})$ denotes the number of inverse-ordered pairs in the permutation $(j_0\cdots j_{n-1})$ (an inverse-ordered pair is a pair whose number on the left side is larger than its number on the right side).

Lemma 2 says that, if the entries of a block matrix over $\mathcal{M}_m(\mathbb{F})$ are pairwise commutative, or equivalently, the matrix is defined over a commutative sub-ring of the matrix ring $\mathcal{M}_m(\mathbb{F})$, we can compute its determinant by computing the determinant of it as a matrix over this sub-ring firstly to obtain a matrix in $\mathcal{M}_m(\mathbb{F})$, and then computing determinant of this resulting matrix. This lemma will help us a lot in our constructions of MDS matrices based on block Vandermonde and Cauchy-like matrices.

2.3   XOR Counts of Matrices over $\mathbb{F}_2$

In 2014, the authors of [19] proposed using XOR count to estimate the implementation cost of cryptographic primitives. The XOR count of a matrix over $\mathbb{F}_2$ is the number of XOR operations of the matrix-vector multiplication, which is called d-XOR count. Afterwards, [5] proposed the idea of reusing intermediate results to decrease the XOR count, resulting a new metric called s-XOR count. In this paper, we use the metric s-XOR count to calculate the implementation cost of a matrix.

Definition 3. *[5] An invertible matrix A has an s-XOR count of t over* $\mathbb{F}_2$, *denoted by* $XOR(A) = t$, *if t is the minimal number such that A can be written as*

$$A = P \prod_{k=1}^{t} (I + E_{i_k, j_k})$$

with $i_k \neq j_k$ *for all k, where* E_{i_k, j_k} *is the matrix with a unique non-zero element 1 at the* $(i_k, j_k$-*th entry,* $k \in \{1, \cdots, t\}$.

As an example, consider

$$\begin{pmatrix} 0 & 1 & 1 \\ 0 & 0 & 1 \\ 1 & 1 & 1 \end{pmatrix} \begin{pmatrix} b_1 \\ b_2 \\ b_3 \end{pmatrix} = \begin{pmatrix} b_2 \oplus b_1 \\ b_1 \\ b_3 \oplus b_2 \oplus b_1 \end{pmatrix}.$$

We can reuse the intermediate result $b_2 \oplus b_1$, so we get its s-XOR count is 2. For the block matrices we consider the form

$$A = \begin{pmatrix} A_{0,0} & A_{0,1} & \cdots & A_{0,n-1} \\ A_{1,0} & A_{1,1} & \cdots & A_{1,n-1} \\ \vdots & \vdots & \vdots & \vdots \\ A_{n-1,0} & A_{n-1,1} & \cdots & A_{n-1,n-1} \end{pmatrix}$$

where $A_{i,j} \in M_m(\mathbb{F}_2)$, $0 \leq i, j \leq n-1$, it can be easily derived that its s-XOR count is

$$XOR(A) = \sum_{i,j=0}^{n-1} XOR(A_{i,j}) + n \times (n-1) \times m. \tag{1}$$

2.4   Cauchy Matrix and Vandermonde Matrix

Cauchy matrix and Vandermonde matrix are two important kinds of special matrices in linear algebra. They both have the feature that their determinants can be represented into nice formulas. If the elements appearing in the formulas are pairwise distinct, then the determinants of them are non-zero. For simplification, we only consider Cauchy and Vandermonde matrices over finite fields here.

Definition 4. *Given* $x_0, x_1, \ldots, x_{n-1} \in \mathbb{F}_{2^m}$ *and* $y_0, y_1, \ldots, y_{n-1} \in \mathbb{F}_{2^m}$, *such that* $x_i + y_j \neq 0$ *for all* $0 \leq i, j \leq n-1$, *the matrix* $C = (c_{i,j}) = \left(\frac{1}{x_i + y_j} \right)$ *is called a Cauchy matrix.*

It is well known that the determinant of C is

$$\det(C) = \frac{\prod_{0 \leq i < j \leq n-1} (x_i + x_j)(y_i + y_j)}{\prod_{0 \leq i,j \leq n-1} (x_i + y_j)}.$$

So if x_i's and y_j's are pairwise distinct for all $0 \leq i, j \leq n-1$, then $\det(C) \neq 0$, i.e. C is non-singular.

It is easy to see that any square sub-matrix of a Cauchy matrix is still a Cauchy matrix, so we have the following proposition.

Proposition 1 (See [11]). *For pairwise distinct $x_i, y_j \in \mathbb{F}_{2^m}$ $(0 \le i, j \le n-1)$, the Cauchy matrix $C = (\frac{1}{x_i + y_j})$ is an MDS matrix.*

Definition 5 (See [29]). *The matrix*

$$V = van(x_0, x_1, \cdots, x_{n-1}) = \begin{pmatrix} 1 & x_0 & x_0^2 & \cdots & x_0^{n-1} \\ 1 & x_1 & x_1^2 & \cdots & x_1^{n-1} \\ \vdots & \vdots & \vdots & \vdots & \vdots \\ 1 & x_{n-1} & x_{n-1}^2 & \cdots & x_{n-1}^{n-1} \end{pmatrix}$$

is called a Vandermonde matrix, where $x_i \in \mathbb{F}_{2^m} (0 \le i \le n-1)$.

It is well known the determinant of V is $\det(V) = \prod_{0 \le i < j \le n-1}(x_i + x_j)$, namely, we have $\det(V) \ne 0$ if and only if all of $x_i (0 \le i \le n-1)$ are distinct.

Proposition 2 (See [20]). *Let $V_1 = van(x_0, x_1, \cdots, x_{n-1})$ and $V_2 = van(y_0, y_1, \cdots, y_{n-1})$ be two $n \times n$ invertible Vandermonde matrices over $\mathbb{F}_{2^m}$ satisfying $x_i \ne y_j, 0 \le i, j \le n-1$. Then $V_1 V_2^{-1}$ is an MDS matrix.*

Proposition 3 (See [29]). *Notations and assumptions are the same with those in Proposition 2 and further assume that $x_i = y_i + r$ for some $r \in \mathbb{F}_{2^m}^*$. Then $V_1 V_2^{-1}$ is an involutory MDS matrix.*

3   MDS Matrices and Involutory MDS Matrices Constructed from Block Matrices

In this section, we construct MDS matrices and involutory MDS matrices from block Vandermonde matrices and block Cauchy-like matrices.

3.1   (Involutory) MDS Matrices from Block Vandermonde Matrices

Before we construct MDS matrices, we first introduce some properties of block Cauchy matrix.

Definition 6. *Let $A_0, A_1, \ldots, A_{n-1}$ and $B_0, B_1, \ldots, B_{n-1}$ be $m \times m$ matrices over $\mathbb{F}_2$ satisfying that $A_i + B_j$ is non-singular for any $0 \le i, j \le n-1$. Then the matrix $C = (\frac{I}{A_i + B_j})$ is called a block Cauchy matrix over $\mathcal{M}_m(\mathbb{F}_2)$.*

Under certain conditions, applying Lemma 2, the determinant of a block Cauchy matrix computed is similar to Cauchy matrix over finite field. We can easily have the determinant of a block Cauchy matrix as follows.

Proposition 4. *Let $A_0, A_1, \ldots, A_{n-1}, B_0, B_1, \ldots, B_{n-1}$ be $m \times m$ matrices over $\mathbb{F}_2$ which are pairwise commutative, satisfying that $A_i + B_j$ is non-singular for any $0 \le i, j \le n-1$. Then the determinant of the block Cauchy matrix $C = (\frac{I}{A_i + B_j})$ is*

$$\det(C) = \frac{\prod_{0 \le i < j \le n-1} \det(A_i + A_j) \det(B_i + B_j)}{\prod_{0 \le i, j \le n-1} \det(A_i + B_j)}.$$

By Proposition 4 and Theorem 1, we can give a construction of MDS matrices with block Cauchy matrices as follows.

Theorem 2. *Assume* $\{A_0, \ldots, A_{n-1}, B_0, \ldots, B_{n-1}\}$ *is a set of* $m \times m$ *matrices over* $\mathbb{F}_2$ *which are pairwise commutative, and the sum of any two elements of it is non-singular. Then the block Cauchy matrix* $C = (\frac{I}{A_i + B_j})$ *is an MDS matrix.*

Proof. From the definition of a block Cauchy matrix, it is obvious any square sub-matrix of it is still a block Cauchy matrix. It is clear from Proposition 4 that all sub-matrices of C are non-singular under the conditions of this theorem.

In [29], the authors construct MDS matrices with Vandermonde matrices over finite fields. In the following we consider the construction of MDS matrices with block Vandermonde matrices.

Definition 7. *The matrix*

$$V = Van(A_0, A_1, \cdots, A_{n-1}) = \begin{pmatrix} I & A_0 & A_0^2 & \cdots & A_0^{n-1} \\ I & A_1 & A_1^2 & \cdots & A_1^{n-1} \\ \vdots & \vdots & \vdots & \vdots & \vdots \\ I & A_{n-1} & A_{n-1}^2 & \cdots & A_{n-1}^{n-1} \end{pmatrix}$$

is called a block Vandermonde matrix, where $A_i \in \mathcal{M}_m(\mathbb{F}_2)(0 \leq i \leq n-1)$.

Now we give the construction of an MDS matrix with two block Vandermonde matrices.

Theorem 3. *Let* $V_1 = Van(A_0, A_1, \cdots, A_{n-1})$ *and* $V_2 = Van(B_0, B_1, \cdots, B_{n-1})$ *be two block Vandermonde matrices, where* $A_i, B_j \in \mathcal{M}_m(\mathbb{F}_2), 0 \leq i, j \leq n-1$, *are commutative and the sum of any two of them is non-singular. Then* $V = V_1 V_2^{-1}$ *is an MDS matrix.*

Proof. Assume the inverse of V_2 is $V_2^{-1} = (S_{i,j})$, where $S_{i,j} \in \mathcal{M}_m(\mathbb{F}_2), 0 \leq i, j \leq n-1$. Then we have

$$V_2 V_2^{-1}[i,j] = \sum_{k=0}^{n-1} S_{k,j} B_i^k = \begin{cases} 0 & i \neq j \\ I & i = j. \end{cases}$$

Let $p_j(X) = \sum_{k=0}^{n-1} S_{k,j} X^k$ be a matrix polynomial. Then we can see that $p_j(X)$ is actually the Lagrange interpolation polynomial, that is,

$$p_j(X) = \sum_{k=0}^{n-1} S_{k,j} X^k = \prod_{\substack{k=0 \\ k \neq j}}^{n-1} \frac{X + B_k}{B_j + B_k}. \tag{2}$$

Therefore, we have

$$V_1 V_2^{-1}[i,j] = \sum_{k=0}^{n-1} S_{k,j} A_i^k = p_j(A_i) = \prod_{\substack{k=0 \\ k \neq j}}^{n-1} \frac{A_i + B_k}{B_j + B_k}. \tag{3}$$

Let $C = ((A_i + B_j)^{-1})$, which is a block Cauchy matrix. Let $D_1 = diag(\prod_{k=0}^{n-1}(A_0 + B_k), \prod_{k=0}^{n-1}(A_1 + B_k), \cdots, \prod_{k=0}^{n-1}(A_{n-1} + B_k))$ and $D_2 = diag(\prod_{\substack{k=1}}^{n-1}(B_0 + B_k)^{-1}, \prod_{\substack{k=0\\k\neq 1}}^{n-1}(B_1 + B_k)^{-1}, \cdots, \prod_{k=0}^{n-2}(B_{n-1} + B_k)^{-1})$ be two block diagonal matrices. Then we have

$$D_1 C D_2[i,j] = \prod_{k=0}^{n-1}(A_i + B_k)(A_i + B_j)^{-1} \prod_{\substack{k=0\\k\neq j}}^{n-1}(B_j + B_k)^{-1} = \prod_{\substack{k=0\\k\neq j}}^{n-1}\frac{A_i + B_k}{B_j + B_k} = V_1 V_2^{-1}[i,j]$$

Since $A_i + B_j, A_i + A_k, B_i + B_k, 0 \leq i,j,k \leq n-1$ and $i \neq k$, are non-singular, we know the block Cauchy matrix C is MDS, and $\det(D_1) \neq 0$ and $\det(D_2) \neq 0$. From Lemma 1, we know $V = V_1 V_2^{-1} = D_1 C D_2$ is an MDS matrix.

Theorem 3 is a direct generalization of the result over finite fields given in [29]. However, our proof technique is quite different from the one used in [29]. In fact, the MDS property is confirmed by computing the branch number in [29]. The main argument is based on the basic fact that the polynomial $p(x) = \sum_{i=0}^{n-1} p_i x^i$ has at most $n - 1$ different roots in any finite field, where $p_i \in \mathbb{F}_{2^m}$. However, following this approach to prove our result of Theorem 3, we will meet some difficulties since it seems we do not have such argument that the polynomial $P(X) = \sum_{i=0}^{n-1} P_i X^i$ has at most $n - 1$ roots, which are $m \times m$ matrices over $\mathbb{F}_2$, where $P_i \in \mathbb{F}_2^m$. Therefore, we provide a new method to prove Theorem 3. The advantage of this approach is that it can reflect the deeper structure of the matrix $V_1 V_2^{-1}$ for two block Vandermonde matrices V_1, V_2.

Based on Theorem 3, we can easily obtain the following theorem.

Theorem 4. *The matrix V of Theorem 3 is an involutory MDS matrix if $B_i = A_i + R$, where $R \in \mathcal{M}_m(\mathbb{F}_2)$ and $R \neq A_i, i = 0, 1, \ldots, n-1$.*

Proof. From (3) and $B_i = A_i + R$, we know

$$V_1 V_2^{-1}[i,j] = \prod_{\substack{k=0\\k\neq j}}^{n-1}\frac{A_i + B_k}{B_j + B_k} = \prod_{\substack{k=0\\k\neq j}}^{n-1}\frac{A_i + R + A_k}{A_j + R + A_k + R} = \prod_{\substack{k=0\\k\neq j}}^{n-1}\frac{B_i + A_k}{A_j + A_k}.$$

Similarly we have

$$V_2 V_1^{-1}[i,j] = \prod_{\substack{k=0\\k\neq j}}^{n-1}\frac{B_i + A_k}{A_j + A_k},$$

which is equal to $V_1 V_2^{-1}[i,j]$. So the matrix $V = V_1 V_2^{-1}$ is an involutory MDS matrix.

We need to notice that the elements used to form the block Vandermonde matrices in the above constructions should be pairwise commutative and the sum of any two of them should be non-singular. It seems difficult to find such kind of elements in the matrix ring. However, we can give a simple way to deal with this problem by considering matrix polynomials. More precisely, we replace $A_i, B_j, R, 0 \leq i, j \leq n - 1$ with $f_i(B), g_j(B), r(B), 0 \leq i, j \leq n - 1$, where $B \in \mathcal{M}_m(\mathbb{F}_2)$ and $f_i(x), g_j(x), r(x) \in \mathbb{F}_2[x]$. Now naturally they are pairwise commutative. In order to ensure that $f_i(B) + f_k(B), g_i(B) + g_k(B), f_i(B) + g_j(B), 0 \leq i, j, k \leq n - 1$ and $i \neq k$, are non-singular, we need the following proposition [35].

Proposition 5 (See [35]). *Let $\mathbb{F}$ be a field, $B \in \mathcal{M}_m(\mathbb{F}), m_B(x)$ be the minimal polynomial of B, and $g(x) \in \mathbb{F}[x]$. Then $\det(g(B)) \neq 0$ if and only if $GCD(g(x), m_B(x)) = 1$, where $GCD(g(x), m_B(x))$ denotes the greatest common divisor of $g(x)$ and $m_B(x)$.*

According to Proposition 5, we construct the desired $A_i, B_j, R, 0 \leq i, j \leq n-1$ of Theorem 4 by replacing them with $f_i(B), g_j(B), r(B), 0 \leq i, j \leq n - 1$, where $B \in \mathcal{M}_m(\mathbb{F}_2)$ and $f_i(x), g_j(x), r(x) \in \mathbb{F}_2[x]$. As an example, we give a 4×4 involutory block MDS matrix over $\mathcal{M}_8(\mathbb{F}_2)$ as follows.

Example 1. Let $f_0(x) = 1, f_1(x) = x^7 + x^3 + x^2 + 1, f_2(x) = x^7 + x^2, f_3(x) = x^3, r(x) = 1$ and $m_B(x) = x^8 + x^7 + x^5 + x^4 + x^3 + x + 1 = (x^4 + x + 1)(x^4 + x^3 + 1)$. Then we obtain an involutory block MDS matrix V by Theorem 4 for

$$B = \begin{pmatrix} 0 & 1 & 0 & 0 & 0 & 1 & 1 & 0 \\ 1 & 1 & 0 & 0 & 1 & 1 & 0 & 1 \\ 1 & 0 & 0 & 0 & 0 & 0 & 0 & 0 \\ 0 & 1 & 0 & 0 & 0 & 0 & 0 & 0 \\ 0 & 1 & 1 & 0 & 1 & 0 & 0 & 1 \\ 1 & 1 & 0 & 1 & 0 & 1 & 1 & 1 \\ 0 & 0 & 0 & 0 & 1 & 0 & 0 & 0 \\ 0 & 0 & 0 & 0 & 0 & 1 & 0 & 0 \end{pmatrix}. \tag{4}$$

The XOR count of V is 160.

It is worth noting that the minimal polynomial $m_B(x)$ in Example 1 is a reducible polynomial. If the minimal polynomial $m_B(x)$ is an irreducible polynomial, then the matrix B will be similar to a matrix representation of an element of the finite field $\mathbb{F}_{2^8}$.

3.2   (Involutory) MDS Matrices from Block Cauchy-Like Matrix

In this section, we will give a new structure to construct MDS matrices and involutory MDS matrices.

Definition 8. *Let $A_0, A_1, \ldots, A_{n-1}$ and $B_0, B_1, \ldots, B_{n-1}$ be $m \times m$ matrices over $\mathbb{F}_2$ satisfying that $I + A_i B_j$ is non-singular for any $0 \leq i, j \leq n - 1$. Then the matrix $T = (\frac{I}{I + A_i B_j})$ is called a block Cauchy-like matrix over $\mathcal{M}_m(\mathbb{F}_2)$.*

We find that block Cauchy-like matrices share some beautiful features with block Cauchy matrices. For example, any square sub-matrix of a block Cauchy-like matrix is still a block Cauchy-like matrix; the determinant of a block Cauchy-like matrix computed is similar to block Cauchy matrix, so we have the following proposition.

Proposition 6. *Let $A_0, A_1, \ldots, A_{n-1}, B_0, B_1, \ldots, B_{n-1}$ be $m \times m$ matrices over $\mathbb{F}_2$ which are pairwise commutative, satisfying that $I + A_i B_j$ is non-singular for any $0 \leq i, j \leq n - 1$. Then the determinant of the block Cauchy-like matrix $T = (\frac{I}{I + A_i B_j})$ is*

$$\det(T) = \frac{\prod_{0 \leq i < j \leq n-1} \det(A_i + A_j) \det(B_i + B_j)}{\prod_{0 \leq i,j \leq n-1} \det(I + A_i B_j)}.$$

Theorem 5. *Assume $\{A_0, \ldots, A_{n-1}, B_0, \ldots, B_{n-1}\}$ is a set of $m \times m$ matrices over $\mathbb{F}_2$ which are pairwise commutative, and $I + A_i B_j, (0 \leq i, j \leq n - 1)$ and $A_i + A_k, B_i + B_k, (0 \leq i < k \leq n-1)$ are non-singular. Then the block Cauchy-like matrix $T = (\frac{I}{I + A_i B_j})$ is an MDS matrix.*

The proof is very simple based on previous arguments, so we omit it here.

In Theorem 4, we use two block Vandermonde matrices to construct involutory MDS matrices. In the following we introduce another novel approach based on two block Vandermonde matrices.

Theorem 6. *Let $V_1 = Van(A_0, A_1, \cdots, A_{n-1})$ and $V_2 = Van(B_0, B_1, \cdots, B_{n-1})$ be two block Vandermonde matrices, where $A_i, B_j \in \mathcal{M}_m(\mathbb{F}_2), 0 \leq i, j \leq n - 1$, are pairwise commutative and $I + A_i B_j, 0 \leq i, j \leq n - 1$, and $A_i + A_k, B_i + B_k, 0 \leq i < k \leq n - 1,$ are non-singular. Then $V^* = V_1(V_2 P)^{-1}$ is an MDS matrix, where $P = \begin{pmatrix} & & I \\ & \cdot^{\cdot^{\cdot}} & \\ I & & \end{pmatrix}$ and I is an $m \times m$ identity matrix.*

Proof. Assume the inverse of V_2 is $V_2^{-1} = (S_{i,j})$, where $S_{i,j} \in \mathcal{M}_m(\mathbb{F}_2), 0 \leq i, j \leq n - 1$. It is easy to see the inverse of P is also P. Then we have

$$(V_1(V_2 P)^{-1})[i, j] = ((V_1 P) V_2^{-1})[i, j] = \sum_{k=0}^{n-1} S_{k,j} A_i^{n-1-k}.$$

Let $p_j^*(X) = \sum_{k=0}^{n-1} S_{k,j} X^{n-1-k}$ be a matrix polynomial. We can see that it can be viewed as the reciprocal polynomial of the polynomial

$$p_j(X) = \sum_{k=0}^{n-1} S_{k,j} X^k = \prod_{\substack{k=0 \\ k \neq j}}^{n-1} \frac{X + B_k}{B_j + B_k}.$$

Thus we have

$$p_j^*(X) = \prod_{\substack{k=0 \\ k \neq j}}^{n-1} \frac{I + B_k X}{B_j + B_k}.$$

This leads to

$$(V_1(V_2 P)^{-1})[i, j] = \sum_{k=0}^{n-1} S_{k,j} A_i^{n-1-k} = p_j^*(A_i) = \prod_{\substack{k=0 \\ k \neq j}}^{n-1} \frac{I + B_k A_i}{B_j + B_k}. \tag{5}$$

Let $D_3 = diag(\prod_{k=0}^{n-1}(I+A_0B_k), \prod_{k=0}^{n-1}(I+A_1B_k), \cdots, \prod_{k=0}^{n-1}(I+A_{n-1}B_k))$ and $D_4 = diag(\prod_{\substack{k=1}}^{n-1}(B_0+B_k)^{-1}, \prod_{\substack{k=0 \\ k\neq 1}}^{n-1}(B_1+B_k)^{-1}, \cdots, \prod_{k=0}^{n-2}(B_{n-1}+B_k)^{-1})$ be two block diagonal matrices. We have

$$D_3TD_4[i,j] = \prod_{k=0}^{n-1}(I+A_iB_k)(I+A_iB_j)^{-1}\prod_{\substack{k=0 \\ k\neq j}}^{n-1}(B_j+B_k)^{-1}$$

$$= \prod_{\substack{k=0 \\ k\neq j}}^{n-1}\frac{I+A_iB_k}{B_j+B_k} = (V_1(V_2P)^{-1})[i,j]$$

Since $I+A_iB_j, 0 \leq i,j \leq n-1$, and $A_i+A_k, B_i+B_k, 0 \leq i < k \leq n-1$, are nonsingular, we know the block Cauchy-like matrix T is MDS by Theorem 5, and $det(D_3) \neq 0$, $det(D_4) \neq 0$. From Lemma 1, we know $V^* = V_1(V_2P)^{-1} = D_3TD_4$ is an MDS matrix.

For a block Vandermonde matrix V, we call the matrix VP a reversed block Vandermonde matrix.

Based on the matrix V^* in Theorem 6, we can also give a construction of involutory MDS matrices.

Theorem 7. *Notations and assumptions are the same with those in Theorem 6. Then $\widetilde{V} = (R^{\frac{n-1}{2}}V^*[i,j])$ is an involutory MDS matrix if $B_i = A_iR$, where $R \in \mathcal{M}_m(\mathbb{F}_2)$, $R \neq 0$ and it is the square of a matrix when n is even.*

Proof. From the definition of $\widetilde{V}$, we have

$$\widetilde{V} = (R^{\frac{n-1}{2}}V^*[i,j]) = diag(R^{\frac{n-1}{2}}, R^{\frac{n-1}{2}}, \cdots, R^{\frac{n-1}{2}})V^*.$$

From Lemma 1, we know $\widetilde{V}$ is an MDS matrix.
From (5) and $B_i = A_iR$, we have

$$\widetilde{V}[i,j] = R^{\frac{n-1}{2}}\prod_{\substack{k=0 \\ k\neq j}}^{n-1}\frac{I+A_iB_k}{B_j+B_k} = R^{\frac{n-1}{2}}\prod_{\substack{k=0 \\ k\neq j}}^{n-1}\frac{I+RA_iA_k}{R(A_j+A_k)} = R^{-\frac{n-1}{2}}\prod_{\substack{k=0 \\ k\neq j}}^{n-1}\frac{I+B_iA_k}{A_j+A_k}.$$

Similarly we can derive

$$\widetilde{V}^{-1}[i,j] = R^{-\frac{n-1}{2}}(V^*)^{-1}[i,j] = R^{-\frac{n-1}{2}}V_2(V_1P)^{-1}[i,j] = R^{\frac{n-1}{2}}\prod_{\substack{k=0 \\ k\neq j}}^{n-1}\frac{I+B_iA_k}{A_j+A_k} = \widetilde{V}[i,j].$$

So the matrix $\widetilde{V}$ is an involutory MDS matrix.

Similar to the approach used in the previous subsections, we can also obtain lightweight involutory MDS matrices based on Theorem 7 using matrix polynomials. We give an example in the following.

Example 2. Let $f_0(x) = 0, f_1(x) = x^3 + x, f_2(x) = x, f_3(x) = x^7 + x^5, r(x) = 1$ and $m_B(x) = x^8 + x^2 + 1 = (x^4 + x + 1)^2$. Then we have an involutory MDS matrix $\widetilde{V}$ by Theorem 7 with

$$B = \begin{pmatrix} 0\,1\,0\,0\,0\,1\,0\,0 \\ 0\,0\,1\,0\,0\,0\,1\,0 \\ 0\,0\,0\,1\,0\,0\,0\,1 \\ 1\,0\,1\,0\,1\,0\,0\,0 \\ 0\,1\,0\,1\,0\,1\,0\,0 \\ 0\,0\,1\,0\,1\,0\,1\,0 \\ 0\,0\,0\,1\,0\,1\,0\,1 \\ 1\,0\,1\,0\,1\,0\,1\,0 \end{pmatrix}. \tag{6}$$

The XOR counts of $\widetilde{V}$ is 151.

In the following we will give some comparisons of our construction with previous constructions in Table 1.

Table 1. Comparison of 4×4 involutory MDS matrices

Elements	Reference	Matrix type	XOR count
$\mathcal{M}_8(\mathbb{F}_2)$	Theorem 4	Block Vandermonde	$64 + 4 \times 3 \times 8 = 160$
$\mathcal{M}_8(\mathbb{F}_2)$	Theorem 7	Permutation of block Vandermonde	$38 + 4 \times 3 \times 8 = 151$
$\mathbb{F}_{2^8}$	[28]	Hadamard	$40 + 4 \times 3 \times 8 = 152$
$\mathbb{F}_{2^8}$	[3]	Hadamard	$80 + 4 \times 3 \times 8 = 176$

Here, we only compare with the lightest results over finite field because we are generalized some constructions of finite field. Note that the 4×4 involutory reversed block Vandermonde MDS matrices over $\mathcal{M}_8(\mathbb{F}_2)$ are lighter than previous constructions over finite field. While it is possible to search for more lightweight involutory reversed Vandermonde MDS matrices over sub-field or non-commutative ring, it requires very different search strategy and it is beyond the scope of our work.

4   Conclusion

In this paper, we construct MDS matrices and involutory MDS matrices directly using block Vandermonde matrices and block Cauchy-like matrices. Some of our results are direct generalizations of previous ones obtained over finite fields, but we give a deeper understandings and simpler proofs. A novel structure named block Cauchy-like matrix which was not considered before in the constructions of MDS matrices is also presented. It is interesting that they also have relationships with block Vandermonde matrices. With the approaches introduced in this paper, we can also obtain some new lightweight involutory MDS matrices.

References

1. Augot, D., Finiasz, M.: Exhaustive search for small dimension recursive MDS diffusion layers for block ciphers and hash functions. In: Proceedings of 2013 IEEE International Symposium on Information Theory (ISIT), pp. 1551–C1555. IEEE (2013)
2. Augot, D., Finiasz, M.: Direct construction of recursive MDS diffusion layers using shortened BCH codes. In: Cid, C., Rechberger, C. (eds.) FSE 2014. LNCS, vol. 8540, pp. 3–17. Springer, Heidelberg (2015). https://doi.org/10.1007/978-3-662-46706-0_1
3. Barreto, P.S., Rijmen, V.: The Khazad legacy-level block cipher. Submission to the NESSIE Project
4. Berger, T.P.: Construction of recursive MDS diffusion layers from Gabidulin codes. In: Paul, G., Vaudenay, S. (eds.) INDOCRYPT 2013. LNCS, vol. 8250, pp. 274–285. Springer, Cham (2013). https://doi.org/10.1007/978-3-319-03515-4_18
5. Beierle, C., Kranz, T., Leander, G.: Lightweight multiplication in $GF(2^n)$ with applications to MDS matrices. In: Robshaw, M., Katz, J. (eds.) CRYPTO 2016. LNCS, vol. 9814, pp. 625–653. Springer, Heidelberg (2016). https://doi.org/10.1007/978-3-662-53018-4_23
6. Blaum, M., Roth, R.M.: On lowest density MDS codes. IEEE Trans. Inf. Theory **45**(1), 46–59 (1999)
7. Cui, T., Jin, C., Kong, Z.: On compact cauchy matrices for substitution permutation networks. IEEE Trans. Comput. **64**(7), 1998–2102 (2015)
8. Daemen, J., Rijmen, V.: The wide trail design strategy. In: Honary, B. (ed.) Cryptography and Coding 2001. LNCS, vol. 2260, pp. 222–238. Springer, Heidelberg (2001). https://doi.org/10.1007/3-540-45325-3_20
9. Guo, J., Peyrin, T., Poschmann, A.: The PHOTON family of lightweight hash functions. In: Rogaway, P. (ed.) CRYPTO 2011. LNCS, vol. 6841, pp. 222–239. Springer, Heidelberg (2011). https://doi.org/10.1007/978-3-642-22792-9_13
10. Guo, J., Peyrin, T., Poschmann, A., Robshaw, M.: The LED block cipher. In: Preneel, B., Takagi, T. (eds.) CHES 2011. LNCS, vol. 6917, pp. 326–341. Springer, Heidelberg (2011). https://doi.org/10.1007/978-3-642-23951-9_22
11. Chand Gupta, K., Ghosh Ray, I.: On constructions of involutory MDS matrices. In: Youssef, A., Nitaj, A., Hassanien, A.E. (eds.) AFRICACRYPT 2013. LNCS, vol. 7918, pp. 43–60. Springer, Heidelberg (2013). https://doi.org/10.1007/978-3-642-38553-7_3
12. Gazzoni Filho, D., Barreto, P., Rijmen, V.: The Maelstrom-0 hash function. In Proceedings of the 6th Brazilian Symposium on Information and Computer Systems Security (2006)
13. Gupta, K.C., Pandey, S.K., Venkateswarlu, A.: On the direct construction of recursive MDS matrices. Des. Codes Crypt. **82**(1–2), 77–94 (2017)
14. Gauravaram, P., et al.: Grøstl a SHA-3 candidate. In: Dagstuhl Seminar Proceedings. Schloss Dagstuhl-Leibniz-Zentrum für Informatik (2009)
15. Guo, Z., Liu, R., Gao, S., Wu, W., Lin, D.: Direct construction of optimal rotational-XOR diffusion primitives. IACR Trans. Symmetric Cryptol. **2017**(4), 169–187 (2017)
16. Gohberg, I., Olshevsky, V.: Complexity of multiplication with vectors for structured matrices. Linear Algebra Appl. **192**, 163–192 (1994)
17. Jean, J., Peyrin, T., Sim, S.M., Tourteaux, J.: Optimizing implementations of lightweight building blocks. IACR Trans. Symmetric Cryptol. **2017**(4), 130–168 (2017). https://doi.org/10.13154/tosc.v2017.i4.130-168

18. Kranz, T., Leander, G., Stoffelen, K., Wiemer, F.: Shorter linear straight-line programs for MDS matrices yet another XOR count paper. IACR Trans. Symmetric Cryptol. **2017**, 188–211 (2017). https://doi.org/10.13154/tosc.v2017.i4.188-211

19. Khoo, K., Peyrin, T., Poschmann, A.Y., Yap, H.: FOAM: searching for hardware-optimal SPN structures and components with a fair comparison. In: Batina, L., Robshaw, M. (eds.) CHES 2014. LNCS, vol. 8731, pp. 433–450. Springer, Heidelberg (2014). https://doi.org/10.1007/978-3-662-44709-3_24

20. Lacan, J., Fimes, J.: Systematic MDS erasure codes based on Vandermonde matrices. IEEE Commun. Lett. **8**(9), 570–572 (2004)

21. Liu, M., Sim, S.M.: Lightweight MDS generalized circulant matrices. In: Peyrin, T. (ed.) FSE 2016. LNCS, vol. 9783, pp. 101–120. Springer, Heidelberg (2016). https://doi.org/10.1007/978-3-662-52993-5_6

22. Li, Y., Wang, M.: On the construction of lightweight circulant involutory MDS matrices. In: Peyrin, T. (ed.) FSE 2016. LNCS, vol. 9783, pp. 121–139. Springer, Heidelberg (2016). https://doi.org/10.1007/978-3-662-52993-5_7

23. Li, C., Wang, Q.: Design of lightweight linear diffusion layers from near-MDS matrices. IACR Trans. Symmetric Cryptol. **2017**(1), 129–155 (2017)

24. Rijmen, V., Barreto, P.: The Anubis Block Cipher. The NESSIE (2000)

25. Rijmen, V., Daemen, J.: The Design of Rijndael: AES. The Advanced Encryption Standard. Springer, Berlin (2002). https://doi.org/10.1007/978-3-662-04722-4

26. Rijmen, V., Daemen, J., Preneel, B., Bosselaers, A., De Win, E.: The cipher SHARK. In: Gollmann, D. (ed.) FSE 1996. LNCS, vol. 1039, pp. 99–111. Springer, Heidelberg (1996). https://doi.org/10.1007/3-540-60865-6_47

27. Shannon, C.E.: Communication theory of secrecy systems. Bell Syst. Tech. J. **28**(4), 656–715 (1949)

28. Sim, S.M., Khoo, K., Oggier, F., Peyrin, T.: Lightweight MDS involution matrices. In: Leander, G. (ed.) FSE 2015. LNCS, vol. 9054, pp. 471–493. Springer, Heidelberg (2015). https://doi.org/10.1007/978-3-662-48116-5_23

29. Sajadieh, M., Dakhilalian, M., Mala, H., Omoomi, B.: On construction of involutory MDS matrices from Vandermond matrices in GF (2^q). Des. Codes Crypt. **2012**(64), 287–308 (2012)

30. Sarkar, S., Syed, H.: Lightweight diffusion layer: importance of Toeplitz matrices. IACR Trans. Symmetric Cryptol. **2016**(1), 95–113 (2016)

31. Schneier, B., Kelsey, J., Whiting, D., et al.: Twofish: a 128-bit block cipher. NIST AES Proposal, vol. 15, p. 23 (1998)

32. Silvester, J.R.: Determinants of block matrices. Math. Gaz. **84**(501), 460–467 (2000)

33. Xiao, L., Heys, H.M.: Hardware design and analysis of block cipher components. In: Lee, P.J., Lim, C.H. (eds.) ICISC 2002. LNCS, vol. 2587, pp. 164–181. Springer, Heidelberg (2003). https://doi.org/10.1007/3-540-36552-4_12

34. Youssef, A.M., Mister, S., Tavares, S.E.: On the design of linear transformations for substitute permutation encryption networks. In: Workshop on Selected Areas of Cryptography 1996, pp. 40–48 (1997)

35. Zhao, R., Zhang, R., Li, Y., Wu, B.: On constructions of a sort of MDS block diffusion matrices for block ciphers and hash functions. Sci. Chin. Inf. Sci. **2016**(59), 99–101 (2016)

Exploiting Preprocessing for Quantum Search to Break Parameters for $\mathcal{MQ}$ Cryptosystems

Benjamin Pring[(✉)]

University of Bath, Bath BA2 7AY, UK
b.i.pring@bath.ac.uk

Abstract. In this paper we re-examine quantum search applied to the Multivariate Quadratic ($\mathcal{MQ}$) hardness problem over the finite field GF(2). This problem is key to the security of a number of proposed *post-quantum* public-key cryptosystems designed to be resistant against attacks from quantum computers and in this paper we give a warning of the dangers of extrapolating parameters based upon the efficiency of quantum search algorithms. Our methods demonstrate that by applying preprocessing to the $\mathcal{MQ}$ problem, we can reduce the computational load on the quantum computer and, in a generalisation of multi-target search for single-targets, improve the efficiency of the basic quantum search oracle for the $\mathcal{MQ}$ problem over GF(2). Our work builds upon the $\mathcal{MQ}$ oracle introduced by Westerbaan and Schwabe [19] and improves it to the extent that it breaks all quantum-resistant security parameters for the Gui cryptosystem [16] proposed by the original authors [15]. Our results hold both in the logical gate model and when the algorithm is fully costed in terms of the Clifford+T universal gate set.

Keywords: Quantum search · Multivariate Quadratic · Cryptography

1 Introduction

The Multivariate Quadratic ($\mathcal{MQ}$) problem, informally that of finding a solution to a system of m degree two equations over a finite field in n variables, is a significant problem in cryptography, given that asymmetric encryption systems can be broken by solving large $\mathcal{MQ}$ systems of equations [4] and that a family of public-key cryptosystems rely upon the hardness of solving this problem [16,18].

These systems of equations are vulnerable to classical algorithms when the system is either underdetermined ($m \ll n$), overdetermined ($m \gg n$) or sparse (they contain equations using far fewer than n variables). The hardest instances of the $\mathcal{MQ}$ problem are therefore thought to be dense systems when $m \approx n$. These algorithms can generally be placed on a spectrum with one extreme being pure Gröbner bases techniques and the other being exhaustive brute force search, which possesses a complexity of $\mathcal{O}(2^{n+2} \log_2 n)$ [7]. Whilst the best known classical Las Vegas methods for solving systems over GF(2) possess an asymptotic

© Springer Nature Switzerland AG 2018
L. Budaghyan and F. Rodríguez-Henríquez (Eds.): WAIFI 2018, LNCS 11321, pp. 291–307, 2018.
https://doi.org/10.1007/978-3-030-05153-2_17

complexity of $\mathcal{O}(2^{0.792n})$ [5], it is thought that these methods only have an advantage over classical exhaustive search when $n > 200$. Whilst classical search is obviously still an effective tool, quantum search (sometimes referred to as Grover's algorithm [12]) offers a further advantage. For a search space of size $N = 2^n$ with M satisfying assignments, the basic *query complexity* of classical search is $\mathcal{O}(\frac{N}{M})$ whilst the quantum query complexity is $\mathcal{O}(\sqrt{\frac{N}{M}})$. This measure of complexity refers *only* to the number of queries the algorithm must make to a *black box oracle*, which upon input of a bitstring of length n, reveals only whether the bitstring is a solution to our system of equations. This black-box oracle can be realised in classical or quantum circuitry via interpreting each bit of the length n bitstring as an assignment to a separate variable and evaluating the system of equations upon this assignment. Once all equations are evaluated, it is easily checked whether the system of equations is satisfied. Quantum search additionally includes a *diffusion step*, the cost of which is often negligible compared to the implementation of the quantum black-box, or *quantum oracle*. The concrete gate count and gate depth for the entire quantum search algorithm to terminate resulting in one of the M items with probability ≈ 1 is given by

$$\frac{\pi}{4} \cdot \sqrt{\frac{N}{M}} \cdot \Big(\text{Cost(Quantum oracle)} + \text{Cost(Diffusion step)}\Big), \qquad (1)$$

where implementing the diffusion step costs $\mathcal{O}(\log_2 N)$. Unfortunately, or fortunately for the security of cryptosystems, this query complexity has been proven to be a lower-bound [8] and so techniques to reduce the total circuit complexity or gate depth cost consist of reducing the cost of the quantum oracle, lowering the query-complexity or accepting a smaller probability of success. We focus on altering the stated problem so that we only obtain partial information concerning the solution. This allows for a reduction of the query complexity at the expensive of increasing the cost of the quantum oracle, resulting in concrete gains.

1.1 Concrete Gains

In order to determine the threat that quantum computing poses to cryptographic standards, a recent trend has been to produce quantum resource estimates for various problems, such as quantum search applied to AES [11], preimage attacks on SHA-2 and SHA-3 [1] and that of Shor's algorithm [17]. Ultimately logical gates must be implemented in a particular architecture and one potential choice is to use the Clifford Gate set and the additional T gate, which together form the universal *Clifford+T* gate set for quantum computation. Unfortunately quantum computers are expected to require a vast amount of error correction and the cost of error correction for T gates is predicted to overwhelm that required for the Clifford gate set. This has led to recent quantum resource estimation papers providing resource counts by counting the T gates and the Clifford gates separately, allowing for an approximation of the difficulty of realising these algorithms in potential near-future architecture. Schwabe and Westerbaan introduced the cryptanalysis of the binary $\mathcal{MQ}$ problem [19] and provided low-qubit algorithms to solve it, but only examined the problem in terms of logical gates. Their work demonstrated that the binary $\mathcal{MQ}$ problem becomes vulnerable to quantum computers in the near future, as important instances can be solved with only hundreds of logical qubits, compared to the thousands required to break other cryptographic problems [1,11,17]. This motivates our paper and our results provide further evidence that this problem may be even easier to solve.

Recently, the classical algorithms of BooleanSolve and XL have been adapted to act as oracles in this manner in conjunction with quantum search and their proposed gate count is lower than that required for quantum search with the oracle of Schwabe of Westerbaan [19]. Whilst these are asymptotically superior algorithms, our methods are based purely search-based, require no such assumptions and may require fewer resources for smaller instances.

1.2   Contributions

We provide a method that can be applied to any binary $\mathcal{MQ}$ oracle that evaluates m equations in n variables and stores their results on m registers to create a *partial search* oracle. This technique defines the quantum search oracle as one which checks whether $n - b$ variables lead to a satisfying solution, where $0 \leq b < n$. Using this strategy increases the cost of the quantum oracle, but allows us to benefit from the drop in query complexity. In order to implement this, we must design a circuit based upon a classical preprocessing stage and after the quantum search terminates, we must locate the remaining b variables. These classical costs will be negligible as b will be small compared to n.

Parameters for post-quantum cryptosystems are often chosen based upon the efficiency of the best-known attack, as was the case with the Gui cryptosystem [15,16] and Schwabe and Westerbaan's $\mathcal{MQ}$ oracle [19] and we highlight that our methods should be viewed as a warning that basic structure can be exploited in quantum algorithms in unexpected ways, so that parameters should be chosen conservatively with regards to the best-known quantum attack.

1.3   Organisation of Paper

Section 2 recaps the basics of quantum circuitry required to cost our algorithm. Section 3 reviews quantum search. In Sect. 4 we review existing binary $\mathcal{MQ}$ oracles and describe our quantum algorithm for partial search. In Sect. 5 we review our results in the Clifford+T gate model. We give our conclusions in Sect. 6.

2   Quantum Algorithms and Circuitry

2.1   Quantum Computing

Both classical computers and quantum computers can be thought of as performing operations upon a register which stores memory. In the case of classical computers, the register contains *bits*, each of which can take on the value 0 or 1 and an n-bit state may be in one of the 2^n possible states $x \in \{0,1\}^n$. In contrast, the register of a quantum computer contains *qubits* and may be in a *superposition* of the 2^n states which a classical register might store. We refer to these 2^n states as the *computational basis states*. By superposition, we mean that the quantum state is

$$|\psi\rangle = \sum_{x \in \{0,1\}^n} \alpha_x |x\rangle, \qquad (2)$$

where $\alpha_x \in \mathbb{C}$ represents the *amplitude* of any particular state $|x\rangle$. Measuring the quantum register results in the collapse of the quantum state into a single n-bit string, where the probability of measuring state $|x\rangle$ is $|\alpha_x|^2$, so that

$$\sum_{x \in \{0,1\}^n} |\alpha_x|^2 = 1. \qquad (3)$$

Quantum algorithms work via manipulating the α_x amplitudes into a state whereby measurement will collapse the superposition into a single bit-string which contains useful information. The manipulation of these amplitudes is performed by means of *quantum gates* which are the quantum analog of classical logic gates whose fundamental property is that they are reversible, in that no information is lost through computing their output. This reversibility allows computations to be uncomputed, but means that arbitrary boolean functions must be implemented with an additional overhead.

Quantum states themselves may be represented as a vector of complex coefficients and quantum gates as unitary matrices—quantum gates therefore act *linearly* upon the quantum states, so that a quantum circuit implementing the boolean function $h : \{0,1\}^n \longrightarrow \{0,1\}$ acts by

$$U_h \sum_{x \in \{0,1\}^n} \alpha_x \cdot |x\rangle = \sum_{x \in \{0,1\}^n} \alpha_x \cdot U_h |x\rangle. \qquad (4)$$

Owing to the property of linearity and the nature of Grover's algorithm, it will be sufficient for our purposes to consider the operation of quantum gates upon the individual computational basis states. At times we will represent algorithms via quantum circuits, which should be read from left (input) to right (output).

Quantum circuits must be reversible and so to implement an arbitrary boolean function we must adapt the function. This can be achieved by defining the quantum circuit to include the input (n bits), output (m bits) and working memory (w bits) so that the boolean function $h : \{0,1\}^n \longrightarrow \{0,1\}^m$ is interpreted as the quantum circuit acting upon $n + m + w$ qubits

$$U_h |x\rangle |y\rangle |0\rangle^w \mapsto |x\rangle |y \oplus h(x)\rangle |g(x)\rangle. \qquad (5)$$

This computation results in a number of so-called "garbage bits", $|g(x)\rangle$, which can be thought of as the end state of the working memory and ancillae bits.

If these garbage bits are not dealt with, they will interfere with the diffusion step of quantum search, which relies upon the register being identical on all places apart from the computational basis states representing the search space $|x_1 \ldots x_n\rangle$. There is a simple procedure to compute $h(x)$ without garbage bits, which requires an extra space of m-qubits. The procedure consists of computing $|x\rangle |y\rangle |g(x)\rangle |h(x)\rangle$ as in (5) and copying the result to the output register before uncomputing U_h.

2.2   Quantum Gates

The logical quantum gates we describe below may be decomposed into their Clifford+T gate components as in (Fig. 2) and we provide further details in Sect. 5. Owing to the linear action of quantum gates, as in (4), it suffices to study the action of these logical gates acting upon an arbitrary computational basis state $|x_1 \ldots x_k\rangle$, where $x_i \in \{0,1\}$. We will wish to implement the boolean functions $\wedge_k$ for $k \geq 2$, $\oplus$ and NOT in quantum circuitry. We briefly review these primitives, provided by the designers of the original binary $\mathcal{MQ}$ oracle [19] (Fig. 1).

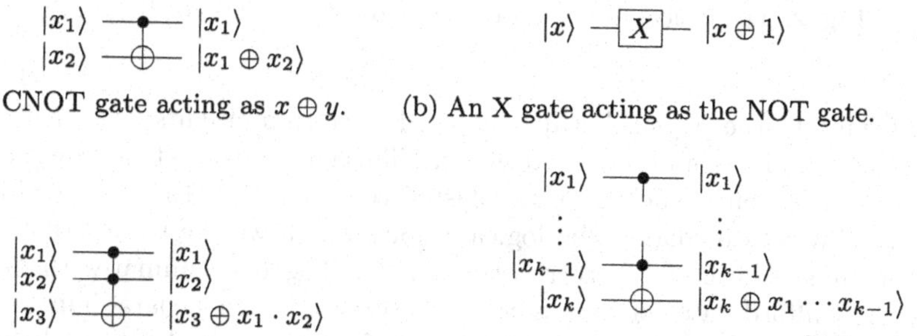

$$|x_1\rangle \quad\text{———}\bullet\text{———}\quad |x_1\rangle$$
$$|x_2\rangle \quad\text{———}\oplus\text{———}\quad |x_1 \oplus x_2\rangle$$

(a) A CNOT gate acting as $x \oplus y$.

$$|x\rangle \quad\text{—}\boxed{X}\text{—}\quad |x \oplus 1\rangle$$

(b) An X gate acting as the NOT gate.

$$|x_1\rangle \quad\text{———}\bullet\text{———}\quad |x_1\rangle$$
$$|x_2\rangle \quad\text{———}\bullet\text{———}\quad |x_2\rangle$$
$$|x_3\rangle \quad\text{———}\oplus\text{———}\quad |x_3 \oplus x_1 \cdot x_2\rangle$$

(c) A Toffoli gate acting as $x \wedge y$.

$$|x_1\rangle \quad\text{———}\bullet\text{———}\quad |x_1\rangle$$
$$\vdots \qquad\qquad \vdots$$
$$|x_{k-1}\rangle \quad\text{———}\bullet\text{———}\quad |x_{k-1}\rangle$$
$$|x_k\rangle \quad\text{———}\oplus\text{———}\quad |x_k \oplus x_1 \cdots x_{k-1}\rangle$$

(d) A k-bit Toffoli gate acting as $\bigwedge\limits_{i=1}^{k-1} x_i$.

Fig. 1. Self-inverse gates for building boolean circuits in quantum circuitry.

As the set $\{\wedge, \oplus, \text{NOT}\}$ is a universal boolean gate set, this collection of quantum gates is sufficient to implement arbitrary boolean circuits. We will allow the X gate to act as a primitive gate for the Clifford gate set, though will treat the Toffoli and k-bit Toffoli gate as logical gates for now, whose construction may be optimised dependent upon the availability of ancillae bits.

The Hadamard gate is used in the *diffusion stage* of quantum search and for the construction of Toffoli and k-bit Toffoli gates—facts we will use only for the gate count. Importantly, the Hadamard gate acts upon the state $|1\rangle$ to create

$$H|1\rangle \mapsto \frac{|0\rangle - |1\rangle}{\sqrt{2}} = |-\rangle, \tag{6}$$

which is used to realise the action of the quantum oracle, covered in Sect. 3.2.

2.3 Universal Gate Sets and Error Correction

Quantum circuits, much like boolean circuits, may be built out of a finite universal gate set. In line with other quantum resource estimation papers [1,11,17], we consider the Clifford+T universal gate set for quantum circuits.

Whilst other universal quantum gate sets are naturally possible, this choice represents a potential, well studied gate set and allows a direct comparison with other literature on quantum resource estimation. The Clifford gate set is generated by the gate set {*Controlled-NOT* (CNOT), *Hadamard* (H) and *Phase* (S)} and, with the addition of the T gate and its inverse $T^\dagger$, allows approximation of any reversible quantum gate. We will work directly with logical CNOT, X, Hadamard, Toffoli and k-bit Toffoli gates, which may be easily converted into Clifford+T gate representations.

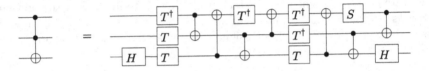

Fig. 2. The logical Toffoli gate decomposed into Clifford+T gates [3,20].

Owing to the physical scales involved, quantum circuits are inherently vulnerable to noise from their environment. This can be corrected for with an error correction scheme, such as that of surface codes [10], though requires many physical qubits to realise the logical qubits which we use to describe quantum algorithms. The cost of error correction for T gates dominates that required for the Clifford gate set and so it has become a common paradigm to separate quantum resource estimations into the Clifford gate complexity and the T gate complexity. We provide gate count and depth in terms of logical gates and T gates, in line with recent quantum resource estimates [1,11,17] in Sect. 6.

3   Quantum Search

3.1 Grover's Algorithm

Search problems may be defined by a boolean function $h : \{0,1\}^n \longrightarrow \{0,1\}$ such that $h(x) = 1$ if and only if x is one of the items that fits our criteria. If h is a black-box (that is, when evaluated upon x it reveals no information other than its output), then the number of classical queries we must make to find one of M items in a search space of size N is clearly $\mathcal{O}(\frac{N}{M})$.

Grover's quantum search algorithm [12] outputs one of M *marked* items we are searching for in a space of size N and, after an initial setup phase to generate the uniform superposition of n-bit strings, consists of iterating a single procedure, the *Grover iteration*, an optimal number of times before measuring

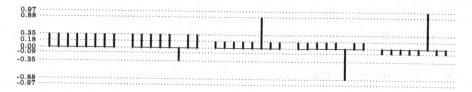

Fig. 3. Evolution of state magnitudes for $N = 8$, $M = 1$ and two iterations.

the quantum register. The Grover iteration is composed of two steps—the call to the *quantum oracle* and the *diffusion step* on n-qubits. In essence, the quantum oracle inverts the phase of the items we are searching for, thereby "marking" them and the diffusion step inverts every amplitude around the *mean* of all amplitudes. There is an optimal lower bound to the number of queries to the quantum oracle for measurement to result in a marked item with probability ≈ 1, which is $\approx \frac{\pi}{4} \cdot \sqrt{\frac{N}{M}}$ [8,12]. Before this bound is reached, the probability of measurement resulting in a marked item will monotonically increase (Fig. 3).

The total cost of executing Grover's algorithm will be the product of the query complexity with the circuit complexity/depth complexity of the quantum oracle and the diffusion step, therefore being

$$\frac{\pi}{4} \cdot \sqrt{\frac{N}{M}} \cdot \Big(\text{Cost(oracle)} + \text{Cost(diffusion step)}\Big). \qquad (7)$$

As we will see, using an efficient exhaustive search method within the quantum oracle simultaneously increases the cost of the oracle whilst both (under mild assumptions and for $M \ll N$) maintaining correctness of the algorithm and reducing the query complexity. As a side benefit, the cost of calling the diffusion step is also mildly reduced. These factors will all lead to real-world gains.

3.2   The Quantum Oracle

Defining the boolean function $h : \{0,1\}^n \longrightarrow \{0,1\}$ to be the function for which $h(x) = 1$ if and only if x is one of the M items we are searching for, the *quantum oracle* is a quantum circuit which acts upon each computational basis by

$$|x\rangle \mapsto \begin{cases} -|x\rangle & \text{if } h(x) = 1 \\ |x\rangle & \text{otherwise.} \end{cases} \qquad (8)$$

Whilst this can be performed in several ways, our method requires the use of the $|-\rangle$ state described by (6), so that the quantum oracle is realised by the action,

$$|x\rangle \left(\frac{|0\rangle - |1\rangle}{\sqrt{2}}\right) \mapsto |x\rangle \left(\frac{|0 \oplus h(x)\rangle - |1 \oplus h(x)\rangle}{\sqrt{2}}\right) = (-1)^{h(x)} |x\rangle \left(\frac{|0\rangle - |1\rangle}{\sqrt{2}}\right). \quad (9)$$

If we ignore the parts of the register which are identical for all computational basis states, then the action of the quantum oracle is as described by (8). Manipulating the phase of the amplitude according to an implemented boolean function is sometimes referred to as *phase kickback*. The resource requirements for the quantum oracle are naturally dependent upon the search problem.

3.3   The Diffusion Step

The *diffusion step* affects the amplitudes via the operation of *inversion around the mean*. Details may be found in the reference section [14], but in essence the diffusion step acts upon the amplitudes of the computational basis states so that

$$\alpha_x \mapsto 2 \cdot \langle \alpha \rangle - \alpha_x \qquad \text{where} \qquad \langle \alpha \rangle = \frac{1}{N} \sum_{x \in \{0,1\}^n} \alpha_x. \qquad (10)$$

As applying the quantum oracle with phase kickback means that the amplitude of all marked items is negative, we have that the amplitude of marked items after the diffusion step is increased and those of unmarked items are decreased. The diffusion step upon n-qubits is commonly implemented by a circuit involving $2n$ Hadamard gates, $2n$ X gates and a single $(n + 1)$-bit Toffoli gate.

4   An $\mathcal{MQ}$ Partial Search Oracle

Definition 1 (The Multivariate Quadratic ($\mathcal{MQ}$) problem). *Given* $f^{(1)}, \ldots, f^{(m)} \in \mathbb{F}_q[x_1, \ldots, x_n]$, *where* $\mathbb{F}_q$ *is the finite field of size q and each equation is of degree two, the Multivariate Quadratic ($\mathcal{MQ}$) problem is to find an* $\bar{x} = (\bar{x}_1, \ldots, \bar{x}_n)$ *with* $\bar{x}_i \in \mathbb{F}_q$ *such that* $f^{(i)}(\bar{x}) = 0$ *for* $i = 1, \ldots, m$.

The *binary* $\mathcal{MQ}$ problem is simply the specialised case when $q = 2$. As a minor modification, we consider the equivalent problem of searching for an $\bar{x} \in \{0,1\}^n$ such that all equations are satisfied when $f^{(i)}(\bar{x}) = 1$ for $i = 1, \ldots, m$. This is trivially obtained via addition of 1 to all equations and allows us to easily output whether all equations are satisfied via one $(m + 1)$-bit Toffoli gate. In all cases we will assume that $n \leq m < 2n$, as efficient algorithms exist when $m \geq 2n$ [13] and when $m < n$, the system can be reduced to this problem [21].

4.1   Previous Work of Schwabe and Westerbaan

Design. We review the original binary $\mathcal{MQ}$ oracle [19], which we will use in our partial search oracle and for comparison. The single equation

$$f^{(k)}(x_1, \ldots, x_n) = \sum_{1 \leq i < j \leq n} a_{i,j}^{(k)} x_i x_j + \sum_{1 \leq i \leq n} b_i^{(k)} x_i + c^{(k)} \qquad (11)$$

with $a_{i,j}^{(k)}, b_i^{(k)}, c^{(k)} \in \{0,1\}$ may be placed into the equivalent representation

$$f^{(k)}(x_1, \ldots, x_n) = c^{(k)} + \sum_{i=1}^{n} y_i^{(k)} x_i \qquad \text{with} \qquad y_i^{(k)} = b_i^{(k)} + \sum_{j=i+1}^{n} a_{i,j}^{(k)} x_j. \qquad (12)$$

This representation is then exploited to compute in a manner that is space efficient and reversible. To load the equation $f^{(k)}$ into the register $|E^{(k)}\rangle$ with the temporary storage qubit $|t\rangle$, the following method may be used. We note $|E^{(k)}\rangle$ and $|t\rangle$ are initialised to $|0\rangle$.

For $i = 1, \ldots, n$

(a) Compute $y_i^{(k)}$ in the $|t\rangle$ register. The linear component involving the variables is computed using CNOTs with $|x_j\rangle$ as the control and $|t\rangle$ as the target when $a_{i,j} = 1$. The addition of $b_i^{(k)}$ is handled via an X gate.

(b) Add $x_i y_i^{(k)}$ to the register $|E^{(k)}\rangle$. This is accomplished via a Toffoli gate with the first control set to $|x_i\rangle$ and the second on the register $|t\rangle$, which holds the value $y_i^{(k)}$ by our previous step.

(c) Perform step (a) again to uncompute $y_i^{(i)}$ from the $|t\rangle$ register.

Once the nonconstant part of the equation is loaded, an X gate may be used to perform addition of the constant $c^{(k)}$ if required. To illustrate this, we convert the equation

$$f^{(k)} = x_1 x_4 + x_2 x_4 + x_1 x_5 + x_3 x_4 + x_3 x_5 + x_4 x_5 + x_1 + + x_3 + x_5 + 1 \quad (13)$$

into

$$y_1^{(k)} = x_4 + x_5 + 1 \qquad y_2^{(k)} = x_4 \qquad y_3^{(k)} = x_4 + x_5 + 1$$
$$y_4^{(k)} = x_5 \qquad\qquad y_5^{(k)} = 1. \qquad\qquad\qquad\qquad (14)$$

which for loading $x_1 y_1^{(k)}$ into the 7th register using the 6th register for temporary storage corresponds to the following circuit (Fig. 4).

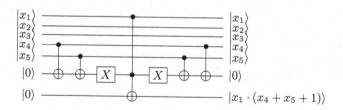

Fig. 4. Addition of $x_1 y_1^{(k)}$ to an equation register via a temporary storage register.

Including the cost of uncomputing the equation registers, this oracle therefore requires $2mn$ Toffoli gates for multiplication, $4mn + 2m$ X gates, $2m(n^2 - n)$ CNOT gates and $n + m + 1$ qubits.

4.2   A Partial Search Oracle

We propose a quantum oracle that can be constructed from *any* circuit that evaluates m binary $\mathcal{MQ}$ equations and stores their result in m registers. For a concrete example we will use the oracle described in Sect. 4.1. Our circuit solves the k-*partial search problem* for the binary $\mathcal{MQ}$ problem, which can simplify solving the initial binary $\mathcal{MQ}$ problem. In essence our strategy is to relax the problem and use preprocessing to reduce the computational load. This can either be optimised towards total gate count or the total number of T gates.

Definition 2 (The k-Partial Search problem). *Given $h : \{0,1\}^n \longrightarrow \{0,1\}$ and a promise that there exist M bit-strings for which $h(x) = 1$, the k-partial search problem is to locate at least k bits of a bitstring for which $h(x) = 1$.*

If we can obtain the first $n - b$ bits of a valid bitstring, then substitution of these values into a system of m equations in n variables will result in a system of m equations in b variables, a far easier problem to solve either classically via either Gröbner bases/XL algorithms [5,9,13] or a classical/quantum search.

4.3   Concept

Our starting point is to view each quadratic equation as a sum of three components, so that the equation $f^{(k)}$ with original representation

$$f^{(k)}(x_1,\ldots,x_n) = \sum_{i=1}^{n}\sum_{j=i+1}^{n} a_{i,j}^{(k)} x_i x_j + \sum_{i=1}^{n} b_i^{(k)} x_i + c^{(k)} \tag{15}$$

is viewed as the equation

$$f^{(k)}(x_1,\ldots,x_n) = g_1^{(k)}(x_1,\ldots,x_{n-b}) + g_2^{(k)}(x_1 \ldots, x_n) + g_3^{(k)}(x_{n-b+1},\ldots,x_n), \tag{16}$$

where

$$g_1^{(k)}(x_1,\ldots,x_{n-b}) = \sum_{i=1}^{n-b}\sum_{j=i+1}^{n-b} a_{i,j}^{(k)} x_i x_j + \sum_{i=1}^{n-b} b_i^{(k)} x_i \tag{17}$$

$$g_2^{(k)}(x_1,\ldots,x_n) = \sum_{i=1}^{n-b}\sum_{j=n-b+1}^{n} a_{i,j}^{(k)} x_i x_j \tag{18}$$

$$g_3^{(k)}(x_{n-b+1},\ldots,x_n) = \sum_{i=n-b+1}^{n}\sum_{j=i+1}^{n} a_{i,j}^{(k)} x_i x_j + \sum_{i=n-b+1}^{n} b_i^{(k)} x_i + c^{(k)}. \tag{19}$$

In this way g_1 consists of the quadratic and linear terms involving only the first $n - b$ variables, g_2 consists of the quadratic terms for which one variable is from the first $n-b$ variables and the second variable is from the last b variables, whilst g_3 consists of the quadratic and linear terms involving the last b variables and the constant term.

Our first observation is that substitution of the last b variables will result in $g_1^{(k)}$ remaining as in (17), whilst $g_2^{(k)}$ will become a linear equation in the first $n-b$ variables and $g_3^{(k)}$ will evaluate to a constant bit. Our circuit design will exploit this, so that our search space for Grover's algorithm will be defined upon the first $n-b$ variables and the oracle itself will evaluate the $g_1^{(k)}$ equations and then implement a method similar to exhaustive search of the final b variables. This exhaustive search is based upon a classical preprocessing step which generates $m \cdot 2^b$ linear equations obtained by substituting $g_2^{(k)} + g_3^{(k)}$ with the last b variables.

The key idea behind our method will be that if b is small, then after the m $g_1^{(k)}$ equations have been evaluated, the addition of the $m \cdot 2^b$ linear equations will be a minor cost compared to that of computing the $g_1^{(k)}$. The use of classical preprocessing reduces this exhaustive search on the final b variables to a circuit using only CNOT gates and $(m+1)$-bit Toffoli gates. This allows us to define the search space on $n-b$ variables and benefit from the drop in query complexity.

4.4   Our Hybrid Classical/Quantum Algorithm

Classical Preprocessing. We first choose an optimal $0 \le b < n$, which can be easily derived for a fixed n and m via the equations in Sect. 5.3. We note that b will be small ($b = 6$ for $n = m = 80$). We then create the m tuples $(g_1^{(k)}, g_2^{(k)}, g_3^{(k)})$ as in (17), (18) and (19). After this is done, we evaluate the $g_2^{(k)}$ and $g_3^{(k)}$ equations upon all 2^b assignments of the last b variables and store the results. We then have m $g_1^{(k)}$ equations, $m \cdot 2^b$ linear $g_2^{(k)}$ equations and $m \cdot 2^b$ $g_3^{(k)}$ constants. So long as b is small, this will be a negligible cost.

The Quantum Oracle. The search space for our oracle is defined upon the first $n-b$ variables. The oracle first loads m registers with the evaluation of the m $g_1^{(k)}(\bar{x}_1, \ldots, \bar{x}_{n-b})$ equations via any circuit design for evaluation of polynomials and then implements 2^b sub-circuits. Each subcircuit performs the addition of the linear equation from the preprocessed $g_2^{(k)}$ and the constant bit from the preprocessed $g_3^{(k)}$ to the respective equation registers and checks whether the equations have been satisfied via an $(m+1)$-bit Toffoli gate wired to the phase flip bit $|-\rangle$. This has the effect of inverting the phase whenever the first $n-b$ variables and the last b variables captured in the m linear equations that the subcircuit implements lead to a satisfying solution together.

The final subcircuit additionally performs uncomputation to leave each equation register in the state $g_1^{(k)}(\bar{x}_1, \ldots, \bar{x}_{n-b})$. The equation registers may then all be uncomputed via the method used to load them with the values $g_1^{(k)}$.

The 2^b subcircuits are equivalent to performing classical exhaustive search upon the 2^b values that the variables $x_{n-b+1}, \ldots, x_n$ may be assigned. Excluding the first and last of these circuits, each circuit must only add the difference between each pre-processed sum, meaning that we use 2^b $(m+1)$-bit Toffoli gates and at most $(2^b + 1) \cdot m(n-b)$ CNOT gates, which can be executed in

$(2^b + 1) \cdot m$ layers with our assumption that $m \geq n$. There will additionally be at most $(2^b + 1) \cdot m$ X gates, which can be performed in $(2^b + 1)$ layers. As the search space is only defined on $n - b$ variables, the entire circuit requires b fewer qubits for the oracle than would otherwise be required.

Classical Postprocessing. It then remains for us to substitute the $n - b$ values obtained from the quantum search bitstring into our system of m equations. We will then have to solve a system of m equations in b variables, but as b will be small, this cost will be negligible compared to the quantum search stage.

Benefits. The major benefit of this method is that for small b we can exploit the smaller query complexity, $\mathcal{O}(2^{\frac{n-b}{2}})$ as opposed to $\mathcal{O}(2^{\frac{n}{2}})$ and the data gathered from the preprocessing stage in an efficient manner. The preprocessing stage helps us via shifting the computational load from the Toffoli gates to CNOT gates, which require less error correction, and $(m + 1)$-bit Toffoli gates, whose T count can be more readily optimised. We will obtain gains so long as the combined benefits of the lower query complexity applied to the entire modified oracle result in a lower gate count to that of the original oracle.

The technique is also embarassingly parallel, in that if we are lucky to have an excess of qubits, then after computing the m values of $g_1^{(k)}$, we can copy these values to m empty registers and execute the 2^b subcircuits in parallel with multiple phase flip bits. Hence any qubits used for parallelism in the circuit for computing $g_1^{(k)}$ need not be idle during the exhaustive search step.

4.5   Cost Analysis

We note that the case $b = 0$ is that of the original evaluation circuit being used as a quantum oracle without modification. As with the case of the original oracle, circuit depth is dependent upon the number of additional qubits available—for reasons of space, we leave discussion concerning circuit depth to an expanded version of this paper, which details the circuit depth of Schwabe and Westerbaan's oracle and provides new $\mathcal{MQ}$ evaluation oracles optimised for depth and T gate count.

Circuit-Size/Depth Complexity. Fixing the cost of a procedure to evaluate m equations of degree d in n variables as Eval(n,m,d), the cost of the application of an $(m + 1)$-bit Toffoli gate as Toff(m+1) and the cost of the diffusion step on n variables as Diff(n), we obtain our maximum advantage approximately when

$$2^{\frac{n-b}{2}} \cdot \left[2 \cdot \text{Eval(n-b,m,2)} + 2^b \cdot \text{Toff(m+1)} + (2^b + 1) \cdot \text{Eval(n-b,m,1)} + \text{Diff(n-b)} \right] \quad (20)$$

is minimised. We examine the exact benefits in terms of Clifford+T gates in Sect. 5 after detailing the cost of the Toffoli and $(k + 1)$-bit Toffoli gates.

4.6   Valid Parameters

We note that if $M > 1$, then multiple solutions may share the same first $n - b$ values. The quantum search procedure described in Sect. 4.4 may then invert the phase an even number of times, which will be equivalent to no phase inversion, hence the quantum search may fail. In reality, we are dealing with the case where $M \ll N$ (often $M \approx 1$), can fix variables in order to obtain this scenario, or even permute the variable indices upon failure and try again. Nevertheless, a negligible chance of failure remains and without a promise that there exists no collision on the first $n - b$ bits of any two solutions, we must view our adaption as a heuristic adaption to quantum search. One option to correct for this would be to increment a counter [19] controlled on the output of the $(m + 1)$-bit Toffoli gate and invert the phase if and only if the counter has been incremented. This would provide for correctness, but the overhead required may destroy our gains.

To obtain gains we must choose an optimal value for b, which is dependent upon the cost of implementing the $(m + 1)$-bit Toffoli gate. The next section reviews the T gate cost metric and the cost of our partial search oracle.

5   Analysis and Impact in the T Gate Model

Up until now we have kept our analysis in the logical gate model. In this section we expand upon Sect. 2.3, briefly justifying the costing of quantum circuitry in terms of separate counts for the Clifford gates and T gates. We provide Clifford+T costs for the Toffoli and k-bit Toffoli gates and provide analysis for partial search oracle. Full details of error correction methods are beyond the scope of this paper and we refer the reader to [1] for a more detailed treatment.

5.1   Error Correction and T Gates

The cost of error correction for T gates dominates that of those from the Clifford gate set. In essence this is because fault-tolerant implementation of the T gate requires a process referred to as *magic state distillation*. This process requires creation of the so-called *magic state*

$$|\Theta\rangle = \frac{|0\rangle + e^{\frac{i\pi}{4}} |1\rangle}{\sqrt{2}}, \tag{21}$$

in an ancillae qubit which is consumed upon application of each T gate [1].

In order for the entire algorithm to be implemented correctly, we must have that each $|\Theta\rangle$ is produced with an error rate of less than or equal to $\frac{1}{T_U}$, where T_U is the total number of T gates used in the entire quantum algorithm [1]. Creation of these magic states may be handled by so-called *magic state distillation factories*, which produce magic states by processing multiple noisy $|\Theta\rangle$ states and distilling them into a single $|\Theta\rangle$ state with less noise. This process may be repeated until a $|\Theta\rangle$ state is produced with the required error threshold. Given that this process is required only for the implementation of T gates, papers which estimate the cost of quantum algorithms include a separation of cost metrics in either in terms of Clifford gates and T gates [1,11], or only Toffoli gates [17].

5.2   Clifford+T Gate Costs for Logical Gates

Our primitive Clifford gates will be {CNOT, S, H, X}. It then remains for us to provide costs for implementations of the logical Toffoli and k-bit Toffoli gates.

The Toffoli and k-bit Toffoli Gates. We will use the construction in Fig. 2 for the Toffoli gate, which requires 7 T gates and 10 Clifford gates.

From (20), it is obvious that the cost of the $(m+1)$-bit Toffoli gate plays a large part in the efficiency of our methods. We detail more efficient constructions in an expanded version of this paper, but choose an older and relatively inefficient construction to demonstrate the gains that can be made by using our partial search method. The k-bit Toffoli gate can be constructed by elementary means using only Toffoli gates and one ancilla qubit, which breaks down into $80k - 240$ Clifford gates and $52k - 168$ T gates [6]. Without using at least one ancilla qubit, the cost of implementing the k-bit Toffoli gate will be $\mathcal{O}(k^2)$ [6,22] and we note that any improvement in circuit design for the multiple controlled Toffoli gate translates has a favourable impact upon the concrete cost of our partial search algorithm. Optimisation programs such as TPar [2] may provide further gains.

5.3   Clifford+T Gate Costs for Our Partial Search Oracle

We provide the cost of the oracle circuits in terms of Clifford gates and in T gates and parameterised by n, m and b, using the original $\mathcal{MQ}$ oracles [19] as our evaluation circuit. This gives us a total cost in terms of Clifford gates of

$$\frac{\pi}{4} \cdot 2^{\frac{n-b}{2}} \cdot \left(2m(n-b)^2 + 23m(n-b) + 2^b(m(n-b) + 81m - 160) + 84n - 160\right) \tag{22}$$

and a total cost in terms of T gates of

$$\frac{\pi}{4} \cdot 2^{\frac{n-b}{2}} \cdot \left(14m(n-b) + 2^b(52m - 116) + 52(n-b) - 116\right). \tag{23}$$

Given a fixed n and m, an optimum value of b can easily be found and from these equations it is clear our gains come from ensuring the drop in the contribution from the query complexity is not outweighed by the term involving 2^b.

6   Impact and Conclusions

Parameter Choice for Quantum Security. The Gui cryptosystem [16] is a proposed *Multivariate Quadratic* public-key signature scheme that is thought to be resistant to quantum attacks. The Gui signature scheme relies upon k (known as the repetition factor) applications of an HFEv- (Hidden Field Equations with Vinegar variables and Minus equations) derived central map, which is a system of m equations in n variables over GF(2). Forging a signature therefore requires inverting the *central map* k times. Whilst parameters for security against only classical attacks were provided in the original paper [16], the authors provide parameters for security against an attack by a quantum computer in a later paper [15], including the parameters for $\lambda = 80, 128$ and 256 bit quantum security (it should require at least 2^λ quantum gates to break the scheme).

For reasons of space, we refer the reader to the original paper for details of Gui, but summarise the number of equations, variables and repetition factor along with the number of gates using the original $\mathcal{MQ}$ oracle and the $\mathcal{MQ}$ oracle with our adaptations in Table 1 below. We use the Clifford+T gate count as we believe this allows for a fair comparison of our methods. Our results still stand (and are in fact better) in a purely logical gate model [19], but we feel that this makes for a fair comparison, given that we use far more multiple controlled Toffoli gates and we do not wish to hide any costs by abstracting this primitive.

Table 1. Number of Clifford+T gates required to break Gui [15, 16].

λ [15]	$n = m$	k	# Gates using [19]	# Gates using our method	b chosen for our method
80	117	2	$2^{80.99}$	$2^{78.38}$	7
128	209	2	$2^{129.40}$	$2^{126.26}$	8
256	457	2	$2^{256.71}$	$2^{252.93}$	10

As is plain from the table above, which can easily be computed by adding formulae (22) and (23), our methods break the proposed quantum-resistant parameters for Gui given in [15]. As our results do not imply any structural weakness in Gui, new parameters for the cryptosystem can easily be chosen to ensure it is secure for the relevant security level, though in doing so we risk again leaving the cryptosystem open to any future optimisations. It is our hope that this demonstrates that when deciding on parameters for a given security level, even when taking into account the best-known quantum algorithms we must be especially conservative, as the structure of the problem may be vulnerable to techniques such as pre processing, which may reduce the overhead of the quantum oracle.

Conclusions. We have demonstrated that techniques such as reformulation of the $\mathcal{MQ}$ problem, classical preprocessing and adaptation of multi-target search techniques for single targets can provide us with concrete improvements which can break the quantum resistant parameters of $\mathcal{MQ}$ cryptosystems. Furthermore we have proved that in relation to the Clifford+T gate set, we can specifically lower the total number of T gates and therefore drastically reduce the amount of error correction required. The techniques described in this paper demonstrate both how quantum algorithms may be optimised and that parameters should be chosen conservatively with regards to best-known quantum algorithms.

Acknowledgements. The author kindly thanks James Davenport and Christophe Petit for their helpful discussions. Benjamin Pring is funded by an EPRSC grant.

References

1. Amy, M., Di Matteo, O., Gheorghiu, V., Mosca, M., Parent, A., Schanck, J.: Estimating the cost of generic quantum pre-image attacks on SHA-2 and SHA-3. In: Avanzi, R., Heys, H. (eds.) SAC 2016. LNCS, vol. 10532, pp. 317–337. Springer, Cham (2017). https://doi.org/10.1007/978-3-319-69453-5_18
2. Amy, M., Maslov, D., Mosca, M.: Polynomial-time T-depth optimization of Clifford+ T circuits via matroid partitioning. IEEE Trans. Comput.-Aided Des. Integr. Circ. Syst. **33**(10), 1476–1489 (2014)
3. Amy, M., Maslov, D., Mosca, M., Roetteler, M.: A meet-in-the-middle algorithm for fast synthesis of depth-optimal quantum circuits. IEEE Trans. Comput.-Aided Des. Integr. Circ. Syst. **32**(6), 818–830 (2013)
4. Bard, G.: Algebraic Cryptanalysis. Springer, New York (2009). https://doi.org/10.1007/978-0-387-88757-9
5. Bardet, M., Faugére, J.C., Salvy, B., Spaenlehauer, P.J.: On the complexity of solving quadratic boolean systems. J. Complex. **29**(1), 53–75 (2013)
6. Barenco, A., et al.: Elementary gates for quantum computation. Phys. Rev. A **52**(5), 3457 (1995)
7. Bouillaguet, C., et al.: Fast exhaustive search for polynomial systems in $\mathbb{F}_2$. In: Mangard, S., Standaert, F.-X. (eds.) CHES 2010. LNCS, vol. 6225, pp. 203–218. Springer, Heidelberg (2010). https://doi.org/10.1007/978-3-642-15031-9_14
8. Boyer, M., Brassard, G., Høyer, P., Tapp, A.: Tight bounds on quantum searching. arXiv quant-ph/9605034 (1996)
9. Courtois, N.T., Patarin, J.: About the XL algorithm over $GF(2)$. In: Joye, M. (ed.) CT-RSA 2003. LNCS, vol. 2612, pp. 141–157. Springer, Heidelberg (2003). https://doi.org/10.1007/3-540-36563-X_10
10. Fowler, A.G., Mariantoni, M., Martinis, J.M., Cleland, A.N.: Surface codes: towards practical large-scale quantum computation. Phys. Rev. A **86**(3), 032324 (2012)
11. Grassl, M., Langenberg, B., Roetteler, M., Steinwandt, R.: Applying grover's algorithm to AES: quantum resource estimates. In: Takagi, T. (ed.) PQCrypto 2016. LNCS, vol. 9606, pp. 29–43. Springer, Cham (2016). https://doi.org/10.1007/978-3-319-29360-8_3
12. Grover, L.K.: A fast quantum mechanical algorithm for database search. In: Proceedings of the 28th Annual ACM Symposium on Theory of Computing, pp. 212–219. ACM (1996)
13. Joux, A., Vitse, V.: A crossbred algorithm for solving boolean polynomial systems. IACR Cryptology ePrint Archive 2017, 372 (2017)
14. Nielsen, M.A., Chuang, I.L.: Quantum Computation and Quantum Information. Cambridge University Press, Cambridge (2010)
15. Petzoldt, A., Chen, M.-S., Ding, J., Yang, B.-Y.: HMFEv - an efficient multivariate signature scheme. In: Lange, T., Takagi, T. (eds.) PQCrypto 2017. LNCS, vol. 10346, pp. 205–223. Springer, Cham (2017). https://doi.org/10.1007/978-3-319-59879-6_12
16. Petzoldt, A., Chen, M.-S., Yang, B.-Y., Tao, C., Ding, J.: Design principles for HFEv- based multivariate signature schemes. In: Iwata, T., Cheon, J.H. (eds.) ASIACRYPT 2015. LNCS, vol. 9452, pp. 311–334. Springer, Heidelberg (2015). https://doi.org/10.1007/978-3-662-48797-6_14

17. Roetteler, M., Naehrig, M., Svore, K.M., Lauter, K.: Quantum resource estimates for computing elliptic curve discrete logarithms. In: Takagi, T., Peyrin, T. (eds.) ASIACRYPT 2017. LNCS, vol. 10625, pp. 241–270. Springer, Cham (2017). https://doi.org/10.1007/978-3-319-70697-9_9

18. Sakumoto, K., Shirai, T., Hiwatari, H.: Public-key identification schemes based on multivariate quadratic polynomials. In: Rogaway, P. (ed.) CRYPTO 2011. LNCS, vol. 6841, pp. 706–723. Springer, Heidelberg (2011). https://doi.org/10.1007/978-3-642-22792-9_40

19. Schwabe, P., Westerbaan, B.: Solving binary $\mathcal{MQ}$ with Grover's algorithm. In: Carlet, C., Hasan, M.A., Saraswat, V. (eds.) SPACE 2016. LNCS, vol. 10076, pp. 303–322. Springer, Cham (2016). https://doi.org/10.1007/978-3-319-49445-6_17

20. Selinger, P.: Quantum circuits of T-depth one. Phys. Rev. A **87**(4), 042302 (2013)

21. Thomae, E., Wolf, C.: Solving underdetermined systems of multivariate quadratic equations revisited. In: Fischlin, M., Buchmann, J., Manulis, M. (eds.) PKC 2012. LNCS, vol. 7293, pp. 156–171. Springer, Heidelberg (2012). https://doi.org/10.1007/978-3-642-30057-8_10

22. Toffoli, T.: Reversible computing. In: de Bakker, J., van Leeuwen, J. (eds.) ICALP 1980. LNCS, vol. 85, pp. 632–644. Springer, Heidelberg (1980). https://doi.org/10.1007/3-540-10003-2_104

Author Index

Printed in the United States
By Bookmasters